AF393726

GENOMIC INSTABILITY AND IMMORTALITY IN CANCER

PEZCOLLER FOUNDATION SYMPOSIA

SERIES EDITOR:
Enrico Mihich, *Roswell Park Cancer Institute, Buffalo, New York*

STANDING PEZCOLLER SYMPOSIA COMMITTEE:
Enrico Mihich (Chairman), *Roswell Park Cancer Institute, Buffalo, New York*
David M. Livingston (Vice-Chairman), *Dana–Farber Cancer Institute, Boston, Massachusetts*
Giuseppe Bernardi, *Pezcoller Foundation, Trento, Italy*
Bruce Chabner, *Massachusetts General Hospital Cancer Center, Boston, Massachusetts*
Carlo M. Croce, *Jefferson Medical College, Philadelphia, Pennsylvania*
Riccardo Dalla Favera, *College of Physicians & Surgeons of Columbia University,*
    New York, New York
Giuseppe Della Porta, *European Institute of Oncology, Milan, Italy*
Thomas Graf, *European Molecular Biology Lab, Heidelberg, Germany*
Giorgio Lenaz, *University of Bologna, Bologna, Italy*
Paolo Schlechter, *Pezcoller Foundation, Trento, Italy*
Claudio Schneider, *LNCIB Area Science Park, Trieste, Italy*
Tadatsugu Taniguchi, *University of Tokyo, Tokyo, Japan*
Fulvio Zuelli, *University of Trento, Trento, Italy*

PROGRAM COMMITTEE:
Enrico Mihich (Chair), *Roswell Park Cancer Institute, Buffalo, New York*
Leland Hartwell (Co-Chair), *University of Washington, Seattle, Washington*
Carol Greider (Co-Chair), *Cold Spring Harbor Laboratory, Cold Spring Harbor, New York*
Garth Anderson, *Roswell Park Cancer Institute, Buffalo, New York*
Margherita Bignami, *Istituto Superiore Sanitá, Rome, Italy*
David M. Livingston, *Dana-Farber Cancer Institute, Boston, Massachusetts*

A Continuation Order Plan is available for this series. A continuation order will bring delivery of each new volume immediately upon publication. Volumes are billed only upon actual shipment. For further information please contact the publisher.

GENOMIC INSTABILITY AND IMMORTALITY IN CANCER

Edited by

Enrico Mihich
Roswell Park Cancer Institute
Buffalo, New York

and

Leland Hartwell
University of Washington
Seattle, Washington

SPRINGER SCIENCE+BUSINESS MEDIA, LLC

Library of Congress Cataloging-in-Publication Data

Genomic instability and immortality in cancer / edited by Enrico
  Mihich and Leland Hartwell.
        p.   cm. -- (Pezcoller Foundation symposia ; 8)
    "Proceedings of the Eighth Annual Pezcoller Symposium on Genomic
  Instability and Immortality in Cancer, held June 17-19, 1996, in
  Trento, Italy"--T.p. verso.
    Includes bibliographical references and index.
    ISBN 978-1-4613-7448-0   ISBN 978-1-4615-5365-6 (eBook)
    DOI 10.1007/978-1-4615-5365-6
    1. Cancer--Genetic aspects--Congresses. 2. Cancer cells-
  -Congresses. 3. Apoptosis--Congresses.   I. Mihich, Enrico.
  II. Hartwell, Leland. III. Pezcoller Symposium on Genomic
  Instability and Immortality in Cancer (1997 : Trento, Italy)
  IV. Series.
    [DNLM: 1. Neoplasms--genetics--congresses. 2. Cell
  Transformation, Neoplastic--genetics--congresses. 3. Cell Cycle-
  -genetics--congresses. 4. Gene Expression--genetics--congresses.
  W1 PE995 v.8 1997 / QZ 202 G3328 1997]
  RC268.4.G476   1997
  616.99'4042--dc21
  DNLM/DLC
  for Library of Congress   97-29967
                                                             CIP

Proceedings of the Eighth Annual Pezcoller Symposium on Genomic Instability and Immortality in Cancer, held June 17–19, 1996, in Trento, Italy

ISBN 978-1-4613-7448-0

©1997 Springer Science+Business Media New York
Originally published by Plenum Press, New York in 1997
Softcover reprint of the hardcover 1st edition 1997

http://www.plenum.com

10 9 8 7 6 5 4 3 2 1

All rights reserved

No part of this book may be reproduced, stored in a retrieval system, or transmitted in any form or by any means, electronic, mechanical, photocopying, microfilming, recording, or otherwise, without written permission from the Publisher

THE PEZCOLLER FOUNDATION

The Pezcoller Foundation was created in 1979 by Professor Alessio Pezcoller (1896–1993) who was the chief surgeon of the S. Chiara Hospital in Trento from 1937 to 1966 and who gave a substantial portion of his estate to support its activities; the Foundation also benefits from the cooperation of the Savings Bank Cassa di Risparmio di Trento e Rovereto.

The main goal of this non-profit foundation is to provide and recognize scientific progress on life-threatening diseases, currently focusing on cancer. Towards this goal, the Pezcoller Foundation awards, every two years, the Pezcoller Prize, recognizing highly meritorious contributions to medical research; it also sponsors a series of annual symposia promoting interactions among scientists working at the cutting edge of basic oncological sciences.

The symposia are held in the Trentino Region of Northern Italy and their scientific focus is selected by Enrico Mihich with the collaboration of an international Standing Symposia Committee. A Program Committee determines the content of each symposium.

The first symposium focused on *Drug Resistance: Mechanisms and Reversal* (E. Mihich, Chairman, 1989); the second on *The Therapeutic Implications of the Molecular Biology of Breast Cancer* (M.E. Lippman and E. Mihich, Co-Chairmen, 1990); the third on *Tumor Suppressor Genes* (D.M. Livingston and E. Mihich, Co-Chairmen, 1991); the fourth on *Cell Adhesion Molecules: Cellular Recognition Mechanisms* (M.E. Hemler and E. Mihich, Co-Chairmen, 1992); the fifth on *Apoptosis* (E. Mihich and R.T. Schimke, Co-Chairmen, 1993); the sixth on *Normal and Malignant Hematopoiesis: New Advances* (E. Mihich and D. Metcalf, Co-Chairmen, 1994); the seventh on *Cancer Genes: Functional Aspects* (E. Mihich and D. Housman, Co-Chairmen, 1995); the ninth on *The Biology of Tumors* (E. Mihich and C. Croce, Co-Chairmen, 1997).

PREFACE

The eighth Annual Pezcoller Symposium, entitled Genomic Instability and Immortality in Cancer, was held in Trento, Italy, June 17–19, 1996 and was focused on the clarification of the mechanisms of genetic instability, a characteristic of neoplastic cells which also determines tumor progression, and immortality consequent to the lack of susceptibility to mechanisms of maturations, senescence and/or apoptosis.

With presentations at the cutting edge of progress and stimulating discussions, this symposium addressed issues related to mutational lability, changes in DNA repair capabilities, gene recombination processes, cell cycle checkpoints and apoptosis, the significance of telomerases in cell immortalization and senescence, and the clinical relevance and exploitation of the phenomena considered.

We wish to thank the participants in the symposium for their substantial contributions and their participation in the spirited discussions that followed. We would also like to thank Drs. Carol Greider, Garth Anderson, Margherita Bignami, and David Livingston, for their essential input as members of the Program Committee, and Ms. A. Toscani for her invaluable assistance. The aid of the Savings Bank Cassa di Risparmio di Trento e Rovereto, and the Municipal, Provincial, and Regional Administrations in supporting this Symposium through the Pezcoller Foundation are also acknowledged with deep appreciation. Finally, we wish to thank the staff of Plenum Publishing Corporation for their efficient cooperation in the production of these proceedings.

Enrico Mihich
Leland Hartwell

CONTENTS

TELOMERES AND CELL DIVISION IN
Drosophila melanogaster

Giovanni Cenci and Maurizio Gatti

Dipartimento di Genetica e Biologia Molecolare
Università di Roma "La Sapienza"
P.le Aldo Moro 5, 00185 Roma, Italy

INTRODUCTION

Telomeres are specialized DNA-protein complexes that protect the ends of linear eukaryotic chromosomes from degradation and incomplete DNA replication. In most organisms telomeres contain arrays of tandem G-rich repeats that are added to the chromosomal termini by a specialized polymerase, called telomerase, which contains an integral RNA template (reviewed by Zakian, 1989). In *Drosophila melanogaster* telomeric DNA is comprised of two LINE-like retrotransposable elements, called HeT-A and TART, which are specifically added to chromosome ends to compensate for replicative DNA loss (for reviews, see Mason and Biessman, 1995; Pardue, 1995). Apart from this difference in the DNA content, *Drosophila* telomeres behave like those of other organisms and are likely to bind similar, evolutionary conserved proteins (Biessman and Mason, 1992). For example, both *Drosophila* and yeast telomeres have heterochromatic properties and share the ability of modulating the expression of adjacent genes, a phenomenon called telomere position effect (Levis et al., 1985; Shore, 1995).

Telomeres are also believed to help organize the architecture of interphase nucleus by interacting with the nuclear envelope (reviewed by Dernburg et al., 1995). A non random organization of the chromosomes within the nucleus was first described by Rabl (1885). Rabl noticed that the telomeres of prophase chromosomes were located at one side of the nucleus, while the centromeres were lying at the opposite side. Because this polarized chromosomal arrangement was similar to that of the preceding anaphase, Rabl suggested that chromosomes maintain their anaphase configuration throughout interphase. Since these early studies, a Rabl orientation of chromosomes has been observed in a wide range of animals and plants; in many cases telomeres were found near the nuclear periphery, suggesting an interaction between these structures and the nuclear envelope (reviewed by Dernburg et al., 1995). However, no clear evidence of a Rabl orientation was found in mouse and human interphase nuclei. Moreover, in these cells telomeres do not appear to be associated with the nuclear envelope (Manuelidis and Borden, 1988; Billia and de Boni, 1991; Vourc'h et al., 1993).

Genomic Instability and Immortality in Cancer
edited by Mihich and Hartwell, Plenum Press, New York, 1997

TELOMERE-TELOMERE ASSOCIATIONS

A number of studies has shown that telomeres not only interact with the nuclear envelope but also with each other (reviewed by Dernburg et al., 1995). Examination of interphase nuclei has shown that in many cases telomeres are tightly clustered at the periphery of the nucleus. Most interestingly, physical connections between homologous and heterologous telomeres of prophase chromosomes have been documented in several organisms (for reviews see Avivi and Feldman, 1980; Dernburg et al., 1995). For example, in favorable preparations from onion and other plant cells, and from the coccid *Chysomphalus ficus,* all the chromosomes are linked through their telomeres to form a continuous chain or "spireme" (Hughes-Schrader, 1957; Wagenaar, 1969; Stack and Clarke, 1973). The telomere-telomere connections observed in onion cells are already present during interphase; they persist in prophase, tend to disappear in metaphase and are no longer detectable during anaphase (Stack and Clarke, 1973). This suggests that the telomeric attachments seen in prophase are a relic of the interphase chromosome organization, and implies that such attachments must form de novo during each cell cycle. The biological meaning of these telomeric associations and the mechanisms underlying this phenomenon are not understood.

Normal mammalian cells do not exhibit telomere-telomere associations (TAs). However such associations have been observed in senescent human fibroblast (Benn, 1976), in cells from patients with ataxia telangiectasia (Hayshi and Schimd, 1975; Taylor et al., 1981), in lymphocytes of one patient suffering from Thiberg-Weissenbach syndrome (Dutrillaux et al., 1978) and in a wide variety of human tumors (for reviews see Hastie and Allshire, 1989; de Lange, 1995). In most of these tumors 20–30% of metaphases showed clear telomeric associations resulting in the formation or either ring or dicentric chromosomes. These aberrant chromosome configurations were generally caused by random association of chromosomes and no tumor-specific patterns of telomeric associations have been described (for reviews see Hastie and Allshire, 1989; de Lange, 1995).

In many cases telomeric associations observed in cancer cells appear to persist during anaphase; they can form anaphase bridges that result in either chromosome breakage or non disjunction (Hastie and Allshire, 1989; de Lange, 1995). Telomeric associations can therefore promote loss of heterozygosity, contributing to tumor development. Moreover, it has been speculated that the frequency of TAs may be particularly high in early stages of carcinogenesis, before telomerase activation has counterbalanced replicative telomere shortening (de Lange, 1995). Despite the potential involvement of TAs in human tumorigenesis, little is known about the molecular mechanisms underlying telomere-telomere attachments. It is generally accepted that the occurrence of TAs is correlated with the short telomere length in cancer cells. However, the finding that some tumor cell lines with particularly short telomeres do not exhibit TAs, indicates that telomeric lenght is not the only determinant of telomeric fusions (Saltman et al., 1993; de Lange, 1995). It is thus likely that specific transacting factors can contribute to the genesis of TAs in tumor cells. Two transacting functions required to prevent telomeric fusions in normal *Drosophila* cells are described in the next section.

Drosophila MUTANTS AFFECTING TELOMERE BEHAVIOR

In the course of genetic screens for mutations affecting the fidelity of chromosome segregation we have identified two genes *[UbcD1* and *pendolino (pen)]* required for proper telomere behavior. Mutations in both genes cause frequent end-to-end associations of chromosomes, a phenomenon that is never seen in wild type cells. Here we describe the phenotypic and molecular characterization of *UbcD1* (Cenci et al., 1997); *pendolino* has been isolated very recently and its cytological and molecular analysis is currently underway (G. Belloni, G. Cenci, P. Dimitri and M. Gatti, unpublished).

We have identified several mutant alleles in the *UbcD1* locus. The strongest alleles are lethal and die either during embryogenesis (null mutations) or at the larval-pupal transition. Milder lesions in the *UbcD1* locus are viable but cause sterility in males. Examination of brain preparation from larvae bearing various combinations of *UbcD1* mutant alleles revealed a common cytological phenotype. Dividing mutant neuroblasts exhibit frequent end-to-end attachments of chromosomes resulting in dicentric and polycentric linear and ring chromosomes. In the strongest *UbcD1* mutants about 40% of the metaphases displayed at least one of such telomeric attachments. Two main types of telomeric associations were distinguished: double telomere associations (DTAs) involving both sister chromatids of each interacting chromosome end, and single telomeric associations (STAs) in which each of the two sister telomeres behaved independently. These two types of chromosomal configurations are genuine telomeric fusions because they are never accompanied by acentric fragments. However, DTAs have the same appearance of mutagen-induced chromosome exchanges generated during G1, whereas STAs are similar to chromatid exchanges induced during S-G2 (Savage, 1970). It is thus likely that DTAs result from telomeric fusions occurred prior to chromosome replication, while STAs are the consequence of telomeric associations occurred in S-G2 (see Fig. 1).

Examination of telomere-telomere attachments produced by *UbcD1* mutations revealed that they involve either homologous or heterologous telomeres, and that all the telomeres of the *D. melanogaster* complement can participate in TAs, although with different frequencies. The pattern of TAs observed in *UbcD1* mutants suggested that actively dividing wild-type brain cells maintain a Rabl orientation during early interphase but progressively lose it as cells proceed through the cell cycle (Cenci et al., 1997).

The analysis of anaphases in *UbcD1* mutant brains revealed the presence of chromatin bridges connecting the two sets of chromosomes migrating to the poles. This finding clearly indicates that the telomeric associations observed in *UbcD1* metaphases persist throughout anaphase. However, several lines of evidence suggest that most telomere-telomere connections are resolved during anaphase without causing chromosome breakage or non disjunction. First, while the frequency of anaphases with bridges was about 39%, the frequency of anaphases displaying lagging acentric fragments was only 3.3%. Second,

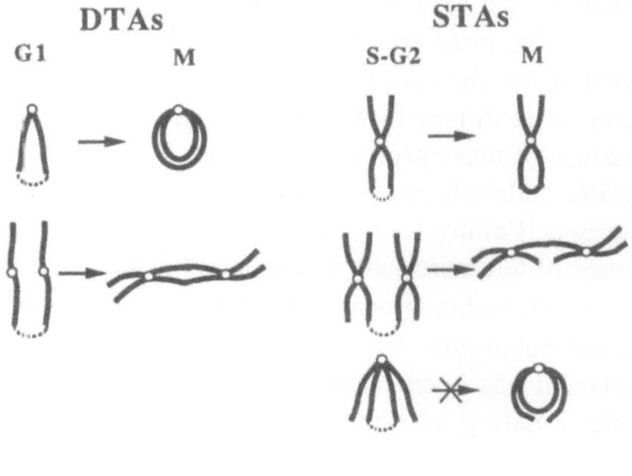

Figure 1. Origin of double telomeric associations (DTAs) and single telomeric associations (STAs) in *UbcD1* mutants. Whereas DTAs rings are rather common, STAs rings are virtually absent. This suggests that the proximity of the opposite telomeres of the same chromosome decreases as cells progress through the cell cycle (see Cenci et al., 1997).

in *UbcD1* mutant brains, in which the frequency of cells with TAs ranged from 10 to 38%, the frequency of hyperploid metaphases (an expected consequence of impaired chromosome segregation) was between 2.4 and 4.3%. Third, in the above-mentioned mutant brains the frequency of segmentally aneuploid cells (i.e. metaphases with a broken chromosome without the corresponding acentric fragment, or metaphases with a normal chromosome complement plus an extra acentric fragment) was also very low, ranging from 0.2 to 1.5%. As detailed in the next section, segmentally aneuploid cells are the expected consequence of the breakage of chromatin bridges during anaphase.

To better define the role of the *UbcD1* locus in telomere behavior we also examined meiotic cell division in *UbcD1* mutant males. Telomere-telomere attachments do also occur during both meiotic divisions of *UbcD1* males. However, in meiotic cells most telomeric connections are not resolved during anaphase and give rise to chromosome breakage. One explanation for the different behavior of TAs during mitosis and male meiosis of *UbcD1* mutants, is that the telomeres of meiotic chromosomes are more tightly connected than those of mitotic chromosomes. An alternative explanation, that we favor, is that meiotic anaphase differs from mitotic anaphase in the mechanics and speed of chromosome movement, and thus in the forces applied across the telomeric connections. For example, if meiotic anaphase were more rapid than the mitotic one, there might not be sufficient time for resolution of the telomere-telomere attachments, leading to chromosome breakage. This is conceivable because the meiotic spindle is much larger than the mitotic spindle and the distance between the separating sets of anaphase chromosomes is higher in meiosis than in mitosis (Cenci et al., 1994; Cenci et al., 1997).

We have cloned the *UbcD1* gene and found that it gives rise to three related transcripts, one specific for the male germline and two expressed throughout development. Isolation and sequencing of cDNAs corresponding to each of these transcripts showed that all three *UbcD1* RNA species result in the same protein product. This polypeptide belongs to the family of the highly conserved ubiquitin conjugating, or E2, enzymes. Database searches revealed that this enzyme had been previously identified in a PCR screen for *Drosophila* homologs of yeast E2 genes (Treier et al., 1992). The UbcD1 protein is strongly homologous (more than 80% of identity) to E2 enzymes of *Saccharomyces cerevisiae, Arabidopsis thaliana, Caenorabditis elegans,* rat and humans (Treier et al., 1992; Girod et al., 1993; Zhen et al., 1993; Sheffner et al., 1994; Jensen et al., 1995; Rolfe et al., 1995; Wing and Jain, 1995). E2 enzymes, together with E1 and E3 enzymes, are responsible for protein targeting by ubiquitination. It has been suggested that E2 enzymes, often in cooperation with E3 enzymes, mediate substrate specificity of ubiquitination. Ubiquitinated proteins are then degraded by a 700 kD protein complex, called the proteasome, which digests the ubiquitin conjugated protein releasing the ubiquitin molecules (for reviews see Jentsch, 1992; Hochstrasser, 1995).

The most straightforward explanation for the presence of TAs in *UbcD1* mutants is that these aberrant telomere-telomere connections result from the failure to degrade *via* the ubiquitin pathway one or more telomere-associated proteins. We do not know the nature of these proteins. However, we propose that during interphase of wild-type *Drosophila* cells telomeres are normally associated, either directly or indirectly through UbcD1 targets. Failure to degrade these proteins by the UbcD1 action would result in the persistence of telomere-telomere associations during mitotic cell division and male meiosis.

D. melanogaster telomeres do not consist of tandemly repeated short sequences like most eukaryotes but contain multiple copies of telomere-specific retrotransposons. However, it has been recently found that the SIR4 protein of yeast telomeres binds the deubiquitinating enzyme UBP3 (Moazed and Johnson, 1996), suggesting that telomeric protein ubiquitination is a general phenomenon. Thus, notwithstanding the uniqueness of *Drosophila* telomere structure, similar functions required for dispersion of chromosome ends prior to metaphase, may occur in the cells of many kinds of organisms.

Drosophila TELOMERES AND CELL CYCLE CHECKPOINT

In most eukaryotic systems telomere loss caused by chromosome breakage activates a checkpoint pathway that results in a prolonged cell-cycle arrest (for reviews see Harwell and Kastan, 1994; Zakian, 1996). These findings were anticipated by the classical studies of Muller (1941) who showed that terminally deleted chromosomes cannot be recovered in the progeny of wild type *Drosophila* males treated with X-rays. However, more recent studies have shown that terminally deleted *Drosophila* chromosomes can be obtained by irradiating females homozygous for the *mu-2* mutation (Mason et al., 1984). These chromosomes, once generated, can be transmitted for many generations in flies that are wild-type for *mu-2*. They do not appear to interfere with the normal progression of the cell cycle even though they keep losing 2–4 bp of terminal DNA at each round of DNA replication (for reviews see Biessman and Mason, 1992; Mason and Biessman, 1995; Pardue, 1995). One interpretation of these findings is that the *mu-2* gene is part of a checkpoint system that senses newly formed chromosome breaks. The broken ends of chromosomes that manage to escape the putative *mu-2* checkpoint may become masked in some unknown manner and fail to activate the cellular pathway that leads to cell cycle arrest.

In the course of our studies on *Drosophila* mutations that cause high frequencies of spontaneuos chromosome breakage (reviewed by Gatti and Goldberg, 1991), we identified a possible checkpoint which eliminates cells with broken chromosomes from larval brains. Examination of larval brains with high levels of chromosome breakage caused either by mutations or by X-rays, revealed that metaphases containing broken chromosomes progress regularly through the next stages of mitosis. However the two daughter cells are unable to complete another cell cycle and fail to progress to the next cell division (Gatti, Cenci and Baker, unpublished results).

The cytological analysis of larval brains carrying the dominant mutation nod^{DTW} (Zhang et al., 1990) suggested that the cell cell cycle arrest caused by chromosome breakage is not related to gene dosage problems, but it is a consequence of an active process that prevents division of cells with broken chromosomes. nod^{DTW} is a temperature sensitive mutation that produces a high frequency of anaphase bridges within a cell cycle after a shift to restrictive temperature. These bridges break during anaphase but the daughter cells that receive broken chromosomes cycle normally and enter mitosis showing the expected pattern of segmental aneuploidy (i.e. cells with broken chromosomes without the corresponding fragment, or cells with a normal chromosome complement plus an extra fragment.) However the daughters produced by these segmentally aneuploid cells are not able to proceed through the cycle and enter the next mitotic division (Gatti and Baker, unpublished results).

Together these observations suggest that in wild-type larval brain cells there is a checkpoint that monitors chromosome integrity at or prior to metaphase. If cells with broken chromosomes are detected, they divide regularly but their daughter cells are prevented to progress to the next cell division. Chromosome breakage that occurs after metaphase (i.e. anaphase) is not sensed until next cell division when it induces a physiological response that leads to the elimination of the damaged cells (Gatti, Cenci and Baker, unpublished results).

REFERENCES

Avivi, L. and Feldman M. 1980. Arrangement of chromosomes in the interphase nucleus of plants. *Hum. Genet.* 55: 281–295.

Benn, P. A. 1976. Specific chromosome aberrations in senescent fibroblast cell lines derived from human embryos. *Am. J. Hum. Genet.* 28: 465–473.

Biessman, H., and Mason J. M. (1992). Genetics and molecular biology of telomeres. *Adv. Genet.* 30: 185–249.

Billia, F., and de Boni U. 1991. Localization of centromeric satellite and telomeric DNA sequences in dorsal root ganglion neurons, *in vitro. J. Cell Sci.* 100: 219–226.

Cenci, G., Bonaccorsi S., Pisano C., Verní F., and Gatti M. 1994. Chromatin and microtubule organization during premeiotic, meiotic and early postmeiotic stages of *Drosophila melanogaster* spermatogenesis. *J. Cell. Sci.* 107: 3521–3534.

Cenci, G., Rawson, R. B., Belloni, G., Castrillon, D.H., Tudor, M., Petrucci R., Goldberg, M.L., Wasserman, S.W. and Gatti M. 1997. UbcD1, a *Drosophila* ubiquitin conjugating enzyme required for proper telomere behavior. *Genes Dev.* 11: 863–875.

de Lange, T. 1995. Telomere dynamics and genome instability in human cancer. In Telomeres (eds E. H. Blackburn and C. W. Greider), pp. 265–293. Cold Spring Harbor Laboratory Press. Cold Spring Harbor, New York, NY.

Dernburg, A. F., Sedat J. W., Cande W. Z., and Bass H. W. 1995. Cytology of telomeres. (eds E. H. Blackburn and C. W. Greider), pp. 295–338.. Cold Spring Harbor Laboratory Press. Cold Spring Harbor, New York, NY.

Dutrillaux , B., Croquett M. F., Viegas-Pequignot E., Aurias A., Coget J., Couturier J., Lejeune J. 1978. Human somatic chromosome chains and rings. *Cytogenet Cell Genet.* 20: 70–77.

Gatti, M. and M. L. Goldberg. 1991. Mutations affecting cell division in *Drosophila. Meth. Cell Biol.* 35: 543–586.

Girod, P. A., Carpenter T. B., van Nocker S., Sullivan M. L., and Vierstra R. D. 1993. Homologs of the essential ubiquitin conjugating enzymes UBC1, 4, and 5 in yeast are encoded by a multigene family in *Arabidopsis thaliana. Plant J.* 3: 545–552.

Hartwell, L. H and Kastan M.B. 1994. Cell cycle control and cancer. *Science* 266:1821–18281.

Hastie, N. D. and Allshire R. C. 1989. Human telomeres: Fusions and interstitial sites. *Trends Genet.* 5: 326–331.

Hayashi, K. and Schmid W. 1975. Tandem duplication q14 and dicentric formation by end-to-end chromosome fusions in ataxia telengiectasia (AT). *Humangenetik* 30: 135–141.

Hochstrasser, M. 1995. Ubiquitin, proteasomes, and the regulation of intracellular protein degradation. *Curr. Opin. Cell Biol.* 7: 215–223.

Hughes-Schrader, S. 1957. Differential polyteny and polyploidy in diaspene coccids (*Homoptera: Coccoidea*). *Chromosoma* 8: 709–718.

Jensen, J. P., Bates P. W., Yang M., Vierstra R. D., and Weissman A. M. 1995. Identification of a family of closely related human ubiquitin conjugating enzymes. *J. Biol. Chem.* 51: 30408–30414.

Jentsch, S. 1992. The ubiquitin-conjugation system. *Ann. Rev. Genet.* 26: 179–207.

Levis, R., Hazelrigg, T. and Rubin G.M. 1985. Effects of genomic position on the expression of transduced copies of the *white* gene of *Drosophila. Science* 219: 558–561.

Manuelidis, L. and Borden, J. 1988. Reproducible compartimentalization of individual chromosome domain in human CNS cells revealed by in situ hybridization and three dimensional reconstruction. *Chromosoma* 96: 397–417.

Mason, J. M. and Biessman H. 1995. The unusual telomeres of *Drosophila. Trends Genet.* 11: 58–62.

Moazed, D. and Johnson A. D. 1996. A deubiquitinating enzyme interacts with SIR4 and regulates silencing in *S. cerevisiae. Cell* 86: 667–677.

Muller, H.J. 1941. Induced mutations in *Drosophila. Cold Spring Harbor Symp. Quant. Biol.* 9: 151–167.

Pardue, M.-L. 1995. *Drosophila* telomeres: another way to end it all. In *Telomeres* (eds E. H. Blackburn and C. W. Greider), pp. 339–370. Cold Spring Harbor Laboratory Press. Cold Spring Harbor, New York, NY.

Rabl, C. 1885. Über Zelltheilung. *Morpholog. Jahrbuch* 10: 214–330.

Rolfe, M., Beer-Romero P., Glass S., Eckstein J., Berdo I., Theodoras A., Pagano M. and Draetta G. 1995. Reconstitution of p53- ubiquitinylation reactions from purified components: The role of human ubiquitin-conjugating enzyme UBC4 and E6-associated protein (E6AP). *Proc. Nat. Acad. Sci.* 92: 3264–3268.

Saltman, D., Morgan, J.L. Cleary, and de Lange T. 1993. Telomeric structure in cells with chromosome end associations. *Chromosoma* 102: 121–128.

Savage, J. R. K. 1970. Sites of radiation induced chromosome exchange. In *Current Topics In Radiation Research,* (ed M. Ebert and A. Howard), 6: 129–194. North-Holland Publishing Company, Amsterdam, London.

Sheffner, M., Huibregtse J. M., and Howley P. M. 1994. Identification of a human ubiquitin-conjugating enzyme that mediates the E6-AP-dependent ubiquitination of p53. *Proc. Natl. Acad. Sci.* 91: 8797–8801.

Shore, D. 1995. Telomere position effects and transcriptional silencing in the yeast *Saccharomyces cerevisiae*. In *Telomeres* (eds E. H. Blackburn and C. W. Greider), pp. 139–191. Cold Spring Harbor Laboratory Press. Cold Spring Harbor, New York, NY.

Stack, S. M. and Clarke C. R. 1973. Differential Giemsa staining of the telomeres of *Allium cepa* chromosomes: observations related to chromosome pairing. *Can. J. Genet. Cytol.* 15: 619–624.

Taylor, A. M. R., Oxford J. M., and Metcalfe J. A. 1981. Spontaneous cytogenetic abnormalities in lymphocytes from thirteen patients with ataxia telagenctasia. *Int. J. Cancer.* 27: 311–319.

Treier, M., Seufert W., and Jentsch S. 1992. *Drosophila UbcD1* encodes a highly conserved ubiquitin-conjugating enzyme involved in selective protein degradation. *EMBO J.* 11: 367–372.

Vourc'h, C. D. Taruscio, Boyle A. L., and Ward D. C. 1993. Cell cycle-dependent distribution of telomeres, centrosomes, and chromosome-specific subsatellite domains in the interphase. *Exp. Cell Res.* 205: 142–151.

Wagenaar, E.B. 1969. End-to-end chromosome attachments in mitotic interphase and their possible significance to meiotic chromosome pairing. *Chromosoma* 26: 410–426.

Wing, S. S. and Jain P. 1995. Molecular cloning, expression and characterization of a ubiquitin conjugation enzyme (E2$_{17kb}$) highly expressed in rat testis. *Biochem. J.* 305: 125–132.

Zakian, V.A. 1989. Structure and function of telomeres. *Annu. Rev. Genet.* 23: 579–604.

Zakian, V.A. 1996. Telomere functions: lessons from yeast. *Trends Cell Biol.* 6: 29–33.

Zhang, P., Knowles, B.A., Goldstein, L.S.B., and Hawley, R.S. 1990. A kinesin-like protein required for distribuitive segregation in *Drosophila. Cell* 62: 1053–1062.

Zhen, M., Heinlein R. , Jones D., Jentsch S. , and Candido E. P. M. 1993. The *ubc-2* gene of *Caenorabditis elegans* encodes a ubiquitin-conjugating enzyme involved in selective protein degradation. *Mol. Cell. Biol.* 13: 1371–1377.

DISCUSSION

Shay: You claim that there is no apoptosis going on. Have you examined interesting molecules like Reaper that have been cloned in *Drosophila* that seem to anticipate cell death? It seems like that would be an interesting molecule to look at to see if the same pathway that works in programmed cell death in *Drosophila* is also applicable in your system.

Gatti: You mean *in vitro*?

Shay: Yes.

Gatti: We did not see any striking evidence of apoptosis in larval brain cells. However, I cannot exclude that cells with broken chromosomes are eliminated by apoptosis. I would like to add that we do not have cell cultures of *UbcD1* cells and that all the experiments I presented have been performed on larval brain cells.

Wahl: If you have telomere associations, are you proposing that in some fashion, in the absence of this protein, that telomeres associate covalently or non-covalently?

Gatti: I believe that telomeres are attached through protein dimers or multimers that bind telomeric DNA. I do not think that they are covalently attached through DNA fibers. This type of attachments would result in DNA breakage at anaphase and would ultimately cause extensive cell death.

Wahl: I am just curious as to how much energy would have to be involved in that protein-DNA interaction to hold together the duplex? Otherwise the association would be fragile and prone to dissociation.

Gatti: I do not know how much energy is involved in the bond. Perhaps what is really important is the speed of anaphase chromosome movement. If this movement is very rapid, as presumably occurs during male meiosis, there might be no sufficient time for resolution of telomere-telomere attachments.

Wahl: And when you get these anaphase bridges do the cells die?

Gatti: No, they do not die because, at least during mitotic division, most bridges are resolved during anaphase.

Nasmyth: You do not particularly see these bridges between sister and non-sister chromatids?

Gatti: We see single chromatid attachments between sisters and non-sisters. Attachments between sisters are more difficult to score given the small size of Drosophila chromosomes. Double telomere associations are easier to score and involve both homologous and heterologous telomeres.

Nasmyth: Are telomeres held together in one place in G2?

Gatti: I do not know. As I said, we only have some evidence that the chromosomes of brain cells maintain a Rabl orientation.

Nasmyth: Yes, but I am trying to get to another question: If you have got a protein holding telomeres together and that needs to be degraded at anaphase, then presumably at some point all the telomeres are held together and then a mechanism is required to dissolve that complex, and your mutant is failing to do that. But the implication of that hypothesis is that there is a structure to be found in wild type cells where all the telomeres are held together in say, G2 or metaphase. What is known about that?

Gatti: This is a very attractive possibility but in situ hybridization experiments performed by Abbie Dernburg and coworkers indicate that telomeres are not tightly clustered during neuroblast interphase. To explain the formation of telomeric associations in *UbcD1* mutants one can envisage that telomeres have a dynamic behavior within a limited area of the interphase nucleus. This would allow telomere-telomere interactions when the UbcD1 protein targets are not properly degraded.

Hartwell: I still want to get back to what Kim Nasmyth was trying to get at, namely, the hypothesis would be that telomeres just have not completed replicating and so you would see then sister-sister association, but you are seeing associations between non-homologous moieties.

Gatti: Yes. Telomeric associations involve both homologous and heterologous telomeres.

Giulotto: I simply wanted to ask you how were these mutants isolated?

Gatti: We performed a large screening for mutants affecting various aspects of *Drosophila* mitosis. We examined larval brain squashes from single P-element induced lethals that die at the larval-pupal transition. The *UbcD1* mutants were isolated on the basis of their cytological phenotype in larval brain cells.

Chapman: I was wondering if it is feasible to look at telomere length and the Ubc D1 mutant to assess if there is a defect in telomere maintenance?

Gatti: Yes, I think it is feasible.

Chapman: Have you done so?

Gatti: No we have not done it.

Stark: I think there is no evidence for a p53 homologue or analogue in *Drosophila*—is that correct? Does anybody know of one? So, if the protein, or homologous protein does not exist I wonder what information there is on homologous function. Do *Drosophila* behave in a way analogous to mammalian cells when, for example, you damage DNA? Do they undergo cell-cycle arrest?

Gatti: You are right. To the best of my knowledge, there is not a p53 homologue in *Drosophila*. However, the fact that *Drosophila* cells undergo both cell cycle arrest and apoptosis, suggests that they have the ability to perform functions analogous to those of p53.

Wahl: I would just like to pursue this idea of the sticky protein a little bit more. Did you bring up the possibility that perhaps this is an anti-recombination protein? That is, the protein associated at the end of the telomere might prevent recombination from occurring at chromosome ends. This may be something that is allowed perhaps during chromosome replication—those ends remain open and available for some period of time before the cycle. The protein must be there to prevent association, and then as chromosome segregation would be affected if this protein is not processed correctly. Perhaps that would enable recombination to occur between ends generating the dicentrics you have described.

Gatti: I think that this is an unlikely possibility. First, because dicentric chromosomes generated by asymmetrical exchanges between telomeres are not expected to be resolved during anaphase. Second, because in *D.melanogaster* there is no evidence of recombination between telomeres.

Hoeijmakers: Concerning the p53 and DNA damage arrest problem, it should be noted that there are also means of DNA damage arrest in the cell which are independent of p53. So even though there may not be a p53 homologue in *Drosophila,* DNA damage arrest may occur in *Drosophila* and many other organisms. Now my question concerns the expression pattern of the Ubc-5D1 gene. Since you find the effects in mitosis and in meiosis, I would like to know whether there is a cell-cycle dependent expression?

Gatti: We have examined the expression pattern of *UbcD1* during *Drosophila* development but we have not analyzed the expression of this gene during the cell-cycle. It would be also very interesting to determine the localization of the UbcD1 protein during the cell cycle. However, we have not yet generated an antibody against this protein.

Hoeijmakers: I see. This brings me to a second question: In mammals, I understand there are more than one homologue of this gene. Is in *Drosophila* also more than one homologue of this gene found?

Gatti: In *Drosophila* there is another E2 enzyme encoded by the *bendless* gene. However, this gene is not highly homologous to UbcD1.

Evan: There is something that I can say that might add to the argument about how *Drosophila* transfuses DNA damage into death, and that is that it has been shown pretty clearly that repair is required more or less for DNA damage-induced death, and that is certainly radiation and drug-induced death. In a Reaper-minus mutant, you can give a thousand times the level of radiation to a *Drosophila* embryo and still not get back up to the normal amount of cell death. Now Reaper is a death domain protein of the TNF receptor class and the implication could well be that DNA damage is actually transduced through a FAS TNF receptor signaling pathway, which we know now, directly recruits the ICE proteosis to trigger cell death. So though it is not known if there is a p53 homologue in *Drosophila* I think the pathway beyond p53 is actually becoming evident.

Gatti: I have no reason to believe the *Drosophila* apoptotic pathway is substantially different from that described in mammals. With respect to *reaper*, it would be very interesting to determine whether this gene is required for the elimination of cells with broken chromosomes.

Hartwell: To come back to this issue of checkpoints, I think from similar data to what you show, you and Bruce argued years ago that there was a G2 to M checkpoint that cells with double strand breaks did not get through, and so my question is could you confirm whether that is correct or not? And then there is the second issue: whether there was a G1 to S checkpoint and that could be examined quite easily in tissue culture cells. The only thing I know is that *Drosophila* does have an MDM2 protein which should act at that stage and I believe p21 has also been found in *Drosophila*.

Gatti: The data I presented about the chromosome breakage checkpoint are essentially the same that Bruce Baker and I obtained some years ago and never published. I only added the results on the frequency of breakage during anaphase, which have been recently obtained by Giovanni Cenci in my laboratory. As to your second question, I should confess that I do not know whether there are data about a G1 to S checkpoint in *Drosophila* cells.

Hartwell: But they do not distinguish G1 from G2, do they?

Gatti: Chromosome breaks produced either during G1 or S-G2, when present at metaphase, cause a cell cycle arrest in the daughter cells which never progress to the next cell division. However, we do not know whether these daughter cells arrest in G1 or in subsequent phases of the cell cycle.

Nasmyth: Going back to what Geoff Wahl was talking about a possibility of recombination being involved, it is not clear from Liz Blackburn's work on *L. lactis* that there is a very strong telomerase independent pathway for generating telomeres which is RAD52 dependent and could be quite a specialized telomeric recombination system. And I am wondering whether you have got a protein in the fly which is shutting that system off, that is telomeres associated and that protein is destroyed at anaphase, because you want to shut off that system when you undergo anaphase and the last thing you want to happen is for you to engage in that recombination; in your mutants the protein is not destroyed at anaphase along the whole of other proteins and then this recombination gets triggered into action and that is what is causing this association. Is there anything that would be consistent with your data?

Gatti: As I said earlier, there is not clear evidence for telomeric recombination in Drosophila. The healing of broken chromosome ends does not appear to occur through recombination but rather through the addition of new HeT-A transposable elements. Moreover, if you induce terminal deletions in mu-2 homozygous females, these chromosomes can remain without telomeres for many generations. This suggests that recombination phenomena involving *Drosophila* telomeres are extremely rare.

Wahl: As George Stark's group has shown, and several other groups as well, one of the consequences of end to end fusions and formation of dicentrics can be amplification, increasing copy number over several cycles. If these descendent cells are viable for several generations, then you should be able to show the formation of large inverted repeats resulting from the recombination subsequent breakage and perhaps another McClintock cycle after that.

Gatti: In which way?

Wahl: Cytogenetically, it should be easy using FISH, for example.

Gatti: I am not sure I have understood your question. Most telomere-telomere attachments are resolved during anaphase and they give rise to very few cells containing chromosome breaks.

Wahl: Perhaps there is a very efficient mechanism for removing the cells that have undergone that process. But there is also a frequency with which cells with these aberrations can "adapt" to go through one more cycles. If you have the system to look for such cells, you should be able to detect the expected cytogenetic abnormalities within a small clone.

Gatti: Unfortunately, you cannot identify cell lineages in squashed preparations of larval brains. Moreover, in Drosophila brain cells broken chromosomes do not undergo a breakage-fusion-bridge cycle. Therefore, I do not expect to detect the DNA amplification phenomena you mentioned earlier.

McEachern: One of the things that seems so unusual about *Drosophila* telomeres is that there is no specific sequence at the very end as opposed to the situation in other eukaryotic cells. It brings up the question of how these ends are kept in a way to distinguish them from DNA damage. And one hypothesis that I have wondered about is whether or not Drosophila telomeres might be capped in a temporal sense, that at a specific time during a cell cycle you might have a protein complex that assembles on the end, on any sort of DNA end, that might be present in cells and that any break that occurs at any other time is then recognized as DNA damage. What might you think of that sort of hypothesis?

Gatti: Actually, *Drosophila* telomeres are mainly composed of HeT-A and TART retrotransposons. These transposable elements may bind telomere-specific proteins. However, I cannot exclude that the same proteins would also bind terminally deleted chromosomes devoid of telomeric DNA. In addition, it is quite possible that these telomeric proteins assemble on telomeres only at specific times of the cell cycle.

Shay: What is the phenotype of mutants and other organisms besides Drosophila homologues? For example, what happens if you knock out the equivalent of Ubc D1 in mice? Do you see the same thing? Also, have you ever taken Ubc D1 and overexpressed it in Drosophila cells? If so, did you see any effects?

Gatti: As far as I know, there are not many mutations in the E2 enzymes of higher eukaryotes, but several E2 mutants have been isolated in yeast. The yeast E2 enzymes with the highest degree of homology with UbcD1 are UBC1, UBC4 and UBC5. Deletion of the gene for one of these enzymes does not cause a clear mutant phenotype; deletion of two of the three renders the cells particularly sensitive to stress; deletion of all three genes is lethal. I do not know whether the mutations in these genes have effects on telomere behavior.

Hartwell: The Mu2 mutant is very interesting because it is transmitting broken ends through mitosis. And so if you look at that mutant with respect to X-ray breaks or other things, do you see transmitting breaks through mitosis?

Gatti: The experiment you are suggesting is very interesting and we will certainly do it.

2

A MOLECULAR CYTOGENETIC VIEW OF CHROMOSOMAL HETEROGENEITY IN SOLID TUMORS

Joe W. Gray, Koei Chin, and Fredric Waldman

Cancer Genetics Program
UCSF Cancer Center
University of California
San Francisco, California 94143-0808

1. INTRODUCTION

The concept that cancers progress, in large part, through accumulation of multiple genetic abnormalities is generally accepted.[1,2,3] The order of progression is best established for colorectal cancer.[3] The development of genetic instability is thought to be a key aspect of tumor progression[4,5] enabling accumulation of the multiple abnormalities required for a complete metastatic phenotype. Several genetic abnormalities that lead to instability have been identified. These involve genes that code for proteins involved in detection and repair of genetic damage,[5] maintenance of telomere length[6] and apoptosis.[7,8] However, the magnitude and type of damage that is accumulated as a result of these defects is only now beginning to be appreciated. The application of molecular cytogenetic techniques such as fluorescence in situ hybridization (FISH[9,10]) and comparative genomic hybridization (CGH[11]) are particularly powerful for the study of damage that results in changes in the structure or number of copies of genomic DNA. FISH allows visualization of specific segments of DNA in single cells and therefore provides information about the variation among cells in the number of copies and the organization of these target sequences. CGH, on the other hand provides a genome-wide overview of changes in relative DNA sequence copy number that occur in the majority of cells of a tumor. Importantly, CGH maps changes onto a normal representation of the genome so that they are easily interpreted. These techniques are particularly useful for studies of DNA sequence copy number changes in spontaneous solid tumors since they do not require cell culture and can be applied to archived tumor material. However, CGH and FISH are not sensitive to genetic events that involve only a small segment of the tumor genome. In addition, CGH is completely insensitive to balanced translocations. Thus, molecular cytogenetic analyses complement other molecular techniques for identification of recurrent aberrations such as analysis of loss of heterozygosity(LOH), differential display and DNA sequence analysis.

Genomic Instability and Immortality in Cancer
edited by Mihich and Hartwell, Plenum Press, New York, 1997

In this paper, we review several recent findings in spontaneous solid tumors made using molecular cytogenetic techniques. Specifically, we describe: (a) the substantial genetic variability that can occur between cells within tumors revealed using FISH, (b) locations of recurrent aberrations in breast and other solid tumors originating in several different organ sites revealed using CGH, (c) genetic events that influence the number and type of aberrations and (d) approaches to identification of novel genes located in regions of recurrent abnormality that may contribute to tumorigenesis or phenotype.

2. DESCRIPTORS OF HETEROGENEITY

2.1. FISH for Assessment of Heterogeneity

FISH is now widely used for assessment of DNA sequence copy number and structural integrity in individual interphase nuclei and metaphase chromosomes.[9,10] With FISH, regions of the genome targeted by a nucleic acid probe are made visible in individual nuclei. The number of signals generated using FISH with appropriately large probes is a quantitative measure of the number of copies of the target DNA sequence copy[11] and can be applied to interphase nuclei from touch preparations,[12] nuclei isolated from paraffin sections[13] and nuclei in intact tissue sections.[14] Furthermore, multi-color FISH allows simultaneous analysis of several different regions of the genome.[15,16] The region of the genome examined is determined by the probe used. Common targets include whole chromosomes, chromosome centromeres and specific DNA sequences. In general, regions larger than ~50 kb must be targeted in order to allow accurate signal enumeration analysis in clinical samples. The number of informative probes is increasing continually as a result of progress of the human genome program. A collection of available probes maintained by the authors can be accessed at http://rmc-www.lbl.gov.

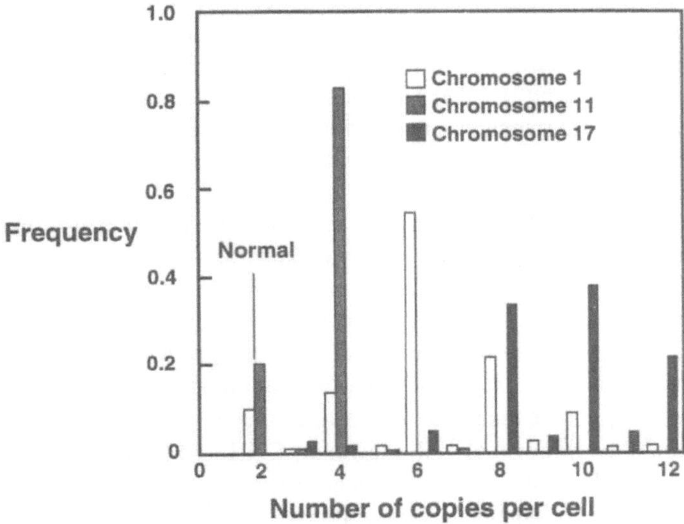

Figure 1. Frequency distribution of the number of signals measured for primary breast tumors after staining using FISH with probes for the centromeres for chromosomes 1, 11, and 17. This tumor shows substantial cell to cell heterogeneity in number of hybridization signals. However, the degree of heterogeneity varies among tumors.

The ability to make specific DNA sequences visible in interphase nuclei makes FISH particularly useful for analysis of DNA sequence copy number heterogeneity in solid tumors. The degree of heterogeneity, measured as variation in the number of hybridization signals per nucleus, varies among tumors and can be considerable in some cases.[12] Figure 1, for example, shows the variation in the number of hybridization signals measured in a primary human breast cancer after FISH with probes to the centromeres of chromosomes 1, 11 and 17. This figure shows results from a tumor with a high variation in copy number. However, other tumors show much less variation. Similar cell-to-cell variability in DNA sequence copy number also occurs in tumor cell lines grown as xenografts in mice. Figure 2, for example, shows DNA sequence copy number distributions measured for five MCF-7 human tumor xenografts after FISH with probes for the centromere of chromosome 1, CMYC and 20q13.2. These distributions were selected from tumors growing at significantly different rates to maximize possible differences in copy number distribution between tumors. The distributions show substantial variability within each tumor but the means of the distributions are nearly the same. Thus, the tumors do not seem to change mean DNA sequence copy number as rapidly as might be expected from the level of heterogeneity measured using FISH.

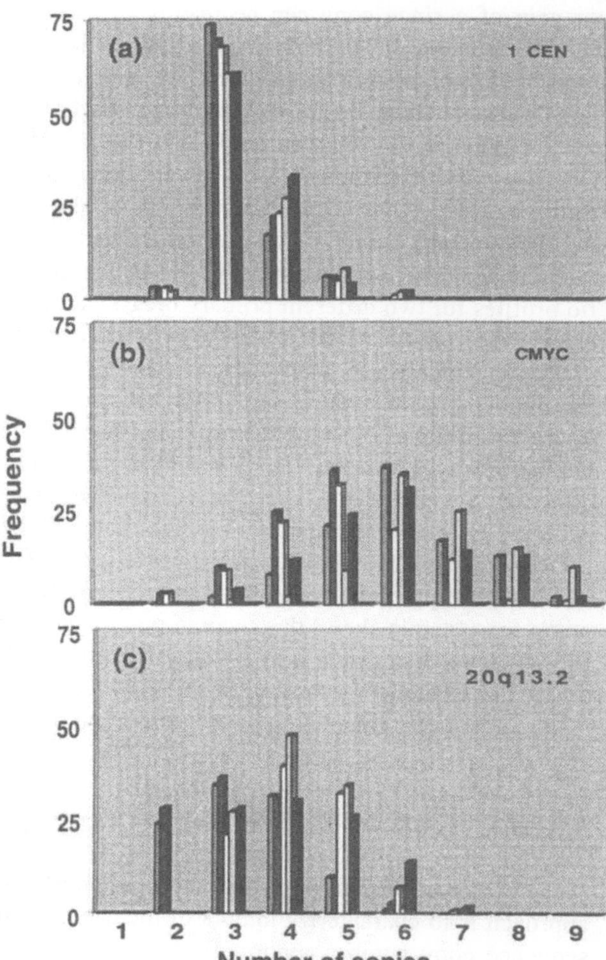

Figure 2. Frequency distributions of the number of signals measured in 5 different MCF-7 tumor xenografts using chromosome specific probes. Panel a. Signals produced using a probe for the centromere for chromosome 1. Panel b. Signals produced using a probe for C-MYC. Panel c. Signals produced using a probe for 20q13.2.

2.2. CGH for Assessment of Average Gene Copy Number

CGH[11] is an in situ hybridization technique in which changes in relative DNA sequence copy number are mapped onto normal metaphase chromosomes. This is accomplished, as illustrated in Figure 3a, by hybridizing differentially labeled DNA from tumor and normal tissues to normal metaphase chromosomes so that the ratio of bound tumor DNA to normal DNA along the normal chromosomes gives an indication of relative copy DNA sequence number in the tumor. Typically, tumor DNA is labeled so it fluoresces green and normal DNA so it fluoresces red. In this case, regions of increased relative DNA content (e.g. regions of trisomy or gene amplification) in the tumor are relatively green and regions of reduced tumor DNA (e.g. regions of monosomy) are relatively red. These regions appear as increases and decreases in green:red fluorescence intensity ratio profiles measured for each chromosomes using digital imaging microscopy. CGH is sufficiently robust that it can be applied to fresh or archived tumor material[17] and only a few hundred to a few thousand cells are needed for analysis.[18] The power of CGH is that it maps changes in relative DNA sequence copy number onto normal chromosomes so the normal genomic locations of DNA sequences that occur at abnormal copy number in tumors can be readily determined. Changes in relative DNA sequence copy number caused by the loss or gain of a singe copy can be readily detected if the involved region is greater than ~10Mb in extent.[19] Much smaller regions of high level amplification can be detected as well.[11] The major weaknesses of CGH are its total insensitively to structural changes and/or abnormalities that do not result in changes in relative DNA sequence copy number.

Figure 3b shows green:red CGH ratio profiles measured for a primary ovarian cancer. The extreme genomic rearrangement that can occur in some tumors is clearly evident in this analysis. Over 20 different regions of abnormality involving almost half of the tumor genome are visible. Of course, the number and genomic location of abnormalities can vary considerably among tumors of the same type. Figure 4, for example, shows CGH ratio profiles for two different primary breast cancers. The two tumors are dramatically different in spectrum of abnormality, even though they are clinically similar. So far, the extent to which these abnormalities translate into altered tumor behavior has not been thoroughly explored. However, differences in relative DNA sequence copy number are likely to influence tumor phenotype and possibly response to therapy if the gene copy number changes translate into differential expression of genes. A few studies point in this direction. Several regions detected as amplified by CGH are associated with altered clinical behavior and/or therapeutic response. Some of these involve known oncogenes such as ERBB2 or CMYC while others involve unknown oncogenes.[11] In addition, the total number of copy number abnormalities is usually associated with clinical outcome. Figure 5, for example shows that ovarian tumors with many abnormalities tend to survive for shorter times than those with few abnormalities.[20] Similar trends have been observed for breast[21] and bladder[22] cancers. Thus, assessment of such abnormalities should be considered as a possible guide to tumor treatment and prediction of future behavior.

3. CHANGES WITH EVOLUTION

CGH has been used to study progression in several solid tumor systems. The general approach is to characterize tumors of increasing grade or stage in order to identify DNA sequence copy number abnormalities (CNAs) that occur at specific periods of development. Cancers analyzed for progression to date include those originating in the breast,

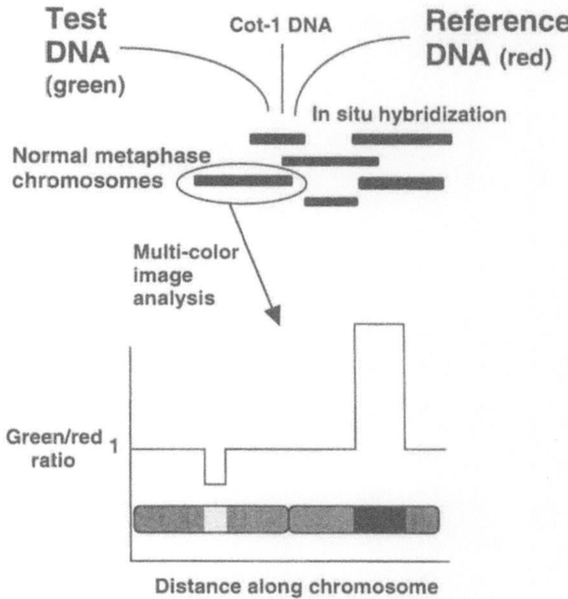

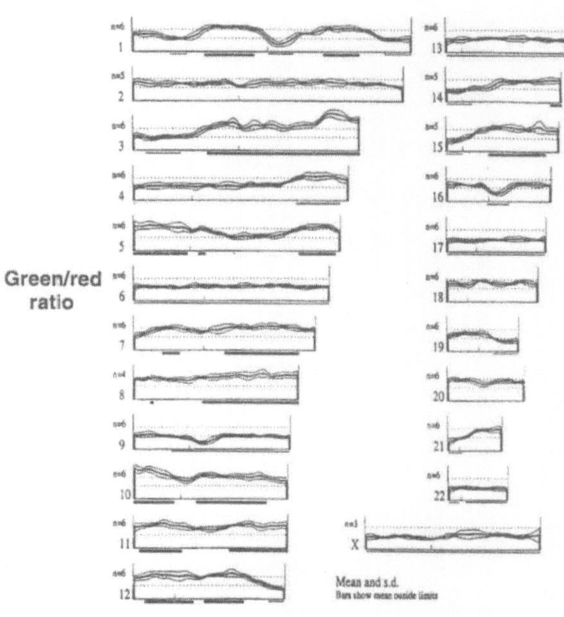

Figure 3. Panel a. Schematic illustration of the principles involved in CGH. Panel b. CGH ratio profiles measured for a primary ovarian tumor. Heavy lines under the profile indicate regions of significantly increased DNA sequence copy number. Light lines under each tumor indicate regions of significantly decreased DNA sequence copy number.

(a)

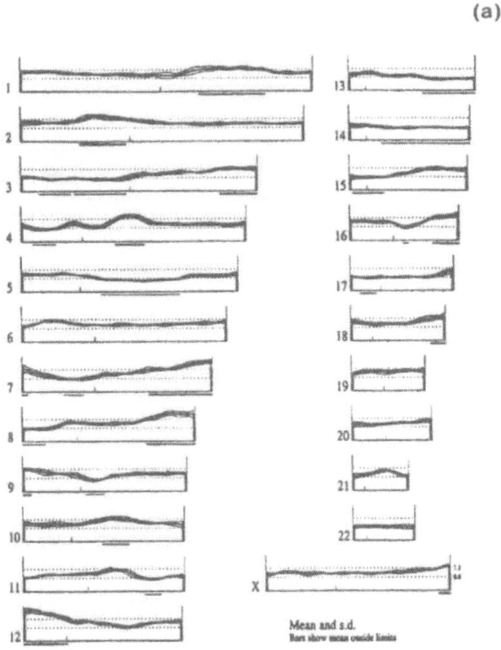

(b)

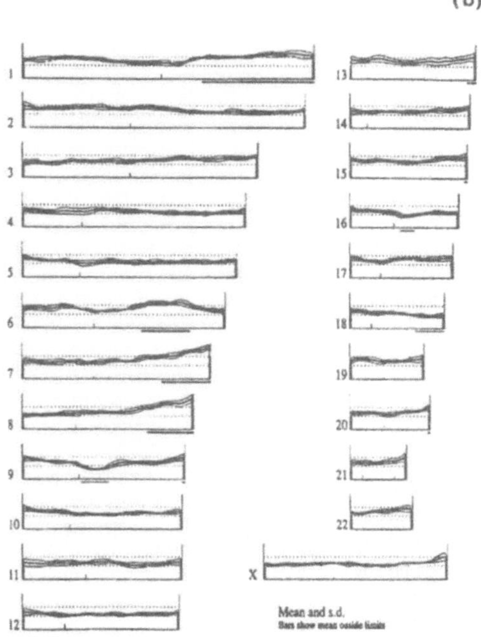

Figure 4. Panels a and b compare CGH ratio profiles measured for two different primary breast tumors.

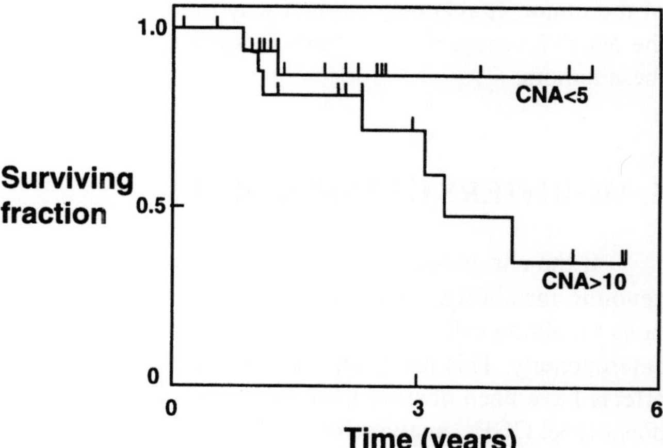

Figure 5. Kaplan-Mayer curve showing survival for ovarian cancer patients whose tumors had <5 or >10 CNAs per tumor. The survival duration of patients whose tumors had many CNAs is significantly shorter than that for patients whose tumors showed few CNAs.

ovary, colon and skin. In ovarian cancer, the number of abnormalities increases with increasing grade and increased copy number at 20q13 and 3q26 appear to be early events and increased copy number at 8q appears to be a late event.[20] Detailed analyses of progression have been carried out in breast cancer where CGH analysis have been carried out on material microdissected from ductal carcinoma in situ (DCIS), primary tumors and metastases (Fredric Waldman, private communication). There does not seem to be a strongly preferred order of occurrence of specific aberrations. Instead, the spectrum of aberrations appears to be similar for all of these stages of the disease although the number of abnormalities does seem to increase with stage. Many of the DCIS lesions show numerous genetic abnormalities suggesting that genetic instability is well established in this stage of many breast tumors. This is supported by analyses of LOH[23] and other genetic analyses[24] of DCIS. Early onset of instability also has been observed in colorectal tumors with CNAs detected in adenomous polyps as well as in invasive cancer.[25] As in breast cancer, the recurrent CNAs in the adenomas are similar to those in colon carcinomas.

The breast cancer studies were conducted pairwise so that the DCIS lesions were from the same patients as the primary tumors. Although some differences appear between DCIS and primary cancers, the majority appear in common suggesting that the genetic divergence between DCIS and invasive cancer usually is modest at the level of resolution possible with CGH. Likewise, the CNAs in primary and metastatic tumors from the same patient are quite similar, again suggesting modest overall chromosomal evolution between primary and metastatic tumors. Slow chromosomal evolution also was observed for MCF-7 xenografts. In fact, CGH analyses of several dozen tumor xenografts were identical in spite of the fact that the tumors grew at significantly different rates. This relative stability is somewhat paradoxical, given the high cell-to-cell variability observed using FISH in primary tumors and the MCF-7 xenografts. The reasons for this seemingly disparate behavior are unknown. However, we speculate that much of the cell-to-cell variability observed using FISH may be due to the continuous production of genetically variant cells that are at a proliferative disadvantage relative to the "stem" cells that populate the majority of cells in tumor. In addition, highly divergent cells do not seem to be responsible for metastases since the CGH karyotypes of metastases are generally similar to those for the primary tumor from which they were derived. Slow growth or accelerated loss of cells that are chromosomally divergent from the stem line may explain the overall slow growth rates observed for many tumors since the genetically divergent cells may comprise the majority of the tumor. It also may explain why the CGH copy number distributions measured for the MCF-7 xenografts are stable in spite of the high chromosomal heterogeneity within these tumors.

4. MODIFIERS OF INSTABILITY

If the chromosomal heterogeneity observed using CGH and FISH is, in fact, due to genomic instability, then genetic events known to affect genetic instability (e.g. aberrations involving cell cycle checkpoints, DNA repair or apoptosis) should affect the level of heterogeneity. This has been found to be the case in several systems. The most dramatic effects have been observed for colorectal tumors displaying the Replication Error (RER) phenotype. CGH[25] analyses of RER tumors and non-RER tumors both show a significant differences in the number of chromosomal abnormalities. RER tumors carry almost no chromosomal abnormalities while those for non-RER tumors show the wide range of CNAs commonly found in many solid tumors. Limited CGH studies of breast tumors show increased numbers of CNAs for tumors carrying p53 mutations compared to tumors with intact p53 function. This difference was also noted in wnt-1 murine mammary tumors arising in mice with varying p53 status.[26] Figure 6 shows that tumors arising in $p53^{+/+}$ mice displayed almost no CNAs while those arising in $p53^{+/-}$ mice carried more. Tumors that eventually lost the second wt p53 allele carried the highest number of abnormalities. The story is not completely clear, however, since tumors arising in $p53^{-/-}$ mice had fewer CNAs. This may be because the $p53^{-/-}$ tumors grew significantly faster than the other tumors so that fewer aberrations might have been accumulated prior to animal sacrifice. However, further study will be required to resolve this issue. In any case, it is becoming clear that the nature of the defects leading to genetic instability may strongly influence the type of genetic damage accumulated by the tumor which may, in turn, affect tumor phenotype and response to therapy.

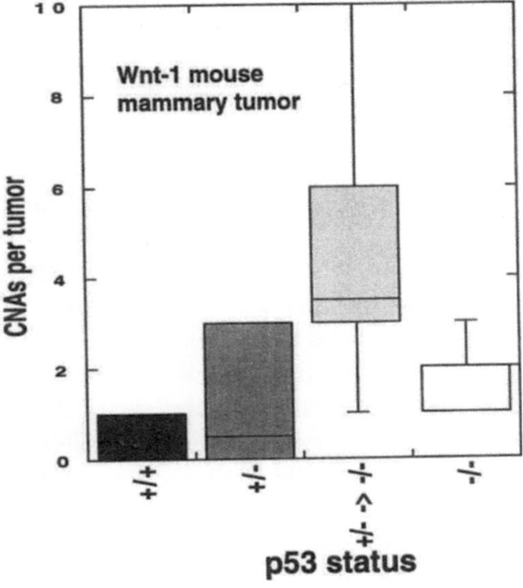

Figure 6. Plot of the number of CNAs measured for murine mammary tumors from wnt-1 transgenic mice crossed with $p53^{-/-}$ mice so that the tumors developed in mice that were homozygous p53(+/+), hemizygous (p53+/-), or nullizygous (p53-/-) for wild type p53. Some tumors in $p53^{+/-}$ mice subsequently lost the other wild type allele so that the tumors became $p53^{-/-}$.

5. CONSISTENT CHROMOSOMAL CHANGES SIGNAL THE LOCATION OF GENES THAT INFLUENCE TUMOR BEHAVIOR

The CGH analyses of the murine p53$^{+/-}$ tumors that eventually lost the wildtype allele of p53 showed recurrent reduced relative DNA sequence copy number on chromosome 11 (the location of p53 in the mouse genome). Presumably, this is because CGH shows the loss of the copy of chromosome 11 carrying the remaining wildtype allele. This suggests that other regions of recurrent CNA revealed by CGH may harbor genes that influence tumor behavior and contribute to tumor evolution. In the wnt-1 tumors, these included decreases in copy number for chromosomes 4, 8, 9, 10, 13, and X.[26] Increases in copy number were much less frequent than decreases in these tumors.

Recurrent CNAs also are found in human tumors. Figure 7, for example, shows regions of frequent CNA in cancers of the ovary.[20] Regions known to harbor genes thought to contribute to the progression of these tumors are detected as recurrent abnormalities in these tumors. These include increased copy number at 8q24 (CMYC), and 17q (ERBB2) and decreased copy number at 17p (p53). However, other regions are recurrently abnormal as well. These include increases in copy number at 3q and decreases in copy number at proximal 17q and 19. In breast cancer, recurrent changes include increases in copy number at 1q, 8q proximal to CMYC, and 20q and decreases in copy number at 11q, 16p, and 16q. These data suggest that some CNAs occur frequently in several different types of human tumors. For example, CNAs involving chromosomes 1, 8, 16 and 17p occurred frequently in both breast and ovarian cancers. A survey of published CGH data on 12 different human tumors confirms that some chromosomes are found to be abnormal much more frequently than others (Schoenberg-Fejzo; private communication). Chromosome regions found to be most frequently increased in copy number include 1q, 3q, 6p, 7, 8q, 13q, 17q and 20q. Those found to be most frequently decreased in copy number include 3p, 6q, 8p, 16 and 17p.

Regions of recurrent abnormality are important in human tumor genetics since they define regions that may harbor genes that contribute to tumorigenesis or tumor phenotype when differentially expressed. Genes in regions of increased copy number are likely to be oncogenes that act in dominant fashion to contribute to tumor phenotype when over expressed as a result of the copy number. However, it is also possible that they encode sequences that act as sinks (i.e. binding sites) for regulatory proteins that normally inhibit the expression of tumor suppressor genes. Alternately, they may encode genes or gene fragments that, when transcribed, act as anti-sense tumor suppressor gene inhibitors. Genes in regions of decreased relative copy number are likely to affect tumor phenotype by contributing to the inactivation of tumor suppressor genes. This notion is consistent with the strong correlation that is observed between loss of heterozygosity and reduced relative copy number,[27] presumably because LOH is usually caused by the physical loss of one copy of the chromosome region under study. However, it is also possible that regions of frequently increased or decreased copy number are simply regions that do not contain genes whose copy number must be tightly controlled to maintain cell viability. This issue can only be resolved through detailed genetic analyses of regions of common copy number abnormality.

6. CHARACTERIZATION AND CLONING OF REGION OF CONSISTENTLY INCREASED COPY NUMBER AT 20q13

We are now analyzing regions of increased copy number in breast in order to define them regions more precisely, identify genes that may contribute to the tumor phenotype

Figure 7. Summary of CGH abnormalities found in 44 primary ovarian cancers. Data are shown chromosome by chromosome. Lines to the left of each ideogram indicate the extent of DNA sequence copy number loss and lines to the right indicate the extent of regions of DNA sequence copy number gain.

when amplified and better understand the mechanisms of amplification. Our analysis of the region of increased copy number at 20q13 is most advanced since the extent of the region is manageable in size. This CNA appears important since it occurs frequently in several tumors including those originating in the breast, ovary, bladder, prostate, and colon. In breast cancers, amplification at 20q13 is associated with poor clinical outcome, increased S-phase and increased tumor grade.[28] This region of the genome also appears to be amplified in bladder epithelial cells transfected with an HPV 16 E7 construct.[29] Most work so far has involved genetic characterization and molecular cloning of the region at chromosome 20q13.2 that is frequently amplified in human breast cancers.

The region of amplification was defined using FISH with a collection of probes mapped along chromosome 20q.[30,31] FISH was used for amplicon mapping since the technique can be applied to the small numbers of cells that can be recovered from archived tumors and yields a quantitative measure of the number of copies of the region of the genome targeted by the probe. Cells were analyzed individually and normal cells admixed in the tumor were ignored during scoring. Figure 8 shows the result of applying this approach to analysis of DNA sequence copy number at several regions along 20q13.2 in the breast cancer cell line BT474Figure 10. FISH analysis of the variation in DNA sequence copy number along chromosome 20 in the breast cancer cell line SK-BR-3. The presence of a highly amplified region at ~20q13.2 (fractional location along chromosome 20 relative to pter = 0.84) is apparent. This same region is increased in copy number in ~10% of human primary breast cancers.[27] Interestingly, this region is frequently co-amplified two other regions on chromosome 20 at 20q11 and 20q13.1.[32] Amplification at 20q13.2 appears to be the most frequent amplification event on chromosome 20. All candidate genes previously mapped to 20q13.2 were excluded based on the fact that they do not map to the most highly amplified region of the amplicon. This suggests that as yet unknown gene(s) is selected by the amplification. Molecular cloning and genetic analysis of this region is now underway.

7. SUMMARY

Studies using FISH and CGH have revealed substantial intra- and inter-tumor heterogeneity in DNA sequence copy number in solid tumors. Regions of recurrent abnormality

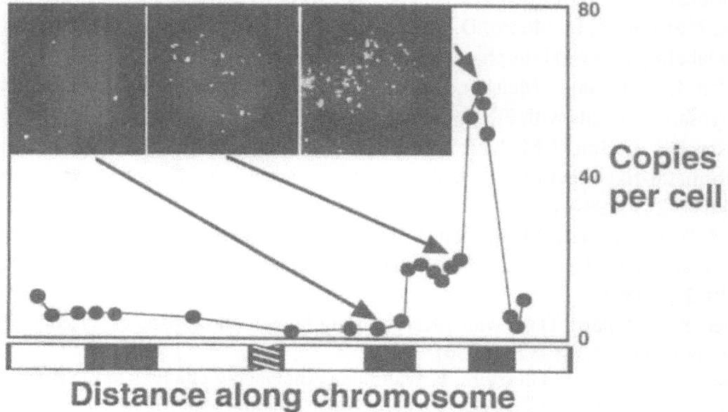

Figure 8. FISH analysis of the variation in DNA sequence copy number along chromosome 20 in the breast cancer cell line BT474. The inserts show hybridization patterns using probes from the regions of chromosome 20 indicated by arrows.

are presumed to harbor genes involved in tumorigenesis or that affect tumor phenotype. The spectrum of these aberrations varies substantially among tumors that appear clinically similar suggesting that the detailed tumor genotype will be an important determinant of tumor behavior and response to therapy. The spectrum of chromosomal aberrations in solid tumors is affected by the type of genetic instability that enable progression. Thus, the mechanism by which genetic instability occur also is important. All of these findings point to the importance of considering the tumor genotype during tumor treatment.

ACKNOWLEDGMENTS

This manuscript was prepared with support from NIH grant CA 58207 and Vysis, Inc.

REFERENCES

1. E.R. Fearon, B. Vogelstein. A genetic model for colorectal tumorigenesis. Cell 61:759–767 (1990)
2. E.R. Fearon. Molecular abnormalities in colon and rectal cancer. In: The Molecular Basis of Cancer (Eds. J. Mendelsohn, P. Howley, M. Israel, L. Liotta). W.B. Saunders and Co. Philadelphia, PA (1995)
3. H.E Varmus, L.A. Godley, S. Roy, I.C. Taylor, L. Yuschenkoff, Y.-P. Shi, D. Pinkel, J.W. Gray, R. Pyle, C.M. Aldaz, et al. Defining the steps in a multistep mouse model for mammary carcinogenesis. Cold Spring Harbor Symposia on Quantitative Biology, 59:491–499 (1994)
4. T.D. Tlsty, A. White, E. Livanos. M. Sage, H. Roelofs, A. Briot, B. Poulose, Genomic integrity and the genetics of cancer. Cold Spring Harbor Symposia on Quantitative Biology. 59:265–275 (1994)
5. R. Fishel, R.D. Kolodner. Identification of mismatch repair genes and their role in the development of cancer. Current Opinion in Genetics and Development, 5:382–395 (1995)
6. C.W. Greider,E.H. Blackburn. Telomeres, telomerase and cancer. Scientific American, 274:92–97 (1996)
7. C.J. Leonard, C.E. Canman, M.B. Kastan. The role of p53 in cell-cycle control and apoptosis: implications for cancer. Important Advances in Oncology,:33–42 (1995)
8. S.J. Martin, D.R. Green. Apoptosis and cancer: the failure of controls on cell death and cell survival. Critical Reviews in Oncology/Hematology, 18:137–153 (1995)
9. D. Pinkel, J. Landegent, C. Collins, J. Fuscoe, R. Segraves, J. Lucas, J.W. Gray. Fluorescence in situ hybridization with human chromosome-specific: detection of trisomy 21 and translocations of chromosome 4. Proceedings of the National Academy of Sciences of the United States of America, 85:9138–9142 (1988)
10. P. Lichter, T. Cremer, J. Borden, L. Manuelidis, D.C. Ward. Delineation of individual human chromosomes in metaphase and interphase by in situ suppression hybridization using recombinant DNA libraries. Human Genetics, 80:224–234 (188)
11. A. Kallioniemi,O.-P. Kallioniemi, D. Sudar, D. Rutovitz, J.W. Gray, F. Waldman, D. Pinkel. Comparative genomic hybridization for molecular cytogenetic analysis of tumors. Science, 258:818–821 (1992)
12. M. Balazs, K. Matsumura, D. Moore, D. Pinkel, J.W. Gray, F.M. Waldman. Karyotypic heterogeneity and its relation to labeling index in interphase tumor cells. Cytometry, 20:62–73 (1995)
13. W. Lee, K. Han, C.P. Harris, L. Meisner. Detection of aneuploidy and possible deletion in paraffin-embedded rhabdomyosarcoma cells with FISH. Cancer Genetics and Cytogenetics, 68:99–103 (1993)
14. C.T. Thompson,P.E. LeBoit, P.M. Nederlof, J.W. Gray. Thick section fluorescence in situ hybridization (FISH). on formalin-fixed, paraffin-embedded archival tissue provides a histogenetic profile. Am. J. Pathology., 144:237–243 (1994)
15. P.M. Nederlof, D. Robinson, R. AbukneshaJ. Wiegant, A.H. Hopman, H.J. Tanke, A.K. Raap. Three-color fluorescence in situ hybridization for the simultaneous detection of multiple nucleic acid sequences. Cytometry, 10:20–27 (1989)
16. M.R. Speicher, S.G. Ballard, D.C. Ward. Karyotyping human chromosomes by combinatorial multi-fluor FISH. Nature Genetics, 12:368–375 (1966)
17. J. Isola, S. DeVries, L. Chu. S. Ghazvini, F. Waldman. Analysis of changes in DNA sequence copy number by comparative genomic hybridization in archival paraffin-embedded tumor samples. American Journal of Pathology, 145:1301–1308 (1994)
18. R.N. Wiltshire, P. Duray, M.L. Bittner, T. Visakorpi, et al. Direct visualization of the clonal progression of primary cutaneous melanoma - application of tissue microdissection and comparative genomic hybridization. Cancer Research 55:3954–3957 (1995)

19. J. Piper, D. Rutovitz, D. Sudar, A. Kallioniemi, O.-P. Kallioniemi, F.M. Waldman, J.W. Gray, D. Pinkel. Computer image analysis of comparative genomic

20. H. Iwabuchi, M. Sakamoto, H. Sakunaga, Y.Y. Ma, M.L. Carcangiu, D. Pinkel, . T. Yang-Feng, J.W. Gray. Genetic analysis of benign, low-grade, and high-grade ovarian tumors. Cancer Research, 55:6172–6180 (1995)

21. J. Isola, O.-P. Kallioniemi, L. Chu, S. Fuqua, S. Hilsenbeck, C.K. Osborne, F.M. Waldman. Genetic aberrations detected by comparative genomic hybridization predict outsome in node-negative breast cancer. American Journal of Pathology 147:905–911 (1995)

22. H. Moch, J. Presti, G. Sauter, N. Buchholz, P. Jordan, M. Mihatsch, F.M. Waldman. Genetic aberrations detected by comparative genomic hybridization are associated with clinical outcome in renal cell carcinoma. Cancer Research, 56:27–30 (1996)

23. P. O'Copnnell, V. Pekkel, S. Fuqua, C.K. Osborne, D.C. Allred. Molecular genetic studies of early breast cancer evolution. Breast Cancer Research and Treatment 32:5–12 (1994)

24. D.S. Murphy, S. Hoare, J. Going, E. Mallon, W. George, S. Kaye, R. Brown, D. Black, W. Keith. Characterization of extensive genetic alterations in ductal carcinoma in situ by fluorescence in situ hybridization and molecular analysis. Journal of the National Cancer Institute, 87:1694–1704 (1995)

25. J. Schlegel, G. Stumm, H. Scherthan, T. Bocker, H. Zirngibl, J. Ruschoff, F. Hofstadter. Comparative genomic in situ hybridization of colon carcinomas with replication error. Cancer Research 55:6002–6005 (1995)

26. L.A. Donehower, L.A. Godley, C.M. Aldaz, R. Pyle, Y.-P. Shi, D. Pinkel, J.W. Gray, A. Bradley, D. Medina, H.E. Varmus. Deficiency of p53 accelerates mammary tumorigenesis in Wnt-1 transgenic and promotes chromosomal instability. Genes and Development, 9:882–895 (1995)

27. M.L. Cher, D. MacGrogan, R. Bookstein, J. Brown, R.B. Jenkins, R.H. Jensen. Comparative genomic hybridization, allelic imbalance, and fluorescence in hybridization on chromosome 8 in prostate cancer. Genes, Chromosomes and Cancer, 1994 Nov, 11(3):153–162 (1994)

28. M.M. Tanner, M. Tirkkonen, A. Kallioniemi, C. Collins, T. Stokke, R. Karhu, D. Kowbel, F. Shadravan, M. Hintz, W.-L. Kuo, et al. Increased copy number at 20q13 in breast cancer: defining the critical and exclusion of candidate genes. Cancer Research, 54:4257–4260 (1994)

29. C. A. Reznikoff, C. Belair, E. Savelieva, Y. Zhai, K. Pfeifer, T. Yeager, K. Thompson, S. DeVries, C. Bindley, M. Newton, G. Sekhon, F. Waldman. Long-term genome stability and minimal genotypic and phenotypic alterations HPV16 E7- but not E6-, immortalized human uroepithelial cells. Genes and Development, 8:2227–2240 (1994)

30. T. Stokke, C. Collins, W.-L. Kuo, D. Kowbel, F. Shadravan, M. Tanner, A. Kallioniemi, O.-P. Kallioniemi, D. Pinkel, L. Deaven, J.W. Gray. A physical map of chromosome 20 established using fluorescence in situ and digital image analysis. Genomics, 26:134–137 (1995)

31. M.M Tanner, M. Tirkkonen, A. Kallioniemi, K. Holli et al. Amplification of chromosomal region 20q13 in invasive breast cancer - Implications. Clinical Cancer Research, 12:1455–1461 (1995)

32. M. Tanner, M. Tirkkonen, A. Kallioniemi, J. Isola, T. Kuukasjarvi, C. Collins, D. Kowbel, X.-Y. Guan, J. Trent, J. Gray, P. Meltzer, O.-P. Kallioniemi. Independent amplification and frequent co-amplification of three non-syntenic regions of the long arm of chromosome 20 in human breast cancer. Cancer Research (in press)

DISCUSSION

Stark: It seems to me that there is a real dilemma here trying to figure out what is important and what is not. It is clear, from what you have shown, that once a cell acquires the ability to tolerate damage that the rate of acquisition of damage is very great. And it could be, let us say, amplification or deletion that you are looking at. There are two major possibilities; one is that a particular region shows up because there is something functionally important in that region, but the other possibility is that a particular region shows up because it happens to get amplified much more frequently than some other region and there is nothing functionally interesting in that region. I just wonder how you are going to get at that. In this situation that you are looking at now, you have got a large number of genes in a region. Maybe one of them is functionally important but, maybe if you are unlucky, that is just a region that likes to get amplified and once it is permitted, it just happens and there is going to be nothing functionally important in that region. How are you going to deal with that problem?

Gray: Obviously that is a very real possibility but I think that the things that allow us to keep going forward are several. One of them, is the strong associations that we see with the clinical behavior, the association that we see with increased S-phase fraction, and the fact that, in a very high frequency, almost all of the E7 transfected cells show that same region of amplification and it is a fairly narrow region. That certainly does not nail it and we are going to have to get to the point where we have cloned the genes in order to be able to answer it definitively. It is a lot of work but I think that it is worth doing. We have to know the answer whether they are biologically important or not.

Stark: But I guess, one thing that might help is correlation of amplification of a region across a wide variety of different kinds of tumors, which you would expect in a region that likes to get amplified. But if there is a functional gene which is required for the development of only a subclass of tumors then you might expect to see tumor restriction and that might be a very important kind of clue.

Gray: We certainly do not see a chromosome 20 amplified in all tumors. We actually picked this one because it was present in a fairly high fraction of them.

Hartwell: I would like to stimulate a discussion here between you and Garth Anderson. I am puzzled. Looking at it from your level of resolution, what one tends to see is large regions of chromosomes which have undergone a change and then when you look at fine resolution within that you find regions, maybe one or two or three, probably megabase size regions that are getting amplified; when Garth looks with PCR primers that are relatively close together in a very much smaller resolution (I may be misspeaking for him but I want to present the dilemma and he can defend himself), it seems like there are very frequent rearrangements, there could be roughly one every hundred KB or so, and I just find those two pictures very different and I do not understand really what is going on.

Anderson: We have been using a technique called inter-SSR PCR where you utilize PCR primers and they are anchored on CA repeats. In other words, you are not amplifying the CA repeat itself but you anchor at the 3 prime end of the CA repeat. Now we have recently been also using CG repeats. We create a sampling fingerprint with about one hundred bands with PCR. Although you use a single primer, you get a representative overall fingerprint of the tumor. What we are seeing is rampant instability in colorectal cancer. We have been looking at early stage polyps in colorectal progression; we have also been doing some stuff with DCIS although, with our results we find that genomic instability in DCIS is relatively uncommon that there are a lot of DCIS's that are very stable. Now, in terms of the anomaly between our results and CGH, clearly we are utilizing a different assay. We are able to pick up events occurring between repeat elements where based on PCR they are forced to be fairly close to one another. But yes, we are picking up evidence that you are getting incredibly abundant mutations in the genome of tumors, utilizing our technique, and I do not know why small events appear to be more prevalent than large events. All I can say right now is that it does look like that is the case, that they are incredibly abundant.

Gray: So following on that, in several, maybe a dozen now, tumors that are amplified on chromosome 20 we have taken that complete physical map that we have built across the region and gone in at as high a resolution as that will permit and mapped out the structure of those amplicons. What we often see is just copy numbers of functions at distance along the genome. There are clearly some amplicons that look like this where we have sampled quite frequently but it is also possible within these things to see some fine structure. Actually MCF7 is the cell line that is particularly prone to look like this. What we have actually focused on in those cases is this region over here which is the one which seems to occur in all of them. We do see some evidence for this kind of high resolution scrambling that you might be picking up. But I think, in general, it is more the exception than the rule. Again we have not done a lot of tumors and I guess one question to ask you is what is the possibility that some of what you are seeing is just inability to amplify some subset of the sequences?

Anderson: Some of that is clearly the case. In other words, if you have a wholesale deletion of an entire chromosome or whatever, you will lose those bands. But where we are picking up new bands we have pretty good evidence that the translocation has occurred within a short region, this type of thing. We have cloned out a number of these bands now and sequenced them and are finding both types of events occurring.

Livingston: You will clone potentially five, six, seven, ten large alleles from these amplified regions. Just this one amplified region (20Q 13.2) will yield you a number of genes. Following up on George Stark's question, if these are not merely signatures of genomic events that are compatible with survival as opposed to tumor progression, what assays would you use to test the possibility that one or more of these amplified alleles contributes to tumor progression?

Gray: Obviously we are bringing up expression systems that will allow us to put in the true amplicon. We have now made clonal libraries of the tumors that contain these complex amplicons, and so the first step is to ask, can we get some sort of a tumorigenic or transformation phenotype with these things? Of course that has the caveat that when you start looking at real solid tumors, proliferation is just one of the endpoints that you can look at and there is angiogenesis and there is adhesion and so on. So I think that being able to do *in vivo* expression of these complex amplicons and look at the whole spectrum of phenotypic changes is going to be what we are driven to. At least on chromosome 20, I am hoping that this really is going to be some sort of a cell cycle phenomenon because of the connection with the increased S phase in the tumors.

Livingston: This increased S phase of those tumor cells in psi 2?

Gray: Yes, in the human tumors.

Livingston: In the lung?

Gray: Yes. This is cytometry data, the question is just what fraction of tumor cells—what is the S phase fraction of the tumor population?

Livingston: And the control for that would be what?

Gray: It is an internal control, in other words we compared the fraction of cells in S phase for those tumors that show increased copy number on 20 with those, for tumors that did not show that amplification. And there is about a two-fold difference.

Livingston: And the degree of contamination of the tumor cell samples by non-tumor cells?

Gray: In most cases those were aneuploid tumors, so we were able to distinguish the tumor from the normal, based just on DNA content.

Stark: I think it is important because everybody may not appreciate this; to have some idea in your mind about what is the basic process that is driving the amplification and how it can generate the heterogeneity. It very likely is that, however they originate, amplifications often progress through bridge breakage fusion cycles and what that means is that in every daughter cell, starting with a clone, there is going to be a different arrangement, there is going to be amplification, there is going to be deletion and that is going to be different in every cell as you go through the envelope progeny that develop from a single initiating event. Amplified regions are going to break in each cell cycle and in each individual cell they are going to break in a different place. The possibility for generating heterogeneity in a situation like this is truly enormous, a single initiating event can just propogate like a chain reaction through the population and generate a tremendous amount of heterogeneity. I think that this can be seen both at the level of gross chromosomal structure and also at the level of fine chromosomal structure as a major mechanism to generate this kind of heterogeneity. In principle there is no big mystery about how a lot of this, at least, can take place.

Gray: O.K. but let me follow up on that a little bit, and talk about the mechanism here. I mean, you postulated this bridge breakage fusion as one way to do that. Generally speaking what we see in the evolution of these things is the amplified sequences end up in complex HSRs that have these multiple sequences present more or less together without too much intervening material. Maybe you would comment on how you see that occurring?

Stark: I think this situation is actually reasonably non-controversial. Clearly in a tumor you are not looking at an early event in the amplification of a particular region so a lot has gone on in very early phases of development that you have no way to examine. And it is really that stage that I am talking about. But I think there is plenty of opportunity for HSRs to be involved either in the early stages of amplification or as late manifestation. And that is a hard issue to resolve as both Geoff and I know.

Wahl: There is no doubt that amplification can be mediated by multiple bridge breakage fusion cycles first pointed out by Barbara McClintock a long time ago, later proposed by Cowell and later shown by Kaufman, by our lab, George's lab and many others. However, analyses of human tumor samples taken directly from patients reveals structures such as shown in this primary breast cancer. These are eccentric fragments, double minute chromosomes that constitute an alternative mechanism of gene amplification. These do not have centromeres but do have replication origins that function like chromosomal replication origins. They replicate once per cycle but they do not segregate equally. When a selective pressure is applied, double minutes containing a gene that gives the cell a survival advantage will accumulate. Now the interesting thing, with regard to what you are saying, is that we have now made artificial constructs *in vitro* where we have put two different double minutes into the same cell and they tend to associate together. We do not understand why that is true, but they associate together. It is at least conceivable that their integration into the chromosome could generate a complex intrachromosomal amplicon derived from disparate regions of the genome.

Gray: Define "associate together". You mean they stick together or they are ligated?

Wahl: We are trying to determine that right now; operationally, in metaphase sporads, two pieces of DNA that have no homology are superimposable by FISH.

Gray: And they are clearly ligated?

Wahl: We do not know if they are ligated together.

Gray: Have you tried pulling fibers out of these things?

Wahl: Not yet, but that may resolve this sort of issue.

Nasmyth: To what extent have you analyzed the status of these amplified regions, that is, are they in such a form that Geoff has suggested, or are they *in situ* or are they ectopic? Because the technique you are using does not tell you anything about that on its own, is that correct?

Gray: In the case of a cell-line that we have analyzed, which is where we can look at this most easily, they are clearly chromosomal, they are ligated. I mean, because they propagate from one cycle to the next, they are retained stably. We have done fiber FISH on these. This is the technique where you take the DNA from the chromosomes, digest to a proteinase, then just pull the fibers out and then hybridize with some probes with the different elements in the amplicon. Within the limitations of that technique, which are great, they do appear to be fairly closely ligated, in other words, 2Q-11 seems to be in relatively close proximity in most cases to Q-13.2 as if they were on the same very close DNA fiber.

Gatti: Everybody here is speaking about the bridge breakage fusion cycle. I read something recently about that and, as far as I am concerned, there is strong evidence that this occurs in *Nature*. Barbara McClintock showed it quite clearly, but to me there is no evidence that something like that occurs in mammalian cells. I know, more or less, that it does not occur in Drosophila. Others have reached the same conclusion that there is no evidence for this cycle in mammalian cells.

Stark: Well I do not know whether you consider dicentric chromosomes to be evidence or not.

Gatti: What is worse, I looked at several different papers about the telomeric fusion in mammals, and lots of nice pictures, but nobody analyzed anaphases. There is no data about anaphases. There are perhaps 30 or 40 published papers and nobody analyzes anaphases in tumor cells.

Stark: So you are complaining that nobody has seen the bridges but the dicentric chromosomes are not sufficient evidence for you of bridge breakage fusion cycles?

Gatti: No.

Stark: Why not? What else can happen to them?

Gatti: What, do you mean by dicentric?

Stark: Yes, if you see dicentric chromosomes.

Gatti: You mean chromosome to chromosome exchanges. In other words, if you have a break, this break is supposed to give rise to fusion between the sister chromatids and then it should become a bridge. That is basically the idea. But nobody has seen the bridges in the anaphase. So that is my impression, that it is difficult to be able to find those bridges.

Bacchetti: Well, I am not sure that bridges have not been analyzed. There are data on tumors where they did see bridges in anaphases. But I do not understand why you do not think that the dicentric and rings that have been observed are not sufficient indication of what would happen afterwards.

Stark: What are they?

Gatti: My impression is that this phenomenon has not been very analyzed.

Bacchetti: You may be right in terms of numbers, that indeed more analysis has been done on metaphases, but again my question, which is also George's question is: what do you expect would happen once dicentrics are formed? What kind of resolution of these structures would you hypothesize?

Gatti: What kind of dicentric are you talking about? The one which involves both systems? Dicentric may be an asymmetric exchange between chromosomes.

Stark: I think I see what you are getting at. Dicentric sister chromatids are not visualized very easily in metaphase spreads. You have to know that there is a connection and if you are not in anaphase between the ends of the sister chromatids. But dicentric sister chromosomes are very readily seen.

Bacchetti: Occasionally you even see sister chromatid fusion, it is much less frequent.

Stark: They are much harder to see.

Hartwell: I think that the issue is not whether a dicentric chromosome would break when it goes through anaphase. The issue is whether a broken chromosome arriving in G1 would replicate, generate two broken sister ends and whether as in other organisms those two broken newly replicated ends would fuse. I do not think that we know that.

Gatti: That is exactly my point. To me there is no evidence for that.

Pinedo: As a clinician I was very intrigued by your tissue analysis versus your cell lines and especially breast cancer. The number of cancer cells are minimal compared to the normal tissue between the tumor. Matrix forms the majority of tumor, with large numbers by fibroblasts and macrophages. How are you going to be able to interpret your data if you have so many normal cells in it? In other tumor types you have a few normal cells, like in renal cancer and other types of cancer, but especially in breast cancer this is not the case. It is very difficult to correlate the MCF-7 data anyway with the cancer—or do you have another way to identify the cancer cells in breast cancer tissue?

Gray: Yes, probably the best way to do this, and we have not done it, but the best way to do this is to take advantage of the fact that from the CGH data, we know within those tumors some of the recurrent high level amplifications. So it is possible for us to mark tumor cells carrying, let us say, amplified c-myc. And then to study heterogeneity of just those cells.

Kolodner: Can we go to your experiment on RER plus and RER minus tumors? How do you define RER plus and RER minus, particularly with regards to the RER minus tumors that are karyotypically stable?

Gray: All of the tumors were RER minus.

Kolodner: But how do you actually do that? Because I think that it is more subtle than the vast literature that has been published on this would lead you to believe.

Gray: I have to confess that we did not do that. Bert Vogelstein did that and so we used his definition.

Anderson: One thing I have always been curious about CGH is: You make the two PCR products, and then you hybridize them always on to a normal metaphase spread. Have you ever gone up the opposite way, in other words, make the tumor metaphase spread? Metaphase is naturally occurring within the tumor; you might pick up greater degrees of heterogeneity or pick up double minutes. In other words, this might give you some sense of instability within the tumor, and within the individual cells where its averaged out for the population and not discernible when you hybridize to a normal metaphase spread.

Gray: You can do that, we have done it a little bit. Generally speaking, we consider the power of the method to be that it maps these changes on to a normal representation of the genome, so we can interpret it. But, it is quite clear that if you do CGH to a metaphase from a population that is amplified, you light up the amplicons. So it can be used that way, we just have not done it very much.

3

CELL CYCLE CONTROL
OF GENETIC STABILITY

Geoffrey M. Wahl,[1] Steven P. Linke,[1,2] Thomas G. Paulson,[1,2]
and Li-chun Huang[1]

[1]Gene Expression Laboratory
The Salk Institute for Biological Studies
La Jolla, California 92037
[2]Department of Biology
University of California San Diego
La Jolla, California 92093

INTRODUCTION

Boveri's deduction that genetic instability is involved in the initiation of tumorigenesis remains one of the most insightful in tumor biology (Boveri, T., 1914). It is remarkable that it came as an extrapolation of his observations on the developmental consequences of multipolar mitoses created by polyspermic fertilization. He noted that the resulting blastomeres exhibited the atypical growth and morphologic features that he had observed in tumor cells. This led him to conclude that the genetic content of each chromosome could not be equivalent and that "a special arrangement for inhibiting [cell division] has been done away with as a consequence of missegregation of chromosomes in cells lacking a bipolar spindle" (as cited in Manchester, K.L., 1995). He recognized that normal growth patterns could never be reestablished once the normal chromosome complement was altered. Thus, Boveri anticipated both the importance of genetic instability in the initiation and progression of cancer and the existence of negative regulators of cell growth, whose loss would result from inappropriate chromosome segregation.

Compelling molecular and cytogenetic data reveal that cancers arise from normal cells through a process of clonal evolution, driven by the effects of multiple genetic alterations that accumulate within the genome of an individual cell. The variable order of their appearance suggests that the cumulative effects of the genetic alterations, rather than a rigid chronology of their occurrence, may be more important in the progression of some cancers (Fearon, E.R., et al., 1987, Fearon, E.R. and Vogelstein, B., 1990). However, at least two barriers must be overcome before mutation-driven tumor progression can occur. First, somatic cells must escape the strictly regulated senescence program that normally limits their growth (Hayflick, L., 1965, Shay, J.W., et al., 1991). Second, controls insuring accurate replication and transmission of chromosomes must apparently be inactivated, since the point mutations, aneu-

Genomic Instability and Immortality in Cancer
edited by Mihich and Hartwell, Plenum Press, New York, 1997

ploidy, and structural chromosome alterations commonly observed in cancer cells arise at far lower rates in normal cells (e.g., see Bhattacharyya, N.P., et al., 1994, Cross, S.M., et al., 1995, Eshleman, J.R., et al., 1995, Livingstone, L.R., et al., 1992, Yin, Y., et al., 1992).

A strong link has been established between the functions of the *p53* gene and the control of genetic stability in humans and rodents. p53 is a classic negative growth regulator of the type envisioned by Boveri. Loss of heterozygosity has been reported in numerous sporadic cancers and inactivation of the signal transduction pathway involving p53 occurs in more than 50% of human cancers (Donehower, L.A. and Bradley, A., 1993, Levine, A.J., et al., 1994). Germline mutation of one p53 allele contributes to the precocious development of multiple types of cancer in Li-Fraumeni patients (Malkin, D., et al., 1990, Srivastava, S., et al., 1990). p53-/- mice are viable, but they develop cancers sooner than normal littermates (Donehower, L.A., et al., 1992). As described below, loss of p53 function in vitro can make cells competent for gene amplification under specific selective conditions, providing a direct link between p53 and the control of this form of genetic instability (Livingstone, L.R., et al., 1992, Yin, Y., et al., 1992). These data establish that p53 is not essential for cell survival, but its loss permits genetic instability to occur.

Checkpoints are biochemical signal transduction pathways designed to insure that a process or product involved in cell cycle progression is completed before a downstream process is initiated (Hartwell, L.H. and Weinert, T.A., 1989). Work over the past decade implicates p53 in at least four cell cycle checkpoints that could affect genetic stability. First, p53 mediates a G0/G1 arrest triggered by various forms of DNA damage, such as double-strand breaks (DSBs) (Huang, L.-C., et al., 1996, Kastan, M.B., et al., 1991, Nelson, W.G. and Kastan, M.B., 1994) and lesions induced by ultraviolet light (Maltzman, W. and Czyzyk, L., 1984). Second, p53 induces a G0/G1 arrest in response to ribonucleotide depletion in the absence of detectable DNA damage (Linke, S.P., et al., 1996). These G1 arrests depend on the ability of p53 to serve as a transcriptional activator of the cyclin-dependent kinase inhibitor p21$^{WAF1/CIP1/SDI1}$ (p21) (Dulic, V., et al., 1994, El-Deiry, W.S., et al., 1994, El-Deiry, W.S., et al., 1993, Harper, J.W., et al., 1993, Linke, S.P., et al., 1996). One important downstream target of p21 kinase inhibition is the retinoblastoma protein (pRb). Mouse embryo fibroblasts deficient in p53, p21, or pRb exhibit altered arrest responses to DNA damage or ribonucleotide depletion (Almasan, A., et al., 1995, Deng, C., et al., 1995, Linke, S.P., et al., 1996). Third, p53 prevents cell cycle progression and DNA re-replication in cells treated with the spindle inhibitors colcemid or nocodazole, implicating p53 in the spindle completion checkpoint (Cross, S.M., et al., 1995). Fourth, p53 also appears to play a role in insuring that the centrosome replicates only once per cycle (Fukasawa, K., et al., 1996). Cells devoid of p53 function synthesize multiple centrioles and spindle poles and undergo DNA re-replication in a single cell cycle, resulting in the aneuploid descendants Boveri predicted would contribute to tumor initiation. While p53 functions in multiple checkpoints to ensure accurate transmission of genetic information, the work described below focuses on the characterization of the p53-mediated G0/G1 arrest induced by ribonucleotide depletion or DSBs.

RESULTS AND DISCUSSION

p53 Induces a G0/G1 Arrest in Response to Ribonucleotide Depletion in the Absence of DNA Damage

Cells with a functional p53 pathway (p53+) arrest in G0/G1 when treated with PALA, a specific inhibitor of de novo UMP biosynthesis (White, A.E., et al., 1994, Yin, Y., et al., 1992). In contrast, cells with a non-functional p53 pathway (p53-) enter S phase when challenged with PALA (White, A.E., et al., 1994, Yin, Y., et al., 1992). This leads to initiation of the replication program without the capacity to generate the precursor pool required to complete S phase and may result in induction of chromatid breaks that can lead to gene amplification and other structural chromosome changes (for discussion see Di Leonardo, A., et al., 1993, Linke, S.P., et al., 1996). While DSBs in normal human diploid fibroblasts (NDF) cause a long-term, senescent-like arrest (Di Leonardo, A., et al., 1994, Little, J.B., 1968), we and others have reported that the PALA-induced G0/G1 arrest can be reversed in the majority of cells by adding the salvage metabolite uridine (Livingstone, L.R., et al., 1992, White, A.E., et al., 1994, Yin, Y., et al., 1992). This observation suggested that p53 may be able to arrest cells in the absence of DNA damage when the levels of specific metabolic intermediates become limiting. Consistent with this inference, we were unable to detect chromosomal aberrations above background in metaphase chromosomes obtained from PALA-treated NDF (Linke, S.P., et al., 1996).

The experiments shown in Fig. 1 were performed to test whether antimetabolite-induced arrest could be initiated in the absence of DNA damage and to define the biochemical specificity of the arrest mechanism. PALA presumably causes DNA damage only in cells proceeding through S phase. Therefore, NDF synchronized in G0 by contact inhibition were released into cycle in the presence of PALA and BrdU and cell cycle progression was analyzed over the next 72 hr. If DNA damage were required for PALA to induce an arrest, cells would have to transit S phase, and incorporate BrdU, to sustain damage before arresting in the next G1. However, Fig. 1 shows that more than 95% of the arrested cells did not enter S phase. Furthermore, addition of uridine to the medium allowed the arrested cells to enter S phase approximately 16 hr later, similar to that observed after cells are released from serum deprivation or contact inhibition without PALA treatment. The time of entry of the first cells into S phase was similar to that observed after release from serum deprivation or contact inhibition without PALA treatment. NDF expressing the HPV16E6 protein (NDFE6), which hastens the degradation of the p53 protein (Scheffner, M., et al., 1990), or immortal cells lacking functional p53 did not arrest when challenged under identical conditions. This demonstrates that the PALA-induced arrest strictly depends on an intact p53 pathway.

The above results indicate that the p53-dependent G0/G1 arrest induced by PALA might result from interference with specific metabolic reactions. Consistent with this idea, we found that another inhibitor of UMP synthesis, pyrazofurin; an inhibitor of CTP synthesis, cyclopentenylcytosine; and two inhibitors of GMP synthesis, tiazofurin and mycophenolic acid, also prevented G0-synchronized NDF from entering S phase in a p53-dependent manner. Treatment with PALA, tetrahydrouridine, and the salvage metabolite cytidine (which specifically depletes UTP pools) also induced a p53-dependent G0/G1 arrest (Fig. 1). In contrast, the dNTP synthesis inhibitors hydroxyurea (HU, see Fig. 1), 5-fluorouracil, and methotrexate (MTX); the AMP synthesis inhibitor alanosine; and the inhibitor of early stage purine synthesis inhibitor 6-methylmercaptopurine riboside (MMPR) elicited arrests independent of p53 status. rNTP pool levels were determined by HPLC to confirm that the antimetabolites depleted the expected pools. Those agents that caused arrest reduced pools to less than 20% of control. Thus, depletion of UTP, CTP, or GTP pools was sufficient to induce a p53-dependent arrest in G0/G1 without progression through S phase.

These observations raise the question of why depletion of other metabolites failed to produce a p53-dependent arrest. Depletion of ATP may not have been effective due to the large size of the endogenous ATP pool and/or the inability of alanosine to reduce it below

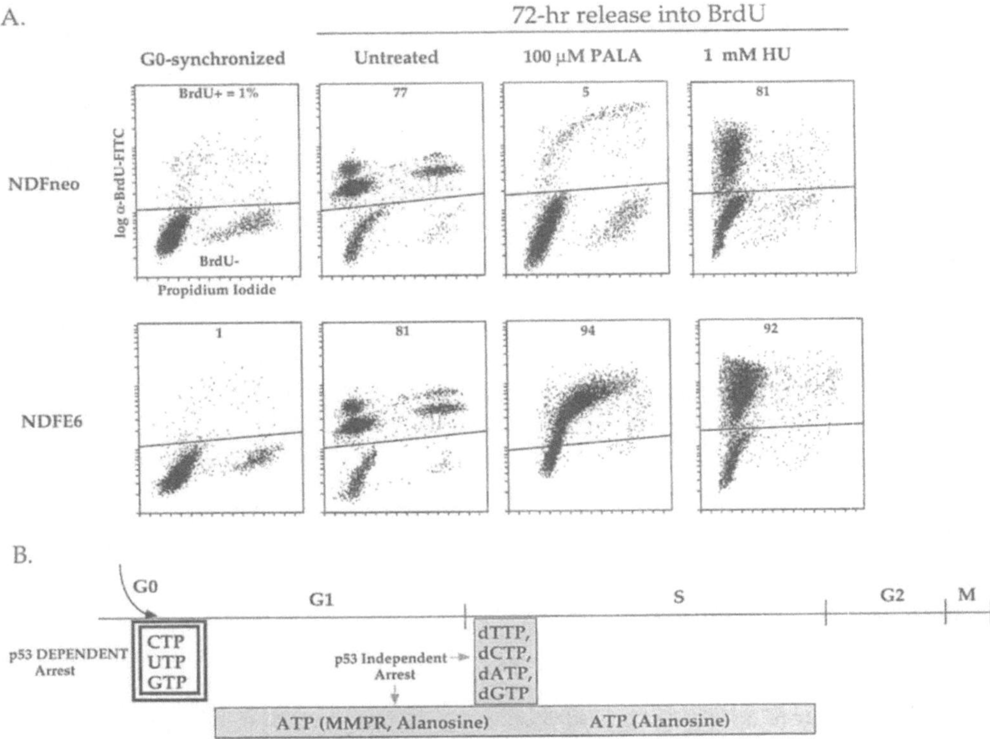

Figure 1. p53 is activated by ribonucletide depletion in the absence of DNA damage. A. Specificity of p53-mediated arrest induced by metabolic inhibitors. NDF infected with a retrovirus encoding the bacterial neomycin phosphotransferase gene (NDFneo, p53$^+$), or encoding the neomycin phosphotransferase gene and the HPV16E6 gene (NDFE6, p53-) were synchronized in G0 by contact inhibition. Cells were released by sub-culture at lower density into medium containing 65 μM BrdU. Cultures were treated with 100 μM PALA (to inhibit de novo UMP biosynthesis) or 1 mM HU (to inhibit production of dNTPs), or they were left untreated. After 72 hr, the percentage of cells able to enter S phase under these conditions was determined by flow cytometry. The percentages shown in the dot plots represent the absolute percentage of BrdU+ events for the the initial G0-synchronized and untreated populations, and the BrdU+ events relative to the untreated control for the PALA and HU samples. B. Summary of the effects of various metabolic inhibitors. Experiments identical to those in "A" were performed using inhibitors of several reactions involved in de novo purine and pyrimidine metabolism. The effects of each drug on rNTP pool levels was determined by HPLC to insure depletion of the expected nucleotides. A summary of the cell cycle positions and p53-dependence of inhibiting the designated nucleotides is shown. These data have been published previously, in part (Linke, S.P., et al., 1996).

approximately 40% of control. On the other hand, while severe depletion of dNTP pools did not induce a G0/G1 arrest, it did arrest cells in early S phase. This demonstrates the specificity of the p53-dependent metabolite sensor, and it implies that p53+ NDF advance through G1 with limiting dNTPs. Arrest triggered by limiting dNTPs apparently requires some progression through S phase, perhaps because DNA synthesis is required to reduce the dNTP levels below the minimum required for chain elongation. Thus, arrest triggered by dNTP depletion may result from reduced activity of DNA biosynthetic enzymes rather than the activation of a specific replication fork progression checkpoint. p53 function is apparently not required for dNTP depletion induced arrest as it occurs equally well in p53+ and p53- cells.

PALA reduced RNA synthesis similarly in p53+ and p53- cells, yet only the former arrested. It is unlikely that inhibition of general mRNA synthesis due to depletion of rNTP precursors is a trigger for p53-dependent arrest, since the RNA polymerase inhibitors α-amanitin and D-ribofuranosylbenzimidazole (DRB) caused a dose-dependent arrest in both p53+ and p53- cells. These data suggest that the ribonucleotide depletion-induced arrest either requires inhibition of the synthesis of specific types of RNA molecules or that another metabolic intermediate(s) is involved. We feel that the latter possibility is less likely, as we have been unable to find evidence of the involvement of intermediates, such as CTP-activated signal transduction molecules or ceramide (which is potentially modulated by UTP levels). Ceramide was an intriguing candidate since it causes pRb dephosphorylation similar to that caused by ribonucleotide depletion (Dbaibo, G.S., et al., 1995, Jayadev, S., et al., 1995). Depletion of dNTP pools would be expected to have a more profound effect on causing DNA damage in G0/G1-phase cells than rNTP depletion due, for example, to inhibition of DNA repair. However, rNTP, but not dNTP, synthesis inhibitors induced a p53-dependent G0/G1 arrest in non-cycling cells. Taken together, these results indicate that the metabolic depletion caused by ribonucleotide synthesis inhibitors, rather than DNA damage, leads to the p53-dependent G0/G1 arrest.

Fig. 2 presents two models by which p53 structure may contribute to its ability to induce arrest in response to ribonucleotide depletion in the absence of DNA damage. p53 has been reported to be associated with two types of RNA molecules. A molecule of 5.8S rRNA has been shown to be covalently attached to the serine residue in the C-terminal CKII phosphorylation site of human, mouse, and rat p53 (Fontoura, B.M., et al., 1992, Samad, A. and Carroll, R.B., 1991; Carroll, B., personal communication). The conservation of the serine modified by CKII and the adjacent three amino acids from all species analyzed thus far suggests that this region may be important for regulating p53 structure and/or function Soussi, T., et al., 1990). 5S rRNA has also been shown to associate with p53 through binding to ribosomal protein L5, which associates with the Mdm2 protein (Marechal, V., et al., 1994). Mdm2 is an important negative regulator of p53 function, presumably due to its ability to bind to the N-terminal transactivation domain and interfere with recruitment of transcriptional co-activators (Chen, C.Y., et al., 1994, Momand, J., et al., 1992, Wu, X., et al., 1993). The model in Fig. 2A proposes that p53 may be sequestered in the cytoplasm due to association with one or both of these rRNAs, possibly because they enable p53 to remain bound to polysomes after translation. Ribonucleotide depletion may disrupt polysome structure, allowing p53 to gain access to the nucleus where it can activate cell cycle inhibitors such as p21. This model does not require removal of the C-terminal 5.8S rRNA for transcriptional activation. Fig. 2B proposes a model involving p53 turnover. As wild-type p53 has a half-life of five minutes or less under normal growth conditions (e.g., Chowdary, D.R., et al., 1994, Scheffner, M., et al., 1993, M. Scheffner, et al., 1990, Yeargin, J. and Haas, M., 1995), it is possible that turnover during ribonucleotide depletion results in the synthesis of a form lacking the C-terminal 5.8S rRNA. This could have several consequences since the C-terminus is highly positively charged, binds to DNA non-specifically, and is involved in forming the p53 tetramers postulated to be the active transcriptional regulator (Ko, L.J. and Prives, C., 1996). Thus, p53 with the C-terminal 5.8S rRNA may not have access to the nucleus, may not bind DNA non-specifically, and may not tetramerize due to the high density of negative charges. The form lacking the 5.8S rRNA may have greater access to the nucleus and/or may form active tetramers more efficiently. Importantly, PALA treatment results in the accumulation of nuclear p53, as predicted by both models (Linke, S.P., et al., 1996). While phosphorylation of the CKII site, binding to C-terminal antibodies, or deletion of

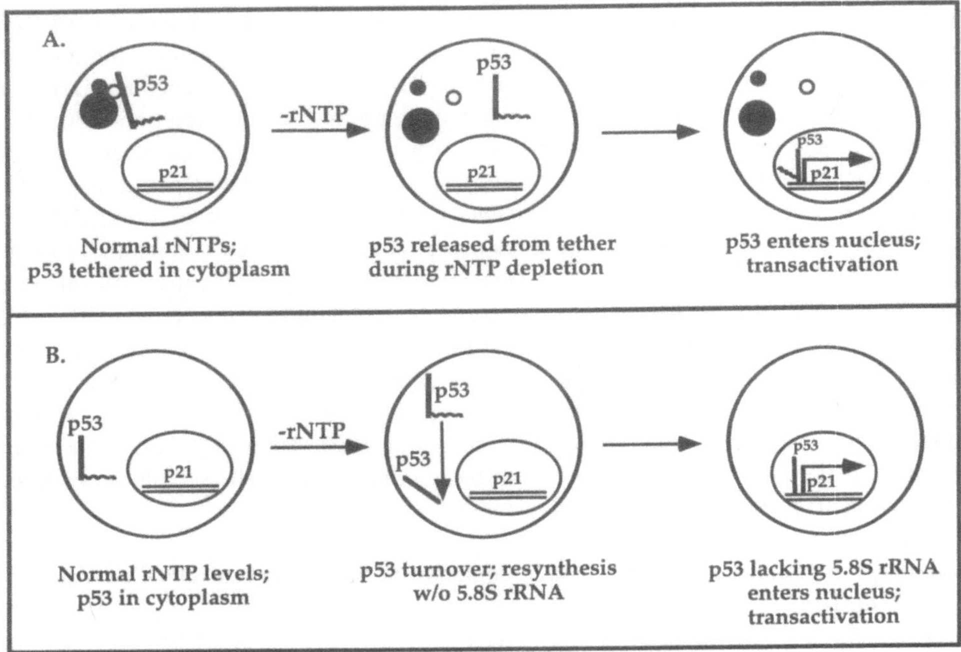

Figure 2. Models for rNTP depletion induced arrest. A. Catch and release model. p53 is proposed to be localized in the cytoplasm in cells growing exponentially. The mechanism of the cytoplasmic tethering is unknown. It may involve association with polysomes (indicated by small and large filled circles representing ribosomal subunits), perhaps through the associations with ribosomal protein L5, 5S rRNA, and/or 5.8S rRNA (squiggly line). rNTP depletion is proposed to disrupt the tether (e.g., by polysome disaggregation), allowing p53 to gain access to the nucleus, whereupon it would transactivate growth inhibiting genes such as p21. B. Release and catch model. This model proposes that rNTP depletion is associated with degradation of p53 containing the 5.8S rRNA, followed by re-synthesis of a form lacking this rRNA. The p53 lacking 5.8S rRNA is able to enter the nucleus, and serve as an effective transcriptional regulator.

the C-terminus activates p53 for sequence-specific DNA binding in vitro and after microinjection (Hupp, T.R., et al., 1992, Hupp, T.R., et al., 1993), p53 with mutations preventing modification of the CKII have also been reported to be transcriptionally active (Fiscella, M., et al., 1994, Hupp, T.R. and Lane, D.P., 1995). The latter observation could be explained by the model in which absence of C-terminal 5.8S rRNA is required for nuclear entry, while the former observations suggest involvement of a second level of control involving negative interaction of the C-terminus with the DNA-binding domain (Hupp, T.R., et al., 1995). It is crucial to test whether ribonucleotide depletion affects p53 association with 5S and 5.8S rRNA, and whether the mutations that prevent RNA association in the N- and C-termini affect the ability of p53 to induce an arrest in response to ribonucletide depletion.

The p53-Mediated Response to Double-Strand Breaks: An Executioner Previously Masquerading as a Guardian

Loss of p53 function clearly contributes to amplification competence, in part because of the deficient metabolite sensor function and ensuing chromatid breakage that can occur under some growth conditions, as described above. However, gene amplification can presumably be initiated by a solitary DSB through formation of a dicentric chromosome or acentric chromosome fragment (e.g., Smith, K.A., et al., 1992, Windle, B., et al., 1991). p53− cells exhibit a frequency of PALA-selected gene amplification that is at least 10^4-fold higher than p53+ cells (Livingstone, L.R., et al., 1992, Tlsty, T.D., 1990, Wright, J.A., et al., 1990, Yin, Y. et al., 1992). We consider it likely that this level of genome protection requires functions in addition to the metabolite sensor capacity of p53.

p53 also induces a G0/G1 arrest in response to various forms of DNA damage. Studies with ionizing radiation, clastogenic chemotherapeutic agents, restriction endonucleases, and nuclear microinjection demonstrate that DSBs trigger such a response (Huang, L.-c., et al., 1996, Kastan, M.B., et al., 1991, Nelson, W.G., and Kastan, M.B., 1994). Similar to the ribonucleotide depletion-induced G0/G1 arrest, the arrest induced by DSBs largely depends on the p53-mediated transcriptional induction of p21 and conversion of pRb into a hypophosphorylated, anti-proliferative form (Deng, C., et al., 1995, Dulic, V., et al., 1994, Linke, S.P., et al., 1996). While other studies indicate that helix-distorting lesions such as those caused by ultraviolet radiation (Maltzman, W., and Czyzyk, L., 1984) and base-pair mismatches or insertion-deletion loops also induce p53-dependent arrest (Lee, S., et al., 1995; Huang, L-c., et al., in preparation), we focus here on the p53-mediated response to DSBs, since they lead to the translocations, deletions, and gene amplification detected in cancer cells.

The relationship between G1 cell cycle arrest and maintaining a low level of structural chromosome rearrangements has been a matter of considerable debate. One very popular model is that low levels of damage activate p53 to induce a transient arrest to allow for repair of the DNA damage, while higher amounts of DNA damage induce apoptosis (Lane, D.P., 1992). Data obtained from asynchronous populations of a human myeloid leukemia cell line, ML-1, have been widely quoted as supporting this model (Kastan, M.B., et al., 1991). A direct prediction of this model is that p53+ cells should be more radioresistant than p53− variants. However, analyses of early passage NDF have shown the opposite. p53+ cells exhibit greater radiosensitivity than isogenic p53− derivatives (Di Leonardo, A., et al., 1994, Di Leonardo, A. et al., 1993, Li, C.Y., et al., 1995). The situation with immortal cell lines, however, is more complex. Some studies suggest that lines with wild-type p53 are more radiosensitive, while others show no clear relationship between p53 status and radiosensitivity (e.g., Li, C.Y., et al., 1995). This complexity is likely due to other genetic changes in cell cycle regulatory or repair factors that are selected during immortalization and extended passage of such cell lines.

The amount of chromosome damage following ionizing radiation should also be lower in p53+ cells if G1 arrest benefits repair. However, we found that ionizing radiation induces similar numbers of anomalies in p53+ and p53− cells (Di Leonardo, A., et al., 1994). One resolution of this observation is provided by analyses of DSB repair in budding yeast, showing that DSBs are repaired very slowly in G1 because the homologous chromosome does not provide an efficient template for repair (Fishman, L.J., et al., 1992, Kadyk, L.C., and Hartwell, L.H., 1992). On the other hand, DSBs are repaired far more efficiently after DNA replication, most likely because the sister chromatid provides a repair substrate that is both identical in terms of DNA sequence and physically closer to the broken end than the homolog (Fishman, L.J., et al., 1992, Kadyk, L.C., and Hartwell, L.H., 1992). Ionizing radiation induces a G2 delay in mammalian cells and in yeast, and it is during this interval that DSBs are repaired most efficiently (Fishman, L.J., et al., 1992, Kadyk, L.C., and Hartwell, L.H., 1992; Kaufmann, W.K., 1995). While p53 overexpression can produce a G2 arrest (Agarwal, M., et al., 1995), it is less certain whether p53 plays a role in G2 arrest under more normal conditions, since p53-deficient primary cells appear to exhibit the same magnitude of G2 delay as normal cells following DNA damage (Paules, R.S., et al., 1995). The observed similarity in the amounts of radiation-induced DNA damage in p53+ and p53− cells is not expected if p53-mediated G1 arrest is required for efficient repair, but it is expected if the majority of DSB repair occurs in S and/or G2 by a checkpoint-repair pathway(s) not requiring p53 function.

An alternative model to explain the genome-protective effect of p53 is that as few as one unrepaired DSB in G1 activates the p53-dependent arrest response to prevent damaged cells from entering further cell cycles. This may be achieved by induction of prolonged or senescent-like arrest or apoptosis, depending on the cell type. This would insure that only those cells lacking DNA damage enter S phase. Several lines of evidence support this model. First, it is clear that p53 induces apoptosis in response to DNA damage in many cell types (Clarke, A.R., et al., 1993, Lowe, S.W., et al., 1993). Second, NDF exposed to ionizing radiation become enlarged within 48 hr and resemble senescent cells as long as three weeks after irradiation (Di Leonardo, A., et al., 1994). Third, p53 and p21 levels are increased and pRb remains hypophosphorylated for at least 96 hr after irradiation (Di Leonardo, A., et al., 1994). These observations are inconsistent with the transient arrest model, but support one of long-term arrest.

Gamma radiation dose curves and microinjection studies in NDF and studies on a temperature-sensitive p53 mutant support the model of one DSB inducing a G0/G1 arrest. Using a cumulative bromodeoxyuridine (BrdU) DNA labeling technique developed in this laboratory, we showed that the ability of G0-synchronized NDF to enter S phase after gamma-irradiation is log-linear in relation to dose and that the curves intersect the *y*-axis at 100% (Di Leonardo, A., et al., 1994; Linke, S.P., et al., in preparation). This suggests that a single event, such as a slowly repaired or irreparable DSB, might prevent progression of cells into S phase. As the dose increases, so might the likelihood that such breaks would occur. The cells that stayed in the initial G0/G1 remained arrested for at least 96 hr. These findings were confirmed and extended by an approach involving nuclear microinjection of DNA into G0-synchronized NDF (Huang, L.-c., et al., 1996). Injection of restriction endonuclease-linearized duplex DNA effectively prevented G0-synchronized NDF from entering S phase after release. In contrast, isogenic NDF expressing HPV16E6 proceeded into S phase when injected with this substrate. Nuclear injection of buffer alone, supercoiled DNA, or circular DNA with a single nick, or cytoplasmic injection of linear DNA had no detectable effects, implying that the microinjection procedure itself does not induce the arrest. We observed an arrest response equivalent to that of serum-deprived cells when each nucleus was injected with at least one linearized molecule. Blunt, 3′-overhang, 5′-overhang, or incompatible ends appeared equivalent in their ability to induce arrest, despite reported differences in repair efficiencies (Morgan, W.F. and Winegar, R.A., 1990). The arrest appeared to be of long duration, since most cells remained in G0/G1 24 to 36 hr post-injection. The microinjection data demonstrate that damaged DNA alone, if present in the nucleus, can trigger a prolonged, p53-dependent arrest. Finally, cells with a thermoregulated p53 arrest in response to the breakage of a single dicentric chromosome only when grown at the permissive temperature (Ishizaka, Y., et al., 1995). These data lead us to propose that as few as one unrepaired/irreparable DNA lesion may trigger the p53-dependent cell cycle arrest pathway to generate a long-term arrest, which would prevent cells with DNA damage from progressing into S phase. This would effectively limit the clonal expansion of cells with broken or dicentric chromosomes, or possibly with one or more short telomeres if they are recognized as unprotected ends.

Two types of experiments indicate that G0 cells, or those in very early G1, may repair some types of damaged DNA poorly (Huang, L.-c., et al., 1996). First, we found that a small primer annealed to a single-stranded DNA circle (i.e., a "large gap") induced arrest. As nicked or circular molecules do not induce arrest, this indicates that cells in G0 do not extend the primer efficiently. In contrast, a circular molecule with a 25-nucleotide (-nt) gap did not induce arrest, indicating that either the gap has to be of sufficient length to activate the arrest mechanism or that the 25-nt gap is repaired so rapidly that it fails to activate the arrest mechanism. Second, we developed a PCR-based assay to determine whether linearized molecules delivered into a G0 or early G1 nucleus could be repaired. We found that molecules injected in G0 remained, apparently intact, for at least 96 hr, and that the ends remained largely or entirely unligated. In contrast, when linearized DNA was injected into nuclei after the cells were released into G1 for at least three hours, PCR bands were observed with sizes consistent with intra- and inter-molecular ligation products. The amount of apparent end-joining increased when the molecules were injected into nuclei of cells released into complete medium for longer times. That is, the amount of product increased when injections were performed in cells that were more advanced in the cell cycle. These data suggest that DNA introduced into nuclei in early G1 by microinjection is ligated inefficiently, which could provide an interval during which the genome could be scrutinized for residual damage from the previous cell cycle.

While some data are incompatible with the model invoking G1 arrest to enable time for repair, the microinjection experiments clearly suggest that a fraction of "broken ends" can be ligated during mid to late G1. On the other hand, the data discussed above are most compatible with a model in which p53-mediated arrest provides a mechanism for cell elimination. One way of resolving these apparently conflicting data is to propose that some lesions can be repaired in G1, while others may be irreparable. As few as one of the latter type may induce a p53 dependent G1 arrest, the most frequent consequence of which is cell elimination through senescence. It remains to be determined whether the naked, linear extrachromosomal DNA introduced by microinjection resembles the possibly irreparable radiation-induced lesions in native chromatin. In order to assess whether results from the repair of microinjected linear plasmid DNA can be extended to the chromatin of bona fide chromosomes, it will be essential to analyze the repair kinetics of a single chromosome break induced as cells enter G1. Experiments to address this point are in progress.

As described above, gamma-irradiation of G0-synchronized NDF produces a prolonged cell cycle arrest, and the fraction of cells that remain arrested increases logarithmically with dose. However, the fraction of cells that enter the first S phase at each radiation dose is significantly greater than the fraction that eventually survive to form macroscopic colonies (Fig. 3; Linke, S.P., et al., in preparation; (Di Leonardo, A., et al., 1993, Li, C.Y., et al., 1995). If NDF with one or more broken chromosomes could enter S phase and remain viable, they should generate variants with chromosome abnormalities such as gene amplification at detectable frequencies. However, the frequency of PALA ressistance is less than 10^{-10} in NDF (Livingstone, L.R., et al., 1992, Tlsty, T.D., 1990, Wright, J.A., et al., 1990, Yin, Y., et al., 1992). These disparities led us to determine whether NDF that entered the first S phase after irradiation were being eliminated in subsequent phases. The cumulative BrdU DNA labeling technique allows quantification of cell cycle phases up to the third S phase after release from G0-synchrony. Fig. 3. shows the ability of one NDF strain (WS1) to escape from the first, second, and third G0/G1 phases following gamma-irradiation. A significant fraction of the cells that entered the first S phase after damage induction presumably generated descendants prone to elimination in subsequent cell cycle phases. After the third G1, the number of cells remaining approximated the number of colonies detected several weeks after irradiation. It is possible that the damage induced in the first cycle produces damage in subsequent cycles through generation of dicentric or

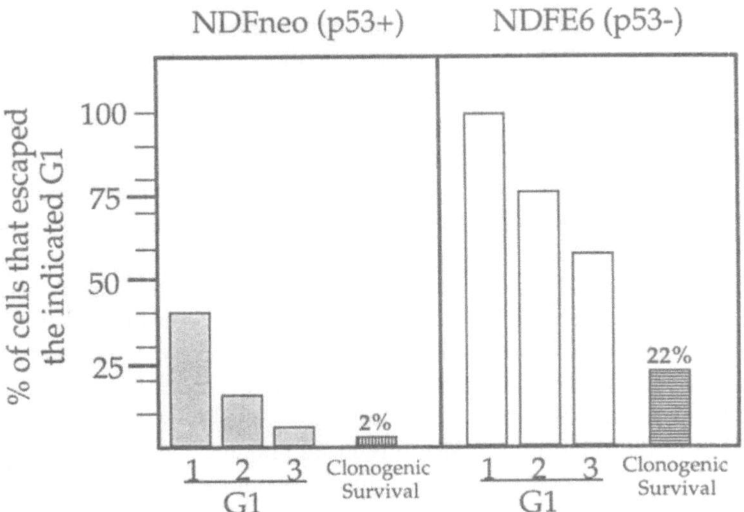

Figure 3. Progressive elimination of p53+ cells following gamma-irradiation. NDFneo or NDFE6 cells were synchronized in G0 by contact inhibition, and then exposed to 4 Gy gamma radiation. The cells were then released into BrdU and the number of cells able to escape the first, second, and third G1 intervals was determined by flow cytometry, as previously described (Linke, S.P., et al., 1996). In each case, the number of cells escaping the indicated G1 is calculated relative to the starting population. In addition, colony forming ability was determined after ten days relative to untreated controls. Note the progressive decline in the number of cells escaping later G1 intervals, and that the decline is steeper in p53+ than p53− cells. The decline in the potential to enter succeeding S phases also diminishes over time.

other abnormal chromosomes that are prone to breakage in later phases (Marder, B.A. and Morgan, W.F., 1993). For example, breakage of a dicentric chromosome in anaphase could trigger the p53-mediated arrest mechanism when the cells enter the next G1. This continuous elimination process would provide one explanation of why NDF fail to undergo gene amplification. This model is also consistent with our observation that irradiated NDF often form very small colonies with very low plating efficiencies (Linke, S.P., and Clarkin, K.C., unpublished observation). Note that p53-deficient NDF also exhibit a diminished ability to escape from the second and third G1 intervals following irradiation, and that the clonogenicty of these p53-deficient cells is also less than the fraction of cells that escape the first G1. The data indicate that a p53-independent mechanism of cell killing is also operating in these fibroblasts. Taken together, the data suggest that p53 helps maintain genetic stability by eliminating cells that have undergone DNA damage rather than enhancing their ability to repair the damage.

The Conditional Nature of Genetic Instability

Cells lacking p53 function should be able to cycle in the presence of chromosome damage. If such damage is generated during the growth of cells in tissue culture, one would expect heterogeneous populations to be generated, the members of which would contain structural and numerical chromosome alterations. If amplification of a region occurred that provided a growth advantage to a cell under a specified condition, a clone would emerge upon appropriate challenge. This is the view of gene amplification suggested by the results of Luria-Delbruck fluctuation analyses performed by several investigators (Kempe, T.D., et al., 1976, Tlsty, T.D., et al., 1989).

Therefore, we were surprised to find that selection of the fibrosarcoma cell line HT1080, which has a defective p53 pathway (Anderson, M.J., et al., 1994), failed to generate variants resistant to the antifolate MTX from more than 10^9 selected cells. In contrast, PALA treatment did not induce cell cycle arrest and resulted in the growth of numerous drug resistant HT1080 clones containing amplified CAD genes. Furthermore, MTX-resistant variants could be obtained if HT1080 cells were allowed to cycle for one or two S phases in the presence of PALA or other drugs that would induce DSBs prior to MTX challenge. The majority of MTX-resistant variants proved to have amplified copies of the DHFR gene (Paulson, T.G., et al., in preparation). This shows that loss of p53 function is necessary, but not sufficient, for gene amplification to occur.

The inability to obtain MTX-resistant variants from HT1080 cells correlated with induction of a late G1 or early S phase arrest by this drug. This is consistent with the data reported above on the metabolite specificity of the p53 ribonucleotide sensor. Both p53+ and p53− cells arrested in late G1 or early S-phase when dTTP pools were depleted by MTX treatment.

Taken together, these results are consistent with the view that the generation of structural chromosome changes, such as gene amplification, require that cells cycle through S phase under conditions that lead to chromatid (or chromosome) breakage. We infer that S-phase progression during nucleotide limitation creates a substrate that is readily broken. We note that p53+ cells that are already in S phase or are committed to entering S phase continue progression through the first S phase during nucleotide limitation, yet they do not seem to accumulate many aberrant chromosomes (Linke, S.P., et al., 1996). This may be explained in two ways. First, perhaps p53 is part of an S-phase replication fork progression checkpoint that limits breakage under nucleotide limiting conditions and/or enhances S phase repair. This is consistent with recent reports suggesting that p53-mediated induction of p21 may inhibit DNA replication without affecting DNA repair (Li, R., et al., 1994). A second possibility is that the p53+ cells have an intact G2 break checkpoint that arrests cells until repair is complete and/or prevents progression of cells with irreparable damage, minimizing the number of cells that enter metaphase with damage. This is supported by data showing a distinct and prolonged G2/M arrest in asynchronous populations of cells treated with nucleotide synthesis inhibitors (Linke, S.P., et al., 1996). In contrast, p53− cells may have lost the G2 damage checkpoint due to the inherent genetic instability of such cells (Paules, R.S., et al., 1995; Kaufmann, W., and Paules, R., personal communication). Alternatively, as p53− cells proceed into S phase with DNA damage or under conditions that can produce damage, activation of the G2 break checkpoint might occur frequently, which would slow cell cycle progression. Cells lacking this checkpoint would, therefore, grow faster and should be highly selected under cell culture conditions. Loss of p53 in some cell lines, and loss of both p53 and the G2 damage checkpoint in others, may explain why related cell lines exhibit different rates of gene amplification (Tlsty, T.D., et al., 1989).

CONCLUSION

These data lead to a more complete understanding of the cell cycle controls that insure genetic stability. p53 senses ribonucleotide limitation and prevents cells from entering S phase under such conditions. This is advantageous, as the bulk of dNTPs are derived by reduction of rNTPs, and rNTPs also serve as precursors of nucleotide-based signal transducers. The availability of a ribonucleotide sensor in early G1, therefore, reduces the probability of entering DNA replication with an insufficient precursor supply. If a cell proceeds through S phase and accumulates damage in excess of the repair capacity, p53 may be activated in the next G1 to eliminate variants with unrepaired or irrepairable damage. While the response to DSBs was discussed here, we have evidence that a p53-dependent G1 arrest is also activated in response to DNA mismatches, insertion-deletion loops, and gaps larger than 30 nucleotides, such as those produced during nucleotide excision repair (Huang, L-c., Clarkin, K., and Wahl, G. M., in preparation). Thus, p53 is activated by the major types of DNA damage created during DNA replication, or by bulky lesions processed by nucleotide excision repair. Should cells escape the G1 arrest induced by p53, they can be eliminated in subsequent cycles by the same mechanism. The system is highly efficient, preventing the propagation of variants that could arise from just a single chromosome break or a small number of more subtle lesions.

Many questions remain to be answered. The mechanisms by which the existence of DNA damage of different types is transmitted to p53 are unknown. It is possible that p53 senses the damage itself through its ability to bind DNA non-specifically (Bakalkin, G., et al., 1994, Jayaraman, J. and Prives, C., 1995, Lee, S., et al., 1995). Alternatively, proteins such as those in the PI-3 kinase family that includes the gene mutated in ataxia telangiectasia (ATM), may transduce and amplify the signal through phosphorylation of p53 (reviewed by Zakian, V.A., 1995). Interestingly, p53 is phosphorylated in vitro in its N-terminal transactivation domain by DNA-PK, a member of the ATM family kinase family which is activated by broken DNA (Gottlieb, T.M. and Jackson, S.P., 1993, Lees-Miller, S., et al., 1992). Mutation of DNA-PK gives rise to one form of severe combined immunodeficiency (scid) in mice (Blunt, T., et al., 1995, Kirchgessner, C.U., et al., 1995). However, we recently found that the G1 damage and metabolite and spindle checkpoint functions of p53 are intact in scid mice, implying that residual DNA-PK activity exists in scid fibroblasts, that DNA-PK is not involved in upstream signal transduction to p53, or that another related kinase can substitute for it in scid cells (Huang, L.-c., et al., 1996). It is also unclear why equivalent activation of p53 by ribonucleotide depletion and DNA damage leads to a reversible arrest in the former case, and apoptosis or senescence in the latter. p53 responds to varied stresses, yet the mechanisms of activation by each remain to be resolved. It remains to be shown whether any of the highly conserved serines, that are in vitro and/or in vivo targets of numerous cell cycle- or metabolically-regulated kinases, contribute to regulation of p53 functions. Finally, we would be remiss if we did not point out the unexplored potential of the p53 metabolite sensor for chemotherapy. p53− cells are exquisitely sensitive to cell killing by certain nucleotide synthesis inhibitors, while at least some normal cells undergo a reversible arrest with such agents. Since p53 is mutated in more than 50% of human cancers, the possibility exists for the development of highly selective chemotherapeutic strategies applicable to a broad range of human neoplasms.

ACKNOWLEDGMENTS

We thank Kristie C. Clarkin for expert technical assistance that contributed to the success of many of the experiments presented in this report. We also acknowledge the participation and helpful suggestions of former members of the lab including: Alexandru Almasan, Aldo Di Leonardo, and Yuxin Yin. We also thank Drs. William Kaufmann and Rick Paules for many helpful discussions and for providing data in advance of publication. This work was supported by the grants from the National Institutes of Health and the G. Harold and Leila Y. Mathers Charitable Foundation. S.P.L was supported, in part, by a fellowship from the H.A. and Mary K. Chapman Charitable Trust. T.G.P was supported, in part, by a Public Health Service Genome Training Grant and the H.A. and Mary K. Chapman Charitable Trust. L-c.H. was supported by the Charles H. and Anna S. Stern Foundation.

REFERENCES

Agarwal, M., Agarwal, A., Taylor, W. and Stark, G., 1995. p53 controls both G2/M and G1 cell cycle checkpoints and mediates reversible growth arrest in human fibroblasts, *Proc Natl Acad Sci U S A* 92:8493–8497

Almasan, A., Linke, S. P., Paulson, T. G., Huang, L. and Wahl, G. M., 1995. Genetic instability as a consequence of inappropriate entry into and progression through S-phase, *Cancer Metas Rev* 14:59–73

Anderson, M. J., Casey, G., Fasching, C. L. and Stanbridge, E. J., 1994. Evidence that wild-type TP53, and not genes on either chromosome 1 or 11, controls the tumorigenic phenotype of the human fibrosarcoma HT1080, *Genes Chromosom Cancer* 9:266–281

Bakalkin, G., Yakovleva, T., Selivanova, G., Magnusson, K. P., Szekely, L., Kiseleva, E., Klein, G., Terenius, L. and Wiman, K. G., 1994. p53 binds single-stranded DNA ends and catalyzes DNA renaturation and strand transfer, *Proc Natl Acad Sci U S A* 91:413–417

Bhattacharyya, N. P., Skandalis, A., Ganesh, A., Groden, J. and Meuth, M., 1994. Mutator phenotypes in human colorectal carcinoma cell lines, *Proc Natl Acad Sci U S A* 91:6319–23

Blunt, T., Finnie, N. J., Taccioli, G. E., Smith, G. C., Demengeot, J., Gottlieb, T. M., Mizuta, R., Varghese, A. J., Alt, F. W., Jeggo, P. A. and Jackson, S. P., 1995. Defective DNA-dependent protein kinase activity is linked to V(D)J recombination and DNA repair defects associated with the murine scid mutation, *Cell* 80:813–823

Chen, C. Y., Oliner, J. D., Zhan, Q., Fornace, A. J., Vogelstein, B. and Kastan, M. B., 1994. Interactions between p53 and MDM2 in a mammalian cell cycle checkpoint pathway, *Proc Natl Acad Sci U S A* 91:2684–8

Chowdary, D. R., Dermody, J. J., Jha, K. K. and Ozer, H. L., 1994. Accumulation of p53 in a mutant cell line defective in the ubiquitin pathway, *Mol Cell Biol* 14:1997–2003

Clarke, A. R., Purdie, C. A., Harrison, D. J., Morris, R. G., Bird, C. C., Hooper, M. L. and Wyllie, A. H., 1993. Thymocyte apoptosis induced by p53-dependent and independent pathways, *Nature* 362:849–852

Cornforth, M. N. and Bedford, J. S., 1993. Ionizing Radiation Damage and Its Early Development in Chromosomes, *Adv in Rad Biol* 17:423–497

Cross, S. M., Sanchez, C. A., Morgan, C. A., Schimke, M. K., Ramel, S., Idzerda, R. L., Raskind, W. H. and Reid, B. J., 1995. A p53-dependent mouse spindle checkpoint, *Science* 267:1353–1356

Dbaibo, G. S., Pushkareva, M. Y., Jayadev, S., Schwarz, J. K., Horowitz, J. M., Obeid, L. M. and Hannun, Y. A., 1995. Retinoblastoma gene product as a downstream target for a ceramide-dependent pathway of growth arrest, *Proc Natl Acad Sci U S A* 92:1347–51

Deng, C., Zhang, P., Harper, J. W., Elledge, S. J. and Leder, P., 1995. Mice lacking p21CIP1/WAF1 undergo normal development, but are defective in G1 checkpoint control, *Cell* 82:675–84

Di Leonardo, A., Linke, S. P., Clarkin, K. and Wahl, G. M., 1994. DNA damage triggers a prolonged p53-dependent G1 arrest and long-term induction of Cip1 in normal human fibroblasts, *Genes Dev* 8:2540–51

Di Leonardo, A., Linke, S. P., Yin, Y. and Wahl, G. M., 1993. Cell cycle regulation of gene amplification, *Cold Spring Harbor Symp. Qaunt. Biol.* 58:655–67

Donehower, L. A. and Bradley, A., 1993. The tumor suppressor p53, *Biochim Biophys Acta* 1155:181–205

Donehower, L. A., Harvey, M., Slagle, B. L., McArthur, M. J., Montgomery, C. J., Butel, J. S. and Bradley, A., 1992. Mice deficient for p53 are developmentally normal but susceptible to spontaneous tumours, *Nature* 356:215–221

Dulic, V., Kaufmann, W. K., Wilson, S. J., Tlsty, T. D., Lees, E., Harper, J. W., Elledge, S. J. and Reed, S. I., 1994. p53- dependent inhibition of cyclin-dependent kinase activities in human fibroblasts during radiation-induced G1 arrest, *Cell* 76:1013–1023

El-Deiry, W. S., Harper, J. W., O'Connor, P. M., Velculescu, V. E., Canman, C. E., Jackman, J., Pietenpol, J. A., Burrell, M., Hill, D. E., Wang, Y., Wiman, K. G., Mercer, W. E., Kastan, M. B., Kohn, K. W., Elledge, S. J., Kinzler, K. W. and Vogelstein, B., 1994. WAF1/CIP1 is induced in p53-mediated G1 arrest and apoptosis, *Cancer Res* 54:1169–74

El-Deiry, W. S., Tokino, T., Velculescu, V. E., Levy, D. B., Parsons, R., Trent, J. M., Lin, D., Mercer, W. E., Kinzler, K. W. and Vogelstein, B., 1993. WAF1, a potential mediator of p53 tumor suppression, *Cell* 75:817–25

Eshleman, J. R., Lang, E. Z., Bowerfind, G. K., Parsons, R., Vogelstein, B., Willson, J. K., Veigl, M. L., Sedwick, W. D. and Markowitz, S. D., 1995. Increased mutation rate at the hprt locus accompanies microsatellite instability in colon cancer, *Oncogene* 10:33–7

Fearon, E. R., Hamilton, S. R. and Vogelstein, B., 1987. Clonal Analysis of Human Colorectal Tumors, *Science* 238:193–7

Fearon, E. R. and Vogelstein, B., 1990. A genetic model for colorectal tumorigenesis, *Cell* 61:759–67

Fiscella, M., Zambrano, N., Ullrich, S. J., Unger, T., Lin, D., Cho, B., Mercer, W. E., Anderson, C. W. and Appella, E., 1994. The carboxy-terminal serine 392 phosphorylation site of human p53 is not required for wild-type activities, *Oncogene* 9:3249–57

Fishman, L. J., Rudin, N. and Haber, J. E., 1992. Two alternative pathways of double-strand break repair that are kinetically separable and independently modulated, *Mol Cell Biol* 12:1292–1303

Fontoura, B. M., Sorokina, E. A., David, E. and Carroll, R. B., 1992. p53 is covalently linked to 5.8S rRNA, *Mol Cell Biol* 12:5145–5151

Fukasawa, K., Choi, T., Kuriyama, R., Rulong, S. and Vande Woude, G. F., 1996. Abnormal centrosome amplification in the absence of p53, *Science* 271:1744–1747

Gottlieb, T. M. and Jackson, S. P., 1993. The DNA-dependent protein kinase: requirement for DNA ends and association with Ku antigen, *Cell* 72:131–42

Harper, J. W., Adami, G. R., Wei, N., Keyomarsi, K. and Elledge, S. J., 1993. The p21 Cdk-interacting protein Cip1 is a potent inhibitor of G1 cyclin-dependent kinases, *Cell* 75:805–816

Hartwell, L. H. and Weinert, T. A., 1989. Checkpoints: controls that ensure the order of cell cycle events, *Science* 246:629–634

Hayflick, L., 1965. The limited in vitro lifetime of human diploid fibroblast strains, *Exp. Cell Res.* 37:614

Huang, L.-c., Clarkin, K. C. and Wahl, G. M., 1996. Sensitivity and selectivity of the p53-mediated DNA damage sensor, *Proc Natl. Acad Sci. U S A* 93:4827–4832

Hupp, T. R. and Lane, D. P., 1995. Two distinct signaling pathways activate the latent DNA binding function of p53 in a casein kinase II-independent manner, *J Biol Chem* 270:18165–74

Hupp, T. R., Meek, D. W., Midgley, C. A. and Lane, D. P., 1992. Regulation of the specific DNA binding function of p53, *Cell* 71:875–86

Hupp, T. R., Meek, D. W., Midgley, C. A. and Lane, D. P., 1993. Activation of the cryptic DNA binding function of mutant forms of p53, *Nucleic Acids Res* 21:3167–74

Hupp, T. R., Sparks, A. and Lane, D. P., 1995. Small peptides activate the latent sequence-specific DNA binding function of p53, *Cell* 83:237–45

Ishizaka, Y., Chernov, M., CM, B. and Stark, G., 1995. p53 dependent growth arrest of ref52 cells containing newly amplified DNA., *Proc Natl Acad Sci U S A* 92:3224–3228

Jayadev, S., Liu, B., Bielawska, A. E., Lee, J. Y., Nazaire, F., Pushkareva, M., Obeid, L. M. and Hannun, Y. A., 1995. Role for ceramide in cell cycle arrest, *J Biol Chem* 270:2047–52

Jayaraman, J. and Prives, C., 1995. Activation of p53 sequence-specific DNA binding by short single strands of DNA requires the p53 C-terminus, *Cell* 81:1021–1029

Kadyk, L. C. and Hartwell, L. H., 1992. Sister chromatids are preferred over homologs as substrates for recombinational repair in Saccharomyces cerevisiae, *Genetics* 132:387–402

Kastan, M. B., Onyekwere, O., Sidransky, D., Vogelstein, B. and Craig, R. W., 1991. Participation of p53 protein in the cellular response to DNA damage, *Cancer Res* 51:6304–11

Kempe, T. D., Swyryd, E. A., Bruist, M. and Stark, G. R., 1976. Stable mutants of mammalian cells that overproduce the first three enzymes of pyrimidine nucleotide biosynthesis, *Cell* 9:541–550

Kirchgessner, C. U., Patil, C. K., Evans, J. W., Cuomo, C. A., Fried, L. M., Carter, T., Oettinger, M. A. and Brown, J. M., 1995. DNA-dependent kinase (p350) as a candidate gene for the murine SCID defect, *Science* 267:1178–83

Ko, L. J. and Prives, C., 1996. p53: puzzle and paradigm, *Genes & Dev.* 10:1054–1072

Lane, D. P., 1992. Cancer. p53, guardian of the genome, *Nature* 358:15–6

Lee, S., Elenbaas, B., Levine, A. and Griffith, J., 1995. p53 and its 14 kDa C-terminal domain recognize primary DNA damage in the form of insertion/deletion mismatches, *Cell* 81:1013–1020

Lees-Miller, S., Sakaguchi, K., Ullrich, S. J., Appella, E. and Anderson, C. W., 1992. Human DNA-activated protein kinase phosphorylates serines 15 and 37 in the amino-terminal transactivation domain of human p53, *Mol Cell Biol* 12:5041–9

Levine, A. J., Chang, A., Dittmer, D., Notterman, D. A., Silver, A., Thorn, K., Welsh, D. and Wu, M., 1994. The p53 tumor suppressor gene, *J Lab Clin Med* 123:817–23

Li, C. Y., Nagasawa, H., Dahlberg, W. K. and Little, J. B., 1995. Diminished capacity for p53 in mediating a radiation- induced G1 arrest in established human tumor cell lines, *Oncogene* 11:1885–92

Li, R., Waga, S., Hannon, G. J., Beach, D. and Stillman, B., 1994. Differential effects by the p21 CDK inhibitor on PCNA- dependent DNA replication and repair, *Nature* 371:534–537

Linke, S.P., Clarkin, K. C., DiLeonardo, A., Tsou, A. and Wahl, G.M., 1996. A reversible p53-dependent G0/G1 arrest induced by rNTP depletion in the absence of detectable DNA damage, *Genes & Dev.* 10:934–947

Little, J. B., 1968. Delayed initiation of DNA synthesis in irradiated human diploid cells, *Nature* 218:1064–5

Livingstone, L. R., White, A., Sprouse, J., Livanos, E., Jacks, T. and Tlsty, T. D., 1992. Altered cell cycle arrest and gene amplification potential accompany loss of wild-type p53, *Cell* 70:923–35

Lowe, S. W., Schmitt, E. M., Smith, S. W., Osborne, B. A. and Jacks, T., 1993. p53 is required for radiation-induced apoptosis in mouse thymocytes, *Nature* 362:847–849

Malkin, D., Li, F., Strong, L. C., Fraumeni, J. F., Nelson, C. E., Kim, D. H., Gryka, G., Bischoff, F. Z., Tainsky, M. A. and Friend, S. H., 1990. Germ line p53 mutations in a familial syndrome of sarcomas, breast cancer and other neoplasms, *Science* 250:1233–1238

Maltzman, W. and Czyzyk, L., 1984. UV irradiation stimulates levels of p53 cellular tumor antigen in nontransformed mouse cells, *Mol Cell Biol* 4:1689–1694

Manchester, K. L., 1995. Theodore Boveri and the origin of malignant tumours, *Trends in Cell Biol.* 5:384–387

Marder, B. A. and Morgan, W. F., 1993. Delayed chromosomal instability induced by DNA damage, *Mol Cell Biol* 13:6667–6677

Marechal, V., Elenbaas, B., Piette, J., Nicolas, J. C. and Levine, A. J., 1994. The ribosomal L5 protein is associated with mdm-2 and mdm-2-p53 complexes, *Mol Cell Biol* 14:7414–7420

Momand, J., Zambetti, G. P., Olson, D. C., George, D. and Levine, A. J., 1992. The mdm-2 oncogene product forms a complex with the p53 protein and inhibits p53-mediated transactivation, *Cell* 69:1237–45

Nelson, W. G. and Kastan, M. B., 1994. DNA strand breaks: the DNA template alterations that trigger p53-dependent DNA damage response pathways, *Mol Cell Biol* 14:1815–1823

Paules, R. S., Levedakou, E. N., Wilson, S. J., Innes, C. L., Rhodes, N., Tlsty, T. D., Galloway, D. A., Donehower, L. A., Tainsky, M. A. and Kaufmann, W. K., 1995. Defective G2 checkpoint function in cells from individuals with familial cancer syndromes, *Cancer Res* 55:1763–73

Samad, A. and Carroll, R. B., 1991. The tumor suppressor p53 is bound to RNA by a stable covalent linkage, *Mol Cell Biol* 11:1598–1606

Scheffner, M., Huibregtse, J. M., Vierstra, R. D. and Howley, P. M., 1993. The HPV-16 E6 and E6-AP complex functions as a ubiquitin-protein ligase in the ubiquitination of p53, *Cell* 75:495–505

Scheffner, M., Werness, B. A., Huibregtse, J. M., Levine, A. J. and Howley, P. M., 1990. The E6 oncoprotein encoded by human papillomavirus types 16 and 18 promotes the degradation of p53, *Cell* 63:1129–36

Shay, J. W., Wright, W. E. and Werbin, H., 1991. Defining the molecular mechanisms of human cell immortalization, *Biochim Biophys Acta* 1072:1–7

Smith, K. A., Stark, M. B., Gorman, P. A. and Stark, G. R., 1992. Fusions near telomeres occur very early in the amplification of CAD genes in Syrian hamster cells, *Proc. Nat. Acad. Sci. U S A* 89:5427–5431

Soussi, T., Caron, de, Fromentel, C and May, P., 1990. Structural aspects of the p53 protein in relation to gene evolution, *Oncogene* 5:945–52

Srivastava, S., Zou, Z., Pirollo, K., Blattner, W. and Chang, E. H., 1990. Germ-line transmission of a mutated p53 gene in a cancer-prone family with Li-Fraumeni syndrome, *Nature* 348:747–749

Tlsty, T. D., 1990. Normal diploid human and rodent cells lack a detectable frequency of gene amplification, *Proc Natl Acad Sci U S A* 87:3132–6

Tlsty, T. D., Margolin, B. H. and Lum, K., 1989. Differences in the rates of gene amplification in nontumorigenic and tumorigenic cell lines as measured by Luria-Delbruck fluctuation analysis, *Proc.Natl.Acad Sci U S A* 86:9441–9445

White, A. E., Livanos, E. M. and Tlsty, T. D., 1994. Differential disruption of genomic integrity and cell cycle regulation in normal human fibroblasts by the HPV oncoproteins, *Genes Dev* 8:666–77

Windle, B., Draper, B. W., Yin, Y. X., O'Gorman, S. and Wahl, G. M., 1991. A central role for chromosome breakage in gene amplification, deletion formation, and amplicon integration, *Genes Dev* 5:160–174

Wright, J. A., Smith, H. S., Watt, F. M., Hancock, M. C., Hudson, D. L. and Stark, G. R., 1990. DNA amplification is rare in normal human cells, *Proc Natl Acad Sci U S A* 87:1791–5

Wu, X., Bayle, J. H., Olson, D. and Levine, A. J., 1993. The p53-mdm-2 autoregulatory feedback loop, *Genes Dev* 7:1126–32

Yeargin, J. and Haas, M., 1995. Elevated levels of wild-type p53 induced by radiolabeling of cells leads to apoptosis or sustained growth arrest, *Curr Biol* 5:423–31

Yin, Y., Tainsky, M. A., Bischoff, F. Z., Strong, L. C. and Wahl, G. M., 1992. Wild-type p53 restores cell cycle control and inhibits gene amplification in cells with mutant p53 alleles, *Cell* 70:937–48

Zakian, V. A., 1995. ATM-related genes: what do they tell us about functions of the human gene?, *Cell* 82:685–7

DISCUSSION

Stark: Perhaps I can start the discussion with what, for me anyway, is a bit of a conundrum; I do not understand your results with methotrexate in terms of your model. Methotrexate does more, of course, than block dNTP synthesis, it blocks one-carbon metabolism and so blocks *de novo* purine biosynthesis. So in methotrexate treated cells, you do not have a synthesis of GTP or ATP, those ribotriphosphates. So methotrexate will starve cells for ribotriphosphates, how come it does not initiate the mechanism that you have specified?

Wahl: Of course that is true but it is concentration dependent. Under our conditions, or at least in these fibroblasts, methotrexate is acting specifically at the level of dTTP as both thymidine and BrdU abrogate the arrest.

Livingston: Do your experiments yield the same results in genetically matched cell lines which are otherwise wild type excepted at the p53 locus?

Wahl: Yes, I understand that caveat. Genetically matched cell lines, I think, is an oxymoron.

Livingston: P53 plus plus versus p53 minus minus.

Wahl: We have looked at Li-Fraumeni cells, which are not genetically matched relative to the heterozygous parent. We have looked at Li-Fraumeni cells in which the p53 transgene has been put back again.

Livingston: What about mouse cells? You know, knockout cells?

Wahl: We have looked at mouse cells, as have Deng and colleagues, and it is quite clear that the PALA is working directly through the p53, p21 pathway as well. Thea Tlsty also looked at $p53^+$ and $p53^-$ mouse cells in her 1992 paper and showed that PALA induced arrest is clearly dependent on p53.

Livingston: It would be interesting to know whether with nucleotides RNTP depletion there is a stop in G_1, and if that is true, if that is p53 sensitive.

Wahl: We have done that. We have not published it but Thea has published it, and also in the controls for the Deng paper on p21, they published it. So it works there as well. They only use PALA. But everything that we found to be true of PALA holds for the other inhibitors (for example, cyclopentenyl cytosine, tetrahydrouridine, etc.), so I believe that is correct.

Livingston: Can you do the experiment plus and minus E6 without BrdU and get the same kind of result? One wonders about BrdU enhancement of nicking and that kind of thing.

Wahl: No, you really have to track where the cells were. We tried to look for nicking in the DNA introduced by BrdU and have been unable to detect it. We are also extraordinarily careful, and keep the lab dark during experiments. Furthermore, the arrest is completely reversible, which would not be expected if damage were induced by drug treatment.

Mihich: There are now some new thymidylate synthase inhibitors which do not have complicating factors; have you had the chance to try those and see in relation to methotrexate and PALA what they are doing?

Wahl: There are several inhibitors that are being made by AGURON and we have analyzed one or two of those. They are specific to thymidine synthetase and they will allow the cell to do basically the same thing as HU does. They will allow the cell to go through G1 and then stop somewhere is S. They do not elicit the G1 arrest of PALA or other rNTP synthesis inhibitors.

Bacchetti: I was wondering whether you had measured ribonucleotides reductase activity in the different cell types? With PALA, p53 minus, p53 plus?

Wahl: We did not measure ribonucleotides reductase but what we did do is use hydroxyurea as a way of inhibiting rNTP to dNTP conversion. We published the flow sort analysis of HU treatment. It shows that HU treat $p53^+$ and $p53^-$ cells stop either close to the G1-S boundary or just inside S-phase.

Bacchetti: But do you think that perhaps depletion of one ribonucleotide may have a different effect on ribonucleotide reductase activity than the hydroxyurea block?

Wahl: Right. We can look at that, I have the paper here. They look to me to be pretty similar. Which is honestly surprising to me, because I felt that if in $p53^-$ cells (maybe David can fill me in on this) as I would have expected that E2F would have been constitutively active. There are other E2F regulators, but I assume that DHFR and TS should have been more abundant in $p53^-$ cells and they should have progressed further into S-phase. Maybe the cells do proceed further, and the level of resolution is insufficient to see it.

Bacchetti: Then I have just a comment. You said that HT1080 cells have wild type p53 but I thought that Eric Stanbridge has shown that there is a mutation in p53 in these cells.

Wahl: HT1080 is very complicated. You have to make sure whom you get it from. Our HT1080 has a wild type p53 gene, but a defective p53 pathway.

Stark: HT1080 as you get them from ATCC, have wild type p53 by sequencing, and it is fully functional, at least as far as inducing p21 and other p53 dependent genes. I will show you some of that later on. Eric Stanbridge's cells have been mutagenized, unfortunately, and as an unwanted consequence, he mutated the p53. He agrees with that. Eric's cells are $p53^-$, everybody else's are $p53^+$.

Bacchetti: Thanks.

Stark: You are welcome.

Wahl: But again there, Silvia, it is very important that, although George said they are wild type for p53, we agree with that, they induce p21, we agree with that. I think we also agree that when you treat them with PALA, they do not arrest. This indicates a defect in the p53 pathway.

Stark: But while cells have a dominant N-RAS mutation, and after all they are human tumors, something has got to be wrong and I think RAS may have something to do with your abnormal phenotype, it would not be an unlikely possibility.

Pinedo: As a follow up to the questions by Dr. Stark, I wonder whether you have looked at gemcitabine's effect. This drug blocks pyrimidine synthesis much closer to CTP than PALA or MTX. Such a study might help to the understanding of the problem.

Wahl: The question is, have we looked at gemcitabine. I cannot remember if we analyzed that drug, but the point you are getting to is a really crucial one. The original experiments that suggested that cells that lose p53, cannot undergo apoptosis. Such cells are likely to be resistant to a whole series of agents. In fact, what I have shown you is the opposite of that; there are specific metabolic pathways that activate p53. In the absence of p53 the cell no longer can respond to drugs that deplete those crucial precursors and it undergoes lethal DNA synthesis. Since p53 is mutated in fifty percent of cancers, I think it is wise to analyze those pathways in greater detail to identify drugs that are currently in clinical trial that might show potential activity against p53 cells.

Nasmyth: I do not know whether you or anyone else here would want to comment on the fact that this G1S control is p53 dependent and, as far as we know, it is p21 dependent. I do not know whether you or anyone else would like to comment on the fact that, thus far, p21 does not seem to have cropped up as a tumor repressor gene nor do mice seem to have particular problems with it.

Wahl: Yes, I think I can comment on that and I think I understand why that is true. In the paper by Deng et al., they commented that the p21$^-$ mice were normal. They exhibited a partial defect in the damage responsive G1 arrest point. They were completely defective in their arrest response to PALA. But, importantly, when they looked at apoptosis in the thymus, it was normal. So my guess is that the apoptotic function of p53 is really the important one with regards to tumorigenesis. And since the p21$^-$ mice have normal apoptosis, my guess is that there are other mechanisms by which apoptosis can be induced. Thus, p21$^-$ deficiency does not abrogate the damage induced apoptotic response, and loss of apoptosis might be what is most highly selected in tumorigenesis. Thus, mutation of p21 would not be highly selected.

Nasmyth: That is my explanation for it. The other one would be the spindle checkpoint that Brian Reid noticed and it would be nice to make the link between p53 and genomic instability, and apoptosis does not give us that.

Wahl: Well it does in the sense that, in the data that I did not go through, one break is enough to activate this G1 arrest point. George Stark has similar data that he published. If, and we do not know this, if one break is enough to initiate apoptosis then you can imagine that a cell, let us say, goes into S-phase with suboptional nucleotides the genome may accumulate breaks. If one or more breaks do not get repaired during the G2 checkpoint, and the cell with damage goes through mitosis and gets back into G1, it will then be eliminated. Alternatively, chromatic fusion may benerate a dicentric chromatid. Dicentric breakage may activate an anaphase, G1 or possibly G2 checkpoint, resulting in cell elimination. It could be that p53 contributes to all of these checkpoints but p21 does not. And I think your point is very well taken because in the p21⁻ mice the spindle checkpoint is normal.

Stark: If I can make a comment, I think it is along the same lines as some of the points that Kim is trying to raise. I will talk about this a little bit this afternoon, but let me just say for now, that it is clear that p53 participates in additional cell cycle checkpoints besides the ones that have been talked about here. It is important for a G2 checkpoint, and I will talk about data later on, showing that it also mediates a checkpoint which monitors the completion of S-phase. In other words if you stop cells in S-phase in a p53 dependent manner, if they are normal, they will not enter mitosis and if they are abnormal with respect to p53 they will enter mitosis. So, p53 is operating at several different checkpoints in addition to the ones that Geoff mentioned on his slide. The other thing I want to mention is that, at least with respect to PALA resistance, CAD gene amplification in human cells is a relatively infrequent mechanism and we have done an extensive analysis of human cells and in quite strong contrast to rodent cells where CAD amplification is the only mechanism that anybody has ever seen, in human cells, it is a very minor mechanism. And so we have the additional issue of having to explain how additional mechanisms of PALA resistance are controlled in human cells in a p53 dependent manner. We understand, at least to a superficial degree, some of these mechanisms in HT1080 cells, for example, one is the generation of 2P chromosome fragments and 2P isochromosomes. These are very common mechanisms of PALA resistance in contrast to CAD gene amplification by the more usual mechanisms. And we also see a very high frequency of cells which overexpress CAD with no gene amplification at all. None of these things happen in normal cells and so we have the problem of trying to understand how a much more complex variety of changes are regulated in human cells in a p53 dependent manner. It is not just amplification.

Wahl: That was the point I was trying to make with the HT1080 experiment with methotrexate. Unscheduled progression through S-phase generate point and other mutations as well as chromosome breaks. The probability of generating these lesions is very, very low in p53⁻ cells. For example, in the micronucleus experiment the cells are treated with PALA, these are asynchronous cells, they all have the same levels of nucleotide depletion and yet in the p53⁺ cells, micronuclei are not generated. By contrast, similarly treated p53⁻ cells generate many micronuclei. That would link p53 function to replicon progression and the formation of breaks.

Hartwell: In the micronuclei experiments, I am just wondering whether micronuclei are useful diagnostic agents as I know you induce them with HU and you see a big increase; but is the basal level enough to be diagnostic for DNA breaks?

Wahl: People have used micronuclei to assay clastongenic as well as aneugenic events for many years and, in fact, it is a classic method for looking at radiation exposure. So it is to some extent useful for that. Micronuclei occur at very low frequency in normal cells in people who are not treated, and they are increased in response to a variety of conditions, such as radiation. I am sure there is literature on drug effects, but I have not analyzed it in detail.

Anderson: Granted that the CAD assay, as in PALAs is not ideal, but it seems like the gene amplification rate you see in the LFS cells, is nowhere near what it is in a lot of tumor lines that you published, or that Thea Tlsty has published. George Stark has published some interesting papers on ras contributing to genomic instability. I am just wondering whether the general assumption is out there, that p53 is the key driver of genomic instability. What is the best controlled experiment where you simply turn-on, turn-off p53? Do you really match the gene amplification rates that you typically see in tumor cells?

Wahl: p53 does not explain all of genetic instability, and I tried to emphasize that poing. Work from Kaufmann and Paules has shown that another very important checkpoint is in G2 when most of the repair of double strand breaks takes place. So there is a long G2 delay in response to double strand breaks. A substantial literature in radiobiology indicates that G2 delay is not under the control of p53, because $p53^-$ cells delay for the same time as $p53^+$ cells. But $p53^-$ cells are genetically unstable, which increases the frequency at which the G2 checkpoint is lost. Cells that have lost the G2 checkpoint may now accumulate chromosome changes at a greater rate. This may explain why Thea Tlsty observed the differences that she reported in cell lines that were related to one another but differed in terms of tumorigenesis. It could be that one cell line had p53 loss, but had the G2 checkpoint intact while another may have had both p53 and the G2 checkpoint inactivated.

Gray: Back to the issue of whether or not the general amplifications that we were discussing earlier are selected for, or just parasitic, or along for the ride. Have you looked at the co-amplification or the simultaneous amplification of other sequences that are known to be amplified in other cells in your PALA treated cells?

Wahl: Well I am glad you asked that question. We are attempting to do such an experiment using comparative genomic hybridzation.

Stark: So the answer to the question is: Do the experiment yourself. Geoff, you have analyzed the effect of PALA in experimentally very convenient systems going from arrested cells to cells that enter cycle, and I am not sure that that is the same thing as answering the question in cells that have not been serum starved and are in-cycle. Do you think there is a way to answer that question about whether or not the same thing is going on there?

Wahl: Well, in fact, Aldo answered that question and it is published in the "Genes and Development" paper that came out in April, and basically the same things happen. The problem, of course, when you are dealing with asynchronous cells is that you can get some damage generated during S-phase. So the experiments are a bit more difficult to interpret. And the reason I chose to present analysis of synchronized cells is because it yields the most unambiguous results. But yes, we have done it both ways in many different cells with the same result.

4

ANEUPLOIDY AND HETEROGENEITY MECHANISMS IN HUMAN COLORECTAL TUMOR PROGRESSION

Walter Giaretti

Laboratory of Biophysics and Cytometry
National Institute for Cancer Research
Genoa, Italy

1. INTRODUCTION

Loss of integrity and cumulative changes in the cellular genome is a hallmark of the multistep process of cancer which progressively leads to uncontrolled cell growth and metastasis (1–3). Genome loss of integrity, genome instability, structural and numerical chromosome aberrations, and aneuploidy are interchangeable concepts and related experimental parameters. They are probably connected to one of the historic hallmarks of cancer (back to the first two decades of this century) that was represented by the observations that tumor cells were characterized by large hyperchromatic nuclei with excess of DNA and by abnormal mitoses (4,5). Genetic alterations appear to be linked to chromosomal aberrations and aneuploidy in cancer cells (6–11), but a cause-effect relationship is still unproven. The large number of genetic and chromosome changes detected by molecular biology, classical cytogenetics and, more recently, by chromosome painting and interphase cytogenetics using fluorescence in situ hybridization (12,13), appears simply too difficult to be set in a cause-effect relationship. An initial preneoplastic genetic event predisposing to more generalized disruption of the genome has been postulated (1). Fragility and instability of the chromosomes may imply a variety of inherited and acquired genetic mechanisms as well as the influence of environmental agents (14–16). Probably, specific genetic changes direct the process of aneuploidization. However, this is not the common view since many scientists believe that aneuploidy is an epiphenomenon of random origin.

2. RESULTS AND DISCUSSION

2.1. Human Colorectal Tumor Progression as a Model for the Study of Aneuploidy Mechanisms

The study of aneuploidization and evolution in human in vivo tumors may be complicated by the fact that parental subclones may be overgrown by successive subclones.

Genomic Instability and Immortality in Cancer
edited by Mihich and Hartwell, Plenum Press, New York, 1997

Aneuploidy evolution is likely to happen in tumors and, therefore, it is important to consider not only advanced malignant lesions, but also their precursors if possible, and to exclude tumors treated by chemical and radiation therapies.

Colorectal cancer represents today one of the most potentially informative systems for the study of the origin and evolution of aneuploidy in humans. In fact, several molecular biology events were discovered to be associated to different stages of tumor progression and, in particular, samples from these stages are easily available. Although *de novo* colorectal cancer has been reported (17), epidemiological, histological, and pathobiological data exist which indicate the dysplastic adenomas as the focal precursors of most human adenocarcinomas of the large bowel (18–22). Histologically, adenomas with mild, moderate and severe dysplasia are intermediate steps, and carcinomas infiltrating the intestinal wall through the muscularis mucosa are the endpoint of the sequence. Adenomas increase in size parallels grade of dysplasia (23), although small adenomas have been reported to regress completely (24). Rate of malignant transformation of adenomas is about 0.25% per year (25) in association with number of polyps, histotype, size, grade of dysplasia and presence of atypical mitoses (26–30). Chromosomal instability in early and late stages of colorectal tumor progression, as well as the link with inherited (31) and acquired (32–35) gene defects and with the growth-regulating functions of stimulatory and suppressor genes (36–39) are essential issues which remain obscure.

Clearly, the study of the chronological appearance of aneuploidy and stemline heterogeneity in human adenomas by sequential biopsies is to be avoided for ethical reasons. According to the present clinical guide lines, colorectal adenomas visible during endoscopy and with an approximate size of 5mm or larger are removed and the patients generally enter follow-up studies with later endoscopic controls. Thus, an interesting experimental possibility remains with the analysis of numerous samples from both benign and malignant lesions of the same tumor type of untreated patients. Knowing that aneuploidy is present in the vast majority (>80%) of the colorectal adenocarcinomas and for the previous reasons, the colorectal adenoma-carcinoma sequence appears to be quite an appropriate model for the study of aneuploidy mechanisms.

2.2. The Unique Contribution of Flow Cytometry

Aneuploidy has been classically approached by chromosome analysis during metaphase. This methodology, however, is often very complex and technically difficult when applied to solid tumors. In addition, classical cytogenetics would not address DNA heterogeneity of cell populations within G0-G1, S, and G2 compartments of the cell cycle.

A unique technique able to address tumor DNA aneuploidy and heterogeneity is DNA content flow cytometry. In addition, the technical facility of "cell/nuclei sorting", that many flow cytometers incorporate, make it uniquely possible to investigate more adequately the underlying molecular genetic mechanisms of the origin and evolution of aneuploidy.

Detection of total DNA changes corresponding approximately to addition or loss of one medium-size human chromosome (±3–4% DNA change) is feasible with this technique using fresh-frozen tissues.

Fresh-frozen specimens for flow cytometry (FCM) measurements were treated as previously detailed (40). Paraffin embedded material was not included because of the inadequacy of the samples for accurate DNA content detection (40). The measurements were taken with a FACS 440 dual laser flow sorter system (Becton-Dickinson, Sunnyvale, CA) that was also used for preparing "sorted nuclei" samples to be submitted to molecular biology. Three parameters were simultaneously measured in list mode acquisition for every individual nucleus, i.e., blue emission (from Dapi), 0°-forward Scatter and 90°-Scatter. The two-parameter combination given by 90°-scatter and 0°-scatter allowed to separate lymphocytes and epithelial nuclei. DNA content of epithelial nuclei was evaluated relatively to DNA content of tissue infiltrating lymphocytes in both adenomas and control mucosa (41). Mixed samples of tissue nuclei and individual specific lymphocytes showed that infiltrating and external lymphocytes superimposed in all cases. Trout erythrocytes and individual-specific normal mucosa were also used as reference DNA standards. Two clear-cut separated DNA peaks were observed in the original DNA content histograms or DNA histogram projections from the above combinations of parameters in the vast majority of the cases. The degree of DNA aneuploidy (DI) was calculated as ratio of mean channel number of epithelial aneuploid G0-G1 peak to mean channel number of peak corresponding to tissue infiltrating G0-G1 lymphocytes and/or control mucosa. Additionally, G0-G1 DNA content in tumor lesions was considered DNA aneuploid when falling outside the DI±3SD range of mucosa of healthy donors with respect to mucosa infiltrating lymphocytes (41). One region of abnormal DI values of special interest was defined as near-diploidy (DI≠1 and DI≤1.4), in comparison with near-triploidy (DI=1.5±0.1), and high aneuploidy (DI>1.6 and DI≠2). DNA tetraploidy (DI=2) was defined at the threshold value above the mean G2+M peak size among the mucosa of healthy donors plus 3SD values and presence of S- and G2M- tetraploid cells. CV values of the G0-G1 peaks and S-phase fraction values were evaluated as reported previously (22,40–43). Sorting of approximately 20.000 nuclei, enriched of the epithelial cell component (to approximately 100% for the well separated DNA aneuploid peaks) was done in order to perform K-ras2 and p53 mutation analysis for DNA content homogeneous subclones.

2.3. DNA Aneuploidy Indices in Colorectal Adenomas and Adenocarcinomas

The DNA index values shown in Fig. 1 (top) represent the cumulative results obtained until October 1996 in our laboratory for human colorectal adenomas and adenocarcinomas investigated by flow cytometry using fresh-frozen material only.

The DNA aneuploidy incidence was approximately 30% in adenomas and 75% in adenomas with early cancer and adenocarcinomas. Figure 1 clearly shows that the aneuploid DNA index values (DI≠1) were highly heterogenous and characterized by a non-random distribution with modal values at DI=0.9, 1.15, 1.50, 1.80, 2.0 and 2.2. A valley was clearly evident at DI=1.3–1.4 which separated near-diploid subclones (peaks A and B) from near-triploid and hypotetraploid (peaks C and D), tetraploid (peak E) and hypertetraploid (peak F) subclones.

DNA aneuploid adenomas (Fig. 1, mid position, and Table 1) were characterized for the majority (72%) by DNA Index values in the near-diploid range (DI≤1.4). Fig. 1 (bottom) and Table 1, for only adenocarcinomas, similarly shows that the majority of aneuploid adenocarcinomas (about 72%) were with higher DNA Indices (DI>1.4).

2.4. Models on the Origin and Evolution of Aneuploidy and Heterogeneity

The fact that abnormal mitoses are commonly observed in both preneoplastic and neoplastic lesions in colon-rectum may suggest that tumor related gene defects and chromosomal rearrangements may increase the probability of errors during mitosis and lead to aneuploid cells by "loss of DNA symmetry" in DNA cell division (22,42–43). This

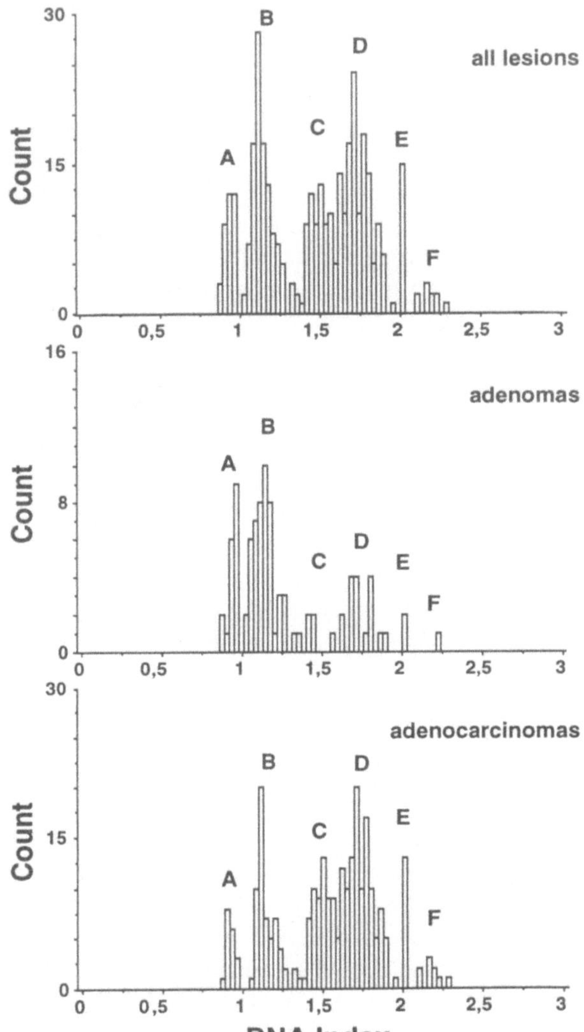

Figure 1. DNA aneuploid Index values (DI≠1, n=370) of intratumor subclones detected by flow cytometry among 340 human colorectal adenomas and 293 adenomas with early cancer and adenocarcinomas. The DI distributions are for both lesions together (top), for adenomas with low and high dysplasia (middle), and for adenomas with early cancer and adenocarcinomas (bottom). A, DNA hypodiploid and B, hyperdiploid near-diploid subclones; C, DNA near-triploid; D, hypotetraploid; E, tetraploid; F, hypertetraploid. Data corresponding to part of these lesions were published previously (22, 40–43, 75).

Table 1. Heterogeneity and relative frequencies of the degree of DNA aneuploidy values DNA Index, DI) as evaluated by flow cytometry among colorectal adenomas and adenocarcinomas (including adenomas with early cancer)[a]

Class	n	A	B	C	D	E	F
All lesions	370	36 (10%)	113 (31%)	64 (17%)	128 (34%)	14 (4%)	15 (4%)
Adenomas	94	18 (19%)	50 (53%)	5 (6%)	17 (18%)	2 (2%)	2 (2%)
Adenocarcinomas	276	18 (6.5%)	63 (23%)	59 (21.5%)	111 (40%)	12 (4.5%)	13 (4.5%)

[a]DI classes were: (**A**) DNA hypodiploid (DI<1); (**B**) hyperdiploid near-diploid (DI≠1 and DI≤ 1.4); (**C**) DNA near-triploid (1.4<DI≤1.6); (**D**) hypotetraploid (1.6<DI<2); (**E**) tetraploid (DI=2); (**F**) hypertetraploid (DI>2). Data corresponding to part of these lesions were published previously (22, 40–43). See also Figure 1.

hypothesis is contrary, however, to the classical view of the stem cell theory, according to which abnormal mitosis would be 'arrested' and would not generate abnormal G1-phase cells (44–48).

An alternative and more commonly accepted view is that aneuploidy in solid tumors is generated by the spontaneous tendency of diploid cancer cells to double their chromosome number and subsequently to loose chromosomes randomly (49). The condition for generating discrete aneuploid peaks, would then be the increased rate of structural abnormalities of the retained chromosomes with possible activation of growth promoting genes and oncogenes (49). This model, though supported by experimental evidence from in vitro and in vivo cancer systems, does not explain the high incidence of near-diploid DNA abnormal stemlines as observed during the colorectal adenoma-carcinoma sequence (22,41–43,50) (see also the two previous paragraphs). These experimental data would favor the alternative model of "loss of DNA symmetry" in abnormal mitoses and generation of near-diploid cells (22,42,43). That abnormal mitoses may complete cell division, albeit during an increased time, and then begin another replication cycle was indicated by recent experimental data (51,52). Due to these experimental data and hypotheses, it is possible that some among the so called "DNA tetraploid" cases, ill defined after measuring DNA content by flow cytometry, may just well indicate a simple increase of the number of mitotic abnormal cells and/or their increased residence time during M. In fact, flow cytometric measurements of the G2 and M phase cells are not separated and, moreover, DNA tetraploid proliferating subclones, characterized by their S-phase and G2+M fractions, are surely rare in several tumor types. In particular, they are rare in early colorectal adenomas where they were detected at the 2% level and approximately 80% among the DNA aneuploid subclones were near-diploid (see section 2.3).

According to the "loss of DNA symmetry" model, simple chromosome losses and gains and other more complex mechanisms such as unbalanced rearrangements (53) may generate two daughter cells with near-diploid aneuploidy. Since the precise chromosome changes are not identifiable with the present technique, the global general terminology "loss of DNA symmetry" in DNA cell division is proposed and used thereafter. "Loss of DNA symmetry" in cell division implies automatically different DNA content in the two daughter cells. DNA content unbalance may be due either to loss of genetic material (i.e., chromosomes, chromosomal fragments, micronuclei) from mitotic cells or to impairment and damage of the complex control mechanisms leading to exact DNA division of normal mitotic cells (54). Chromosome loss may be non-random, smaller size and relative C-band size chromosomes being correlated with higher rate of displacement in methaphase (55).

The experimental data shown in Fig. 1 strongly suggest DNA content evolution from adenomas to adenomas with early cancer and adenocarcinomas by tetraploidization of near-diploid (particularly hypodiploid) subclones. Lower incidence of hypodiploidy in comparison with near-diploid hyperdiploidy may be due to the fact that a hypodiploid postmitotic G1 cell, except when loosing suppressor genes, has a low probability to survive since loss of DNA may impair the synthesis of proteins necessary for its growth. The present concept of hypodiploidy as an early step of tumor aneuploidization/evolution certainly fits with the increasingly recognized importance of chromosome loss (suppressor genes), particularly in solid tumors such as colon cancer and breast cancer.

Figure 2 shows a schematic view of the concept of "loss of DNA symmetry" in cell division for the generation of two DNA index near-diploid cells (with approximately DI= 0.9 and 1.1) and two cells with DI=1.7 and 1.9 after evolution of a hypodiploid cell with DI=0.9 by endoduplication to an abnormal mitosis with DI=3.6. The low incidence of DI values above tetraploidy appears to be associated to a lower probability of tetraploidiza-

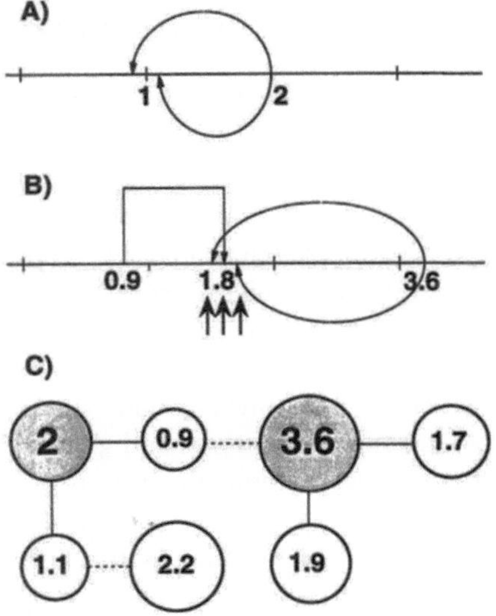

Figure 2. A model of DNA aneuploidization and heterogeneity in human colorectal cancer as adapted from previous publications (22, 42–43). Gray color is for mitotic cells, white for G1 cells; numbers are DNA index (DI) values. A): "loss of DNA symmetry" of an abnormal mitosis with DI = 2; B): endoreduplication of a hypodiploid cell (DI = 0.9) and "loss of DNA symmetry" of its abnormal mitosis with DI = 3.6; C): example of G1 cells (or subclones) with DI = 0.9, 1.1, 1.7, 1.9, and 2.2 generated according to the mechanisms illustrated in A) and B).

tion of the near-diploid hyperdiploid cells (1<DI≤1.4) with respect to the hypodiploid ones. The generation of the near-triploid peaks (DI=1.5±0.1) appears more difficult to be explained with the present hypotheses (see also next section 2.5).

Further experimental work is clearly necessary to better understand the relationship of "delayed" mitoses and abnormal G1-postmitotic cells which may potentially generate near-diploid subclones. It is of notice, however, that colorectal adenomas show massive presence of atypical mitoses (30) and that the existence of near-diploid tumors is well established in the cytogenetic and DNA flow cytometric literature. Within the subgroup of near-diploid colorectal tumors, non-random structural abnormalities have been reported as well as simple chromosome gains and losses (55–59). Recent studies on colorectal cancer have also shown that an increasing number of specific molecular lesions are associated with increased DI values (60). The present aneuploidization/evolution model appears also in agreement with colorectal carcinomas data of patients with verified cancer family syndrome, characterized by early age tumor, good prognosis, and DNA aneuploidy in the near-diploid region for the vast majority (90%) of cases (61), and with other data on sporadic cancer of colon-rectum (60,62), tumor lesions of breast (53), and Barrett's esophagus and adenocarcinomas (63).

On the other hand the model based on the diploid-tetraploid transition (49) is supported by data from both in vitro and in vivo cancer systems analyzed with both classical cytogenetics and DNA flow cytometry. In particular, in germ cell tumor of the testis and in prostate cancer, tetraploidization of DNA diploid cells appears to be an early event, and subsequent loss of chromosome is well documented (64–68). However, in colorectal cancer, it appears unlikely that the combined process of diploid-tetraploid shift and chromosome loss, would generate near-diploid subclones before generating, for example, hypotetraploid subclones.

2.5. A Link between DNA Aneuploidy and K-ras Mutations

Ras oncogenes were shown to be capable to transform mouse NIH-3T3 fibroblasts (69) and induce malignant transformation in Chinese hamster lung cells (70). Moreover, transfected ras oncogenes were reported to enhance genetic instability and predominantly integrate in aberrant chromosomes (71). The report that high expression of the p21(K-ras) mutated protein correlates with an increased rate of abnormal mitoses in NIH-3T3 cells also suggests an intriguing association between the ras protein and genetic instability (72). Mutations of the ras oncogene have been widely investigated in the human colorectal adenomas and carcinomas, and shown to consist manly of single base substitutions of the Kirsten-ras oncogene (K-ras) in codons 12 and 13 (21,34,35,73). It is interesting to observe that the frequency of K-ras mutations in human colorectal adenomas is approximately 50% and that these early preneoplastic tumor lesions are also characterized by frequent atypical mitoses (30) and DNA aneuploidy (22, 41–43, 50).

We have recently reported a relationship between the mutationally activated p21(K-ras) protein and DNA aneuploidization using mouse NIH-3T3 cells transfected with the human K-ras mutated oncogene as a model system in vitro. In particular, we have compared the frequency of abnormal mitoses, the cloning efficiency and the development of DNA aneuploid subclones from both control and K-ras transfected cells (74; see also previous section 2.3).

We found that abnormal mitoses, mainly characterized by lagging chromosomes in prometaphase or anaphase, had a significantly higher frequency in the transfected with respect to control cells (p=0.0001). The generation of subclones was screened by limiting dilution experiments followed by cell expansion. Cloning efficiency was much higher (p=0.0001) for the K-ras transfected cells with 858/2112 (41%) successful subclones with respect to controls that provided 564/2592 (22%) subclones. DNA flow cytometry of nuclei from randomly selected subclones was done in order to evaluate DNA Index values. We found 9 out of 100 DNA aneuploid subclones (with DNA Index values = 0.8; 1.1; 1.1; 1.4; 1.7; 1.9; 1.9; 2.0; 2.0) generated by the K-ras transfected cells versus 1 out of 100 for the controls (p=0.009). Overall, these data indicated that high expression of the mutationally activated human K-ras product in NIH-3T3 cells was associated with abnormal mitoses, increase of cloning efficiency and DNA aneuploidization. In particular, approximately 50% of the DNA aneuploid subclones generated during the limiting dilution experiments were found to be near-diploid (74).

We have also examined the possible association of K-ras2 mutations with DNA ploidy using human sporadic colorectal adenomas. The first 58 adenomas investigated were previously reported and indicated a strong association of G---C/T K-ras2 transvertions with DNA near-diplod aneuploidy(41). Table 2 represents an updated analysis with 115 adenomas.

Table 2. Association of G---C/T K-ras2 transversions with DNA near-diplod aneuploidy in 115 human colorectal adenomas (see text and also a previous study [41])

K-ras2 classes	DI (p=0.001)	
	DI=1	DI≠1
Wild type	53 (62%)	33 (38%)
G---A transitions	10 (71%)	4 (29%)
G---C/T transvertions	2 (13%)	13 (87%)

Notes. DI= DNA Index. DNA diploidy is with DI=1, and DNA aneuploidy is with DI≠1. Among 17 K-ras2 mutated DNA aneuploid subclones 16 were in the near-diplod range (DI≠1 and 2; DI≤1.4).

Table 2 shows a statistically significant association between K-ras2 status, subdivided in three classes, i.e., wild-type, transition mutations (G-A) and transversion mutations (G-C/T) and DNA index (DNA diploidy and aneuploidy) (p=0.001). It is interesting to note that among 17 K-ras2 mutated DNA aneuploid subclones only one was with DNA Index outside the near-diplod range (DI≠1 and 2; DI≤1.4). K-ras2 transitions were associated with DNA diploidy at 72%, transversions with DNA aneuploidy at 87%, whereas DNA aneuploidy occurred in adenomas without mutations at 38%.

A supplementary piece of information was recently reported by comparing DNA index values obtained from 97 intratumor sectors among 19 colorectal adenocarcinomas and their K-ras2 mutations (75). The main results of this study have indicated that K-ras2 mutations were intratumor heterogeneous in a subgroup of 1/3 of these tumors suggesting that the inactivation of K-ras2 may also be a late event of the colorectal carcinogenesis and that sampling of the tumor may be a critical step for molecular biology of K-ras2 (75). These data also suggested a low incidence of K-ras2 mutations among DNA near-diploid aneuploid subclones (75). This last finding was now confirmed with 133 adenocarcinomas (87% of the near-triploid subclones had wild type K-ras2). We may therefore conclude that onset of DNA near-triploid subclones appears to be independent from K-ras2 activation and to indicate a different tumor history and different aneuploidy mechanisms.

2.6. A Link between DNA Aneuploidy and p53 Mutations

Recently, literature reports suggested that p53 mutations may be directly related to aneuploidy since the wild-type p53 would prevent aneuploidy by causing apoptosis of cells with abnormal genomes (76,77).

At difference with other preneoplastic or malignancy predisposing lesions showing a correlation between aneuploidy and p53 inactivation like in Barrett's esophagus (78), colorectal adenomas were only rarely found to be p53 mutated. Since DNA aneuploidy is already present in approximately one third among colorectal adenomas (see also section 2.2), it appears that p53 independent pathways of aneuploidization exist during the colorectal adenoma-carcinoma sequence. This interpretation became reinforced in our hands after performing experiments of "nuclei sorting" by FACS of the epithelial DNA aneuploid component for a certain number of colorectal adenocarcinomas that clearly showed no p53 mutations in the exons 5–8 in approximately half of the cases. A second preliminary intriguing observation was that when both DNA aneuploidy and p53 inactivation were present, heteroduplexes did not seem to be present, suggesting a role of the inactivation of the p53 tumor suppressor gene for specific DI subclone selection. As previously demonstrated (see section 2.2), DI values above near-diploidy with a prevalence of hypotetraploid subclones are a characteristics of colorectal adenocarcinomas. Thus, a possibility exists that p53 inactivation is associated in colorectal adenocarcinomas with hypotetraploidy but not with near-triplody. This possibility on the role of p53 is not merely a speculation but was based as well on our recent preliminary data not yet published (Fronza G., Monaco R. and Giaretti W., unpublished observations). Rare K-ras mutations among near-triploid subclones in colorectal adenocarcinomas was also recently reported (75, see also section 2.5).

2.7. A Link between Aneusomy and 1p36 Deletions

Recent data obtained in our laboratory by fluorescence in situ hybridization (FISH) showed that 1p deletions (using probes to the 1p36 region) are at a relatively high frequency (approximately 40%) in colorectal adenomas (79) supporting the hypothesis that 1p deletions represent an early event in the colorectal tumorigenesis (80). Since literature studies have shown, using cytogenetic (81–83), flow cytometric (22, 40–43, 50) and FISH methodologies (84) that aneuploidy is already present in colorectal adenomas, we have tested the hypothesis that a link exists between 1p deletions and numerical aberrations of chromosomes 1, 7, 17 and 18 obtained by FISH and DNA aneuploidy obtained by multi-parametric flow cytometry. This was not previously investigated, to our knowledge. These data (Di Vinci et al., in preparation), that I anticipate shortly here, essentially showed a strong association of 1p deletions with the investigated aneusomies and DNA aneuploidy in the near-diploid region.

The most appealing hypothesis that may help interpreting these data appears that genes located on the 1p34–36 region could play a role in the maintenance of a normal numerical chromosomal set. This hypothesis is supported by literature findings suggesting that the deregulation of a human 58 kDA protein kinase, codified by a gene mapped on 1p36 region, increases the frequency of mitotic abnormalities and affects the rate of chromosomal numerical aberrations (85,86). A further hypothesis may be the presence of genes in the human 1p36 region regulating methylation. In fact, a "methylation modifier gene "was located in mouse chromosome 4 region syntenic to the human 1p36 band (87). Defects of DNA methylation have been observed in colorectal tumors already prior to progression toward the malignant state (88) and seem to correlate with chromosomal instability (89,90) and with the induction of mitotic disturbances.

3. CONCLUSION

The multiple accumulation of genetic and karyotypic alterations in solid tumor is extremely complex and, so far, these events have not allowed a clear understanding of the processes of tumor DNA aneuploidization and stemline heterogeneity. It is likely that different types of mechanisms of aneuploidization become sequentially operative at different times during the natural history of a neoplasm and that some of these mechanisms become mutually interactive. These mechanisms are likely to be of genetic origin (8,9) but epigenetic phenomena (21) and cell-cell interactions between clonal subpopulations of tumor cells within a neoplasm (92) have also been suggested.

The fact that aneuploidy values, as measured by flow cytometry (more correctly known as DNA Index values, abbreviated DI) do not show a flat distribution for most human tumors, but characteristic peaks as for colorectal lesions (see sections 2.3 and 2.4), suggest that aneuploidy should not be an epiphenomenon of random origin.

Probably, specific genetic changes direct the process of aneuploidization and of the later process of DNA tumor heterogeneity. Even when the tumor changes appear epigenetic or caused by external environmental factors, as for example for the process of methylation of DNA (21), it may well be that the effect of these epigenetic factors are "modulated" by specific genes (or even individual specific susceptibility-genes) and altered by these gene changes (85). Previous studies of our group have investigated if a link exists between DNA aneuploidy and K-ras mutations that were both detected in approximately one half among colorectal adenomas (41; see also section 2.6).

A highly statistically significant association of K-ras2 status with DNA ploidy could be observed. It is of notice that K-ras2 transitions were associated with DNA diploidy (72%) transversions with DNA near-diploid aneuploidy (87%), and DNA aneuploidy occurred in adenomas without K-ras2 mutations at 38%.

One possible interpretation of these findings is that a specific carcinogenic process in the colon of human patients favors the generation of adenomas with both K-ras transversions and aneuploidy and that a different carcinogenic process leads preferentially to adenomas with K-ras transitions without affecting DNA ploidy. This interpretation agrees with experiments performed in rats which indicate that different chemicals lead to different K-ras2 mutations.

Alternatively the fact that, among the adenomas investigated for both K-ras and DNA index, the vast majority of K-ras transversions were found among the DNA near-diploid aneuploid cases suggests that K-ras2 transversions may be related to abnormal mitoses and near-diploid aneuploidy in a cause-effect relationship. This is presently not proven. The observation that abnormal mitoses in colorectal adenomas are massively present (30) and that high expression of the p21ras protein was found to correlate with increased rate of abnormal mitosis in NIH3T3 cells (72,74) and the generation of a prevalence of near-diploid subclones in vitro (74) may, however, reinforce this hypothesis.

This suggested link between the ras protein and aneuploidy might be depending on other ras functions different from the classic signal transduction functions of ras, as for example those that regulate the formation of stress fibers, focal cell adhesion and cytokinesis as recently clarified (91–93).

The role of p53 in generating aneuploidization in human colorectal adenomas is less firm than with other precursor lesions like, for example, Barrett's esophagus (8,9,78). Inactivation of p53 was, in fact, only rarely detected in colorectal adenomas and p53 inactivation is considered a late-event of the colorectal tumor progression (21,34,35,56).

The link of other early genetic events of the colorectal carcinogenesis with DNA aneuploidy-heterogeneity, like the APC gene and mismatch repair genes, is also far from being understood. For mismatch repair genes, in particular, it is curious to observe that there is a clear DNA diploid predominance, as evaluated by flow cytometry, among hereditary non-polyposis colorectal carcinomas and that aneuploidy is mainly in the near-diploid region.

Recent literature data have shown that 1p deletions, as detected by FISH methodologies, represent an early event of the colorectal carcinogenesis (79) and that they are associated with aneusomy and near diploid aneuploidy (Di Vinci et al., in preparation). Whether or not specific deleted genes (possibly in the 1p34–36 region) play a role in directing the process of aneuploidization in the early steps of the colorectal carcinogenesis is not yet proven.

In conclusion, the processes of DNA aneuploidization and heterogeneity, that characterize around 80% of the human epithelial tumors, are not yet entirely clarified. It appears,however, that they may not be dismissed as epiphenomena and that specific defects in cell cycle checkpoints may be responsible for the genomic instability of cancer cells (8, 9). A tentative hypothetical scheme that tries to include some among the important genetic events of the colorectal tumor progression with the processes of DNA repair, apoptosis, aneuploidization, DNA heterogeneity and DNA subclone selection is shown in Fig. 3. It is likely that the understanding of these mechanisms will improve rapidly and that, when more precisely understood, may lead to a better understanding of the pathogenesis, regression, and progression of human epithelial malignancies.

ACKNOWLEDGMENTS

The author thanks the colleagues of his laboratory who have contributed to the research reviewed here including E. Geido, R. Orecchia and E. Zeraschi for flow cytometry

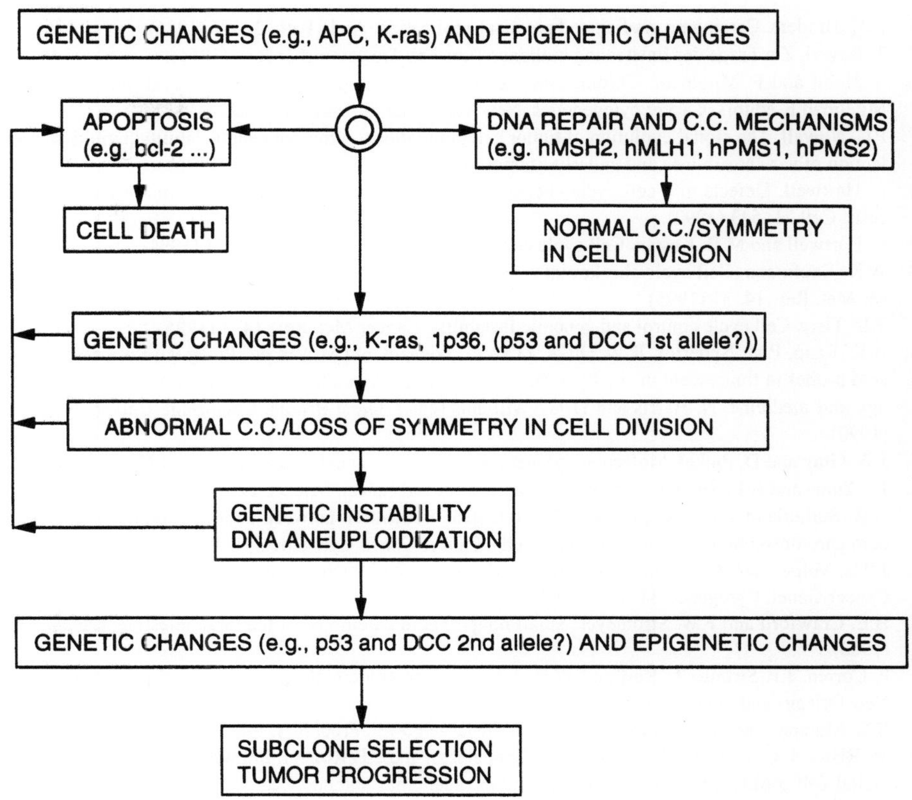

Figure 3. Hypothesized mechanisms of DNA aneuploidization and heterogeneity and association with apoptosis, DNA repair, subclone selection and tumor progression during the colorectal adenoma-carcinoma sequence as adapted from a previous publication (43). (C.C.=cell cycle.)

and sorting; A. Rapallo, N. Pujic and A. Sciutto for the K-ras analysis; A. Di Vinci , E. Infusini and C. Peveri for FISH analysis; G. Fronza and P. Campomenosi (Laboratory of Mutagenesis, National Institute for Cancer Research, Genoa and Chair of Genetics, University of Genoa, Prof. A. Abbondandolo) for the p53 analysis; H. Aste (Gastroenterology, National Institute for Cancer Research, Genoa), F.P. Rossini (Gastroenterology, Ospedale S. Giovanni Vecchio, Turin). I also thank my friends E. d'Amore (Pathology, University of Padova), M. Risio (Pathology, Ospedale of Alba) and R. Monaco (Pathology, Ospedale Cardarelli, Naples) for providing tumor samples and for helpful collaboration and stimulating discussions.

This work was supported by the National Research Council (CNR) grant n. 96.00708.PF39 ACRO and by the Italian Association for Cancer Research "Tumori Colorettali Ereditari".

REFERENCES

1. J. Cairns. The origin of human cancer. Nature 289: 353 (1981).
2. R.A. Weinberg. Oncogenes, antioncogenes, and the molecular bases of multistep carcinogenesis. Cancer Res. 49: 3713 (1989).
3. J.M. Bishop. The molecular genetics of cancer. Science 235: 305 (1987).

4. A.C. Broders. Carcinoma grading and practical application. Arch. Path. 2: 376 (1926).

5. T. Boveri. Zur Frage der Entstehung maligner Tumoren. Gustav Fisher Jcna (1914).

6. S. Heim and F. Mitelman. Chromosomal abnormalities in specific disorders: solid tumors. In: Heim S, Mitelman F, editors. Cancer cytogenetics. New York; Alan R. Liss Inc, p. 227 (1987).

7. P.C. Nowell. The clonal evolution of tumor cell populations. Acquired genetic lability permits stepwise selection of variant sublines and underlies tumor progression. Science 194: 23 (1976).

8. L. Hartwell. Defects in a cell cycle checkpoint may be responsible for the genomic instability of cancer cells. Cell 71: 543 (1992).

9. L. Hartwell and M.B. Kastan. Cell cycle control and cancer. Science 266: 1821 (1994).

10. W.K. Kaufmann. Cell cycle checkpoints and DNA repair preserve the stability of the human genome. Cancer Met. Rev. 14: 31 (1995).

11. T.D. Tlsty. Cell cycle control and genomic instability. Cancer Met. Rev. 14: 1 (1995).

12. A.K. Raap, P.M. Nederlof, R.W. Dirks, J.C.A.G. Wiegant and M. van der Ploeg. Use of haptenize nucleic acid probes in fluorescent in situ hybridization. In: In situ hybridization: application to developmental biology and medicine. N. Harris and D?G? Williams (eds.). Great Britain, Cambridge University press p. 33 (1990).

13. J.W. Gray and D. Pinkel. Molecular cytogenetics in human cancer diagnosis. Cancer 69: 1536 (1992).

14. J.J. Yunis and A.L. Soreng. Constitutive fragile sites and cancer. Science 226: 1199 (1984).

15. G.R. Sutherland and R.N. Simmers. No statistical association between common fragile sites and nonrandom chromosomal breakpoints in cancer cells. Cancer Genet. Cytogenet. 31: 9 (1988).

16. J.P.G. Volpe. Genetic instability of cancer: why a metastatic tumor is unstable and benign tumor is stable. Cancer Genet. Cytogenet. 34: 125 (1988).

17. B.E. Crawford and F.W. Stromeyer. Small nonpolypoid carcinoma of the large intestine. Cancer 51: 1760 (1983).

18. P. Correa, J.P. Strong, A. Reif and W.D. Johnson. The epidemiology of colorectal polyps. Prevalence in New Orleans and international comparison. Cancer 39: 2258 (1977).

19. B.C. Morson. The polyp-cancer sequence in the large bowel. Proc. Roy. Soc. Med. 67: 451 (1974).

20. M. Risio, S. Coverlizza, A. Ferrari, G.L. Candelaresi and F.P. Rossini. Immunohistochemical study of epithelial cell proliferation in hyperplastic polyps, adenomas and adenocarcinomas of the large bowel. Gastroenterology 94: 899 (1988).

21. E.R. Fearon and B. Vogelstein. A genetic model for colorectal tumorigenesis. Cell 61: 759 (1990).

22. W. Giaretti, S. Sciallero, S. Bruno, E. Geido, H. Aste and A. Di Vinci. DNA flow cytometry of endoscopically examined colorectal adenomas and adenocarcinomas. Cytometry 9: 238 (1988).

23. M.J. O'Brien, S.J. Winawer, A.G. Zauber, B. Diaz, L.S. Gottlieb, J. Bond et al. The national polyp study: determinants of high grade dysplasia in colorectal adenomas. Gastroenterology 98: 371 (1990).

24. G. Hoff, A. Foerster, M.J. Vatn, J. Savar and S. Larsen. Epidemiology of polyps in the rectum and colon. Recovery and evolution of unresected polyps 2 years after detection. Scand. J. Gastroenterol. 21: 853 (1986).

25. T.J. Eide. Risk of colorectal cancer in adenoma bearing individuals within a defined population. Int. J. Cancer 38: 173 (1986).

26. J.V. Selby, G.D. Friedman, C.P. Jr. Quesenberry and S.N. Weiss. A case-control study of screening sigmoidoscopy and mortality from colorectal cancer. New Engl. J. Med. 326: 659 (1992).

27. B.C. Morson, I.M.P. Dawson, J.R. Jass, A.B. Price and G.T. Williams. Morson & Dawson's gastrointestinal pathology. Third Edition. Blackwell Scientific Publications. Oxford, p. 577 (1990).

28. T. Muto, H.J.R. Bussey and B.C. Morson. The evolution of cancer of the colon and rectum. Cancer 36: 2251 (1975).

29. Y.S. Kim and S.H. Itskowitz. Carbohydrate antigen expression in the adenoma-carcinoma sequence. In: Basic and clinic perspectives of colorectal polyps and cancer. G. Jr. Steele, R. Burt, S.J. Winawer and J.P. Karr (eds.) New York, Alan R. Liss Inc., p. 241 (1988).

30. C.A. Rubio. Atypical mitosis in colorectal adenomas. Path. Res. Pract. 187: 508 (1991).

31. L.A. Cannon-Albright, M.H. Skolnick, D.T. Bishop, R.G. Lee and R.W. Burt. Common inheritance of susceptibility to colonic adenomatous polyps and associated colorectal cancers. New Engl. J. Med. 319: 533 (1988).

32. E. Solomon, R. Voss, V. Hall, W.F. Bodmer, J.R. Jass, A.J. Jeffreys, et al. Chromosome 5 allele loss in human colorectal carcinomas. Nature 328: 616 (1987).

33. G.C. Burmer and L.A. Loeb. Mutations in the K-ras2 oncogene during progressive stages of human colon carcinoma. Proc. Natl. Acad. Sci. 86: 2403 (1989).

34. B. Vogelstein, E.R. Fearon, S.R. Hamilton, S.E. Kern, A.C. Preisinger, B.A.M. Leppert, et al. Genetic alterations during colorectal-tumor development. New Engl. J. Med. 319: 525 (1988).

35. B. Vogelstein, E.R. Fearon, S.E. Kern, R.H. Stanley, A.C. Preisinger, Y. Nakamura, et al. Allelotype of colorectal carcinomas. Science 244: 207 (1989).

36. G.L. Nicolson. Tumor cell instability, diversification, and progression to the metastatic phenotype: from oncogenes to oncofetal expression. Cancer Res. 47: 1473 (1987).

37. D.J. Slamon. Proto-oncogenes and human cancers. New Engl. J. Med. 317: 955 (1987).

38. S.H. Friend, T.P. Dryja and R.A. Weinberg. Oncogenes and tumor-suppressing genes. New Engl. J. Med. 318: 618 (1988).

39. R. Muschel and L.A. Liotta. Role of oncogenes in metastases. Carcinogenesis 9: 705 (1988).

40. W. Giaretti, M. Danova, E. Geido, G. Mazzini, S. Sciallero, H. Aste et al. Flow cytometric DNA index in prognosis of colorectal cancer. Cancer 67: 1921 (1991).

41. W. Giaretti, N. Pujic, A. Rapallo, S. Nigro, A. Di Vinci, E. Geido and M. Risio. K-ras2 G-C and G-T transversions correlate with DNA aneuploidy in colorectal adenomas. Gastroenterology 108: 1040 (1995).

42. W. Giaretti and L. Santi. Tumor progression by DNA flow cytometry in human colorectal cancer. Int. J. Cancer 45: 597 (1990).

43. W. Giaretti. A model of DNA aneuploidization and evolution in colorectal cancer. Lab. Invest. 71: 904 (1994).

44. H.F. Stich and H.D. Steele. Content of tumor cells. Quantitative genetic analysis of tumor progression. Cancer Met. Rev. 4: 173 (1962).

45. J.C. Fardon and J.E. Prince. A comparison of the ratios of metaphase to prophase in normal and neoplastic tissues. Cancer Res. 12: 793 (1952).

46. T.C. Hsu and P.S. Moorhead. Chromosome anomalies in human neoplams with special reference to the mechanisms of polyploidization and aneuploidization in HeLa strain. Ann. NY Acad. Sci. 63: 1083 (1956).

47. S. Makino. Further evidence favoring the concept of the stem cell in ascites tumors of rats. Ann. NY Acad. Sci. 63: 818 (1956).

48. S. Makino. Neoplasia. In: Human Chromosomes. Igaku Shoin, Tokio, p. 429 (1987).

49. S.E. Shackney, C.A. Smith, B.W. Miller, D.R. Burholt, K. Murtha, H.R. Giles, et al. Model for the genetic evolution of human solid tumors. Cancer Res. 49: 3344 (1989).

50. H.F. Van den Ingh, G. Griffioen and C.J. Cornelisse. Flow cytometric detection of aneuploidy in colorectal adenomas. Cancer Res. 45: 3392 (1985).

51. W.C. Dooley, D.C. Allison and J. Robertson. Discrepancies among the metaphase, telophase, and the G0/G1 and G2 DNA peaks of heteroploid cell lines. Cytometry 12: 99 (1991).

52. W.C. Dooley and D.C. Allison. Non-random distribution of abnormal mitoses in heteroploid cell lines. Cytometry 13: 462 (1992).

53. B. Dutrillaux, M. Gerbault-Seureau, Y. Remvikos, B. Zafrani and M. Prieur. Breast cancer genetic evolution: I. Data from cytogenetics and DNA content. Breast Cancer res. Treat. 19: 245 (1991).

54. T.J. Mitchison. Mitosis: basic concepts. Curr. Op. Cell Biol. 1: 67 (1989).

55. J.H. Ford. Chromosome dysplacement hypothesis. In: B.K. Vig and A.A. Sandberg (eds.). Aneuploidy, incidence and etiology. Alan R. Liss, New York (1987).

56. M. Muleris, R.J. Salmon and B. Dutrillaux. Cytogenetics of colorectal adenocarcinomas. Cancer Genet. Cytogenet. 46: 143 (1990).

57. A.C. Williams, S.J. Harper,C.J. Marshall, R.W. Gill, R.A. Mountford and C. Paraskeva. Specific cytogenetic abnormalities and K-ras mutation in two new human colorectal adenoma derived cell lines. Int. J. Cancer 52: 785 (1992).

58. G. Bardi, B. Johansson, N. Pandis, N. Mandahl, J.E. Bak, C. Lindstrom, et al. Cytogenetic analysis of 52 colorectal carcinomas: non-random aberration pattern and correlation with pathologic parameters. Int. J. Cancer 55: 422 (1993).

59. S. Nakamura, J. Goto, M. Kitayama and I. Kino. Application of the crypt-isolation technique to flow-cytometric analysis of DNA content in colorectal neoplasms. Gastroenterology 106: 100 (1994).

60. G.J.A. Offerhaus, E.P. DeFeyter, C.J. Cornelisse, K.W.F. Tersmette, J. Floyd, S.E. Kern, et al. The relationship of DNA aneuploidy to molecular genetic alterations in colorectal carcinoma. Gastroenterology 102: 1612 (1992).

61. M. Kouri, A. Laasonen, J.P. Mecklin, H. Järvinen, K. Franssila and S. Pyrhönen. Diploid predominance in hereditary nonpolyposis colorectal carcinoma evaluated by flow cytometry. Cancer 65: 1825 (1990).

62. G.I. Meling, R.A. Lothe, A.L. Børresen, C. Graue, S. Hauge, O.P.F. Clausen, et al. The TP53 tumour suppressor gene in colorectal carcinomas. II. Relation to DNA ploidy pattern and clinicopathological variables. Br. J. Cancer 67: 93 (1993).

63. M.J. Mckinley, D.R. Budman, D. Grueneberg, R.L. Bronzo, G.S. Weissman and E. Kahn. DNA content in Barrett's esophagus and esophageal malignancy. Am. J. Gastroenterol. 82: 1012 (1987).

64. A. Zimmermann and F. Truss. The prognostic power of flow-through cytophotometric DNA determinations for testicular diseases. Anal. Quant. Cytol. 2: 247 (1980).

65. E. Thorud, O.P.F. Clausen and T. Abyholm. Fine needle aspiration biopsies from human testes evaluated by DNA flow cytometry. In: O. Lareum, T. Lindmo and E. Thorud (eds.). Flow Cytometry vol. IV. Oslo, Universitetsforlaget, p. 175 (1981).

66. P. Pfitzer, P. Gilbert, G. Roly and K. Vyska. Flow cytometry of human testicular tissue. Cytometry 3: 116 (1982).

67. D.P. Evenson and M.R. Melamed. Rapid analysis of normal and abnormal cell types in human semen and testis biopsies by flow cytometry. J. Histochem. Cytochem. 31: 248 (1983).

68. B. Tribukait. DNA flow cytometry in carcinoma of the prostate for diagnosis, prognosis and study of tumor biology. Acta Oncologica 30: 187 (1991).

69. S. Pulciani, E. Santos, A.V. Lauver, L.K. Long, S.A. Aaronson and D. Barbacid. Oncogenes in solid human tumours. Nature 300: 539 (1982).

70. D.A. Spandidos and N.M. Wilkie. Malignant transformation of early-passage rodent cells by a single mutated human oncogene. Nature 310: 469 (1984).

71. J.E. de Vries, F.H.A.C. Kornips, P. Marx, F.T. Bosman, J.P.M. Geraedts and J. Kate. Transfected c-Ha-ras oncogene enhances karyotypic instability and integrates predominantly in aberrant chromosomes. Cancer Genet. Cytogenet. 65: 35 (1993).

72. N. Hagag, L. Diamond, R. Palermo and S. Lyubsky. High expression of ras p21 correlates with increased rate of abnormal mitosis in NIH3T3 cells. Oncogene 5: 1481 (1990).

73. J.L. Bos, E.R. Fearon, S.R. Hamilton, M. Verlaan de Vries, J.H. van Boom, A.J. van der Eb and B. Vogelstein. Prevalence of ras gene mutations in human colorectal cancers. Nature 327: 293 (1987).

74. S. Nigro, E. Geido, E. Infusini, R. Orecchia and W. Giaretti. Transfection of human mutated K-ras in mouse NIH-3T3 cells is associated with increased cloning efficiency and DNA aneuploidization. Int. J. Cancer 67: 1 (1996).

75. W. Giaretti, R. Monaco, N. Pujic, A. Rapallo, S. Nigro and E. Geido. Intratumor heterogeneity of K-ras2 mutations in colorectal adenocarcinomas. Association with degree of DNA aneuploidy. Am.J. Pathology 149:1 (1996).

76. D.P.Lane. p53, guardian of the genome. Nature 358:15 (1992).

77. M.S. Greenblatt, WP. Bennet, M. Hollstein, C.C. Harris. Mutations in the p53 suppressor gene: clues to cancer etiology and molecular pathogenesis. Cancer Res. 54 (1994).

78. P.C. Galipeau, D.S. Cowan, C.A. Sanchez, M.T Barrett, M.J. Emond, D.S. Levine, P.S. Rabinovitch and B.J. Reid. 17p (p53) allelic losses, 4N (G2/tetraploid) populations, and progression to aneuploidy in Barrett's esophagus. Prc. Natl. Acad. Sci. USA 93 (1996).

79. A. Di Vinci, E. Infusini, C. Peveri, M. Risio, F.P. Rossini and W. Giaretti. Deletions at chromosome 1p by fluorescence in situ hybridization are an early event in human colorectal tumorigenesis. Gastroenterology 111: 102 (1996).

80. G. Bardi, N. Pandis, C. Fenger, O. Kronborg, L. Bomme and S. Heim. Deletion of 1p36 as a primary chromosomal aberration in intestinal tumorigenesis. Cancer Res. 53: 1895 (1993).

81. C.A. Griffin, S. Lazer, S.R. Hamilton, F.M. Giardiello, P. Long, A.J. Krush and S.V. Booker. Cytogenetic analysis of intestinal polyps in polyposis syndromes: Comparison with sporadic colorectal adenomas. Cancer Genet. Cytogenet. 67: 14 (1993).

82. M. Longy, R. Saura, F. Dumas, J.F. Leseve, L. Taine, J.F. Goussot and P. Couzigou. Chromosome analysis of adenomatous polyps of the colon. Cancer Genet. Cytogenet. 67: 7 (1993).

83. L. Bomme, G. Bardi, N. Pandis, C. Fenger, O. Kronborg and S. Heim. Clonal karyotypic abnormalities in colorectal adenomas: clues to the early genetic events in the adenoma-carcinoma sequence. Genes Chromosom. Cancer 10: 190 (1994).

84. J. Herbergs, A.P. de Bruine, P.T.J. Marx, M.I.J. Vallinga, R.W. Stockbrügger, F.C.S. Ramaekers, J.W. Arends and A.H.N. Hopman. Chromosome aberrations in adenomas of the colon. Proof of trisomy 7 in tumor cells by combined interphase cytogenetics and immunohistochemistry. Int. J. Cancer 57: 781 (1994).

85. B.A. Bunnell, L.S. Heath, D.E. Adams, J.M. Lahti and V.J. Kidd. Increased expression of a 58-kDa protein kinase leads to changes in the CHO cell cycle. Proc. Natl. Acad. Sci. USA 87: 7467 (1990).

86. J.M. Lahti, J. Xiang, L.S. Heath, D. Campana and V.J. Kidd. PITSLRE protein kinase activity is associated with apoptosis. Mol. Cell. Biol. 15: 1 (1995).

87. P. Engler, P. Haasch, C.A. Pinkert, et al. A strain-specific modifier on mouse chromosome 4 controls the methylation of independent transgene loci. Cell 65: 939 (1991).

88. S.E. Goelz, B. Vogelstein, S.R. Hamilton and A.P. Feinberg. Hypomethylation of DNA from benign and malignant human colon neoplams. Science 228: 187 (1985).

89. A. Almeida, N. Kokalj-Vokac, D. Lefrancois, E. Viegas-Péquignot, M. Jeanpierre, B. Dutrillaux and B. Malfoy. Hypomethylation of classical satellite DNA and chromosome instability in lymphoblastoid cell lines. Hum. Genet. 91:538 (1993).

90. H.M. Foss, C.J. Roberts, K.M. Claeys and E.U. Selker. Abnormal chromosome behavior in neurospora mutants defective in DNA methylation. Science 262: 1737 (1993).

91. H. Stopper, C. Körber, D. Schiffmann and W.J. Caspary. Cell-cycle dependent micronucleus formation and mitotic disturbances induced by 5-azacytidine in mammalian cells. Mut. Res. 300: 165 (1993).

92. G. Poste, J. Doll and I.J. Fidler. Interactions between clonal subpopulations affect the stability of the metastatic phenotype in polyclonal populations of B16 melanoma cells. Proc. Natl. Acad. Sci. 78: 6226 (1981).

93. A. Hall. Small GTP-binding proteins and the regulation of the actin cytoskeleton. Annu. Rev. Cell Biol. 10: 31 (1994).

94. C.D. Nobes and A. Hall. Rho, rac, and cdc42 GTPases regulate the assembly of multimolecular focal complexes associated with actin stress fibers, lamellipodia, and filopodia. Cell 81: 53 (1995).

95. D.A. Larochelle, K.K. Vithalani and A. De Lozanne. A novel member of the rho family of small GTP-binding proteins is specifically required for cytokinesis. J. Cell Biol. 133: 1321 (1996).

DISCUSSION

Wahl: I just wanted to make sure I understood this, that in the progression you find that there are aneuploid tumors but that these are p53+, is that correct?

Giaretti: The methodology is based on flow cytrometric sorting of DNA aneuploid subclones in colorectal adenocarcinomas. So you have 100% DNA aneuploid cells and then you do the mutation. And what we found is that approximately half of the cells were mutated, the other ones are not. But when mutation and aneuploid are associated, then it seems that the heteroduplex cells are not present, it is almost all homoduplex. So it appears to be deletion of the other allele or both are mutated.

Wahl: Can you rule out the idea that aneuploid tumors that appear to have a wild type p53 genef actually have a defective p53 gene or a defective pathway?

Giaretti: Yes. We have been doing also immunohistochemistry and in some 20% more, we found aberrant protein production. So we have another 20% aberration in p53 but we do not know if this was a mutation in another codon or if it was just overexpression.

Wahl: There are other ways than mutating p53 to inactivate the pathway. Functional assays of the pathway would be needed to assess its integrity.

Giaretti: Yes, these are functional assays.

Shay: Dr. Gray might want to respond to this as well. It seems if we accept that there are p53 independent mechanisms of gene amplification, perhaps what we should be looking for is very early stage cancer, such as preneoplasia. For example, gains or losses, deletions or amplifications in preneoplastic situations where we know there is wild type p53. A good example would be in preneoplastic lesions of the lung, where there are microsatellite alterations 3pLOH that precedes carcinoma *in situ*. Thus, you might be able to apply your type of techniques to try to narrow down regions of amplification. It may, in fact, be important in this whole cascade, leading to gene amplification to determine alterations that are p53 independent.

Giaretti: The incidence of K-RAS mutation in colorectal adenomas is about 50%. But even aberrant crypts were found to be mutated in K-RAS. So, I think that what you suggested perhaps is just K-RAS.

Stark: I do not know what evidence there is for p53-independent amplification, I think it is the other way around. That p53 regulates much more than just amplification and many other forms of abnormality are regulated in a p53 dependent matter.

Gray: Just one other comment on that in terms of the early lesions. It was not until we started very recently looking at those early lesions that we knew the extent of the abnormalities that were there and I think that in light of that, what you are suggesting is very interesting.

Evan: Can I ask for some clarification here, because one of the most intriguing things about p53, it seems to me, is that you can knock out the gene in every cell in an animal. Yet when you actually look, almost all the cells in the animal, you get a normal animal coming out which is really only mildly predisposed to cancer, that is, it gets tumors after three or four months, whereas a normal mouse gets tumors after three or four years and the tumors are still clonal. But almost all the cells in that animal that you can look at, are apparently karotypically normal. So what does that tell us? I am asking for guidance here. What does that tell us about the spontaneous rate of major chromosomal changes within somatic cells? It argues, presumably, that either the cells are forming and becoming abnormal all the time and they are being removed by some arcane mechanism or it is just not happening.

Stark: Perhaps it is telling us that the rate is low.

Wahl: I think that I would agree with George. The rate is low. Cellular homeostasis is obviously a very complex system as we have seen, p53 appears to be involved in a huge number of events, from sensing nucleotide metabolism to spindle integrity. Apparently, all those things must occur with very high fidelity. Even when p53 is inactivated, aneuploidy, re-replication and structural chromosome changes do not occur at high frequencies unless the cell is put under "stressing conditions" (e.g., incubation with spindle and rNTP synthesis inhibitions).

5

p53-DEPENDENT SIGNALING IN RESPONSE TO DNA DAMAGE OR ARREST OF DNA SYNTHESIS AND ITS ROLE IN CELL CYCLE CONTROL

Munna L. Agarwal, William R. Taylor, and George R. Stark

Department of Molecular Biology
Research Institute
The Cleveland Clinic Foundation
Cleveland, Ohio 44195

INTRODUCTION

p53 has a vital role in the normal cellular response to DNA damage or other stress (Michalovitz *et al.,* 1990; Lowe *et al.,* 1993; Vogelstein and Kinzler, 1992). p53 function is lost during the evolution of a diverse group of human malignancies (Hollstein *et al.,* 1991; Greenblatt *et al.,* 1994). Homozygous inactivation of p53 predisposes mice to neoplasia and hemizygous inactivation of p53 in humans is the basis of the hereditary cancer syndrome first described by Li and Fraumeni (Donehower *et al.,* 1992; Malkin *et al.,* 1990; Li and Fraumeni, 1969). The tumor suppressor function of p53 is likely to be based on its ability to mediate cell cycle arrest or apoptosis. Overexpression of a temperature-sensitive mutant allele of p53 in rodent fibroblasts causes them to arrest mainly in G1 at a permissive temperature (Michalovitz *et al.,* 1990). Using dexamethasone-regulated p53, Lin *et al.* (1992) showed that human glioblastoma cells arrest late in G1 when p53 is active. In addition to G1 arrest, regulation by p53 of the other major cell cycle checkpoint, G2/M, has recently been reported in two different systems (Stewart *et al.,* 1995; Agarwal *et al.,* 1995).

The ability of p53 to regulate cell cycle progression under different circumstances probably depends on its ability to act as a sequence-specific DNA binding protein capable of transcriptionally activating genes containing an appropriate enhancer element (Farmer *et al.,*1992; Pietenpol *et al.,* 1994; Dulic *et al.,* 1994). The p53-regulated gene encoding p21/waf1 is important in carrying out the p53-induced cell cycle arrest program (Dulic *et al.,* 1994). p21 is a potent inhibitor of cyclin-dependent protein kinases, which drive cell cycle progression (El-Deiry *et al.,* 1993; Harper *et al.,* 1993; Xiong *et al.,* 1993) and a p53-dependent increase in p21 is observed in response to DNA damage (Macleod *et al.,* 1995).

Genomic Instability and Immortality in Cancer
edited by Mihich and Hartwell, Plenum Press, New York, 1997

The p53 protein accumulates when DNA is damaged (Maltzman and Czyzyk, 1984; Tishler *et al.*, 1993; Fritsche *et al.*, 1993). The protein normally turns over rapidly, and regulated inhibition of its degradation is an important factor in its accumulation in response to damage (Kastan *et al.*, 1991). Also, Mosner *et al.*(1995) have suggested that the synthesis of p53 protein is inhibited at the translational level in G_0-arrested cells, through the binding of p53 to the 5′ region of its own mRNA. The p53 protein requires activation in order to function as a transcription factor. In particular, the C-terminal region negatively regulates the ability of p53 to bind to DNA, and this inhibition can be relieved artificially by antibodies that bind to this region (Hupp *et al.*, 1995). *In vivo,* the activation step probably requires phosphorylation of one or more serine residues in the C-terminal region.

Genetic Analysis of Mammalian Signaling Pathways

In order to reveal the as yet unknown pathways leading from DNA damage or arrest of DNA synthesis to the activation and induction of p53, we have used a genetic approach which has already been highly successful in helping to reveal the signaling pathways that govern responses to interferon-α and interferon-γ (Darnell *et al.*, 1994). The human fibrosarcoma cell line HT1080 was transfected with constructs in which interferon-regulated DNA elements control expression of either the drug-selectable marker guanine phosphoribosyl transferase or the cell-surface marker cd2. Cells with varying levels of cd2 expression can be separated by using a fluorescence-activated cell sorter (FACS). Following extensive chemical mutagenesis with the intercalating agent ICR191, a collection of mutant cell lines was obtained which failed to respond to either or both interferons. As determined by cell-cell fusions, all the mutant lines are recessive and comprise eight different complementation groups. A combination of functional complementation with normal DNA and the use of cDNAs corresponding to components of the pathways that were identified independently in biochemical experiments has revealed that each complementation group lacks a specific signaling component: a receptor subunit, a receptor-associated tyrosine kinase (JAK) or a transcription factor subunit (STAT).

Mutant Cell Lines Defective in Inducing Active p53

Funk *et al.* (1992) isolated GGACATGCCCGGGCATGTCC, a 20-base-pair palindromic sequence which binds tightly to p53. This sequence, when placed upstream of a basal promoter, drives strong p53-dependent induction of reporter genes. Higher levels of p53-dependent induction were observed when the consensus sequence was combined with fragment A, a 33-base pair element isolated from a human genomic clone (Kern *et al.*, 1992).

We tested the status of p53 in HT1080 cells, used previously in the interferon work. Western analysis and direct sequencing of the coding region of the cDNA revealed that the p53 of HT1080 cells is wild-type, inducible, and fully functional as a transcriptional activator. HT1080 cells grow without undergoing apoptosis or cell-cycle arrest even when wild-type p53 is expressed at a high level from a constitutive promoter. HT1080 cells were transfected with p53conA.cd2 and one of the resulting clones, H10, was used for all subsequent studies.

H10 cells were exposed to γ-radiation, UV-radiation or adriamycin, stained for cd2, and analyzed for growth. We could not find conditions in which cd2 was well induced and that allowed appreciable cell survival using agents that damage DNA. Since the use of DNA-damaging agents was not possible, we examined other means of inducing p53.

Treatment of H10 cells for 48 hr with either aphidicolin or N-(phosphonacetyl)-l-aspartate (PALA), which arrest DNA synthesis by inhibiting DNA polymerase directly or by starving cells for pyrimidine nucleotides, respectively, induces p53 protein and cd2 expression strongly. Treatment with high concentrations of PALA for 48 hr induced a dramatic increase in the cell surface expression of cd2, with little or no effect on cell viability upon removal of the drug.

Independent pools of H10 cells were subjected to four rounds of treatment with the chemical mutagen ICR191, to approximately 70% lethality each time. The mutagenized cells from one pool were treated for 48 hr with a very high concentration of PALA, stained with an antibody to cd2, and sorted. The least responsive 3–5% of the cells were recovered and grown again. The entire process was carried out a total of four times, with gradual enrichment for cells that failed to induce cd2 in response to PALA (Fig. 1). After the fourth sort, individual colonies were grown from single cells. One such clone, PD1A, is completely defective in cd2 induction in response to all the agents tested except adriamycin, where a partial response is retained. Western analysis showed normal induction of p53 protein in clone PD1A. In order to eliminate the possibility of a mutation in p53 itself, we sequenced the entire coding region of p53 mRNA from PD1A cells and found it to be unaltered.

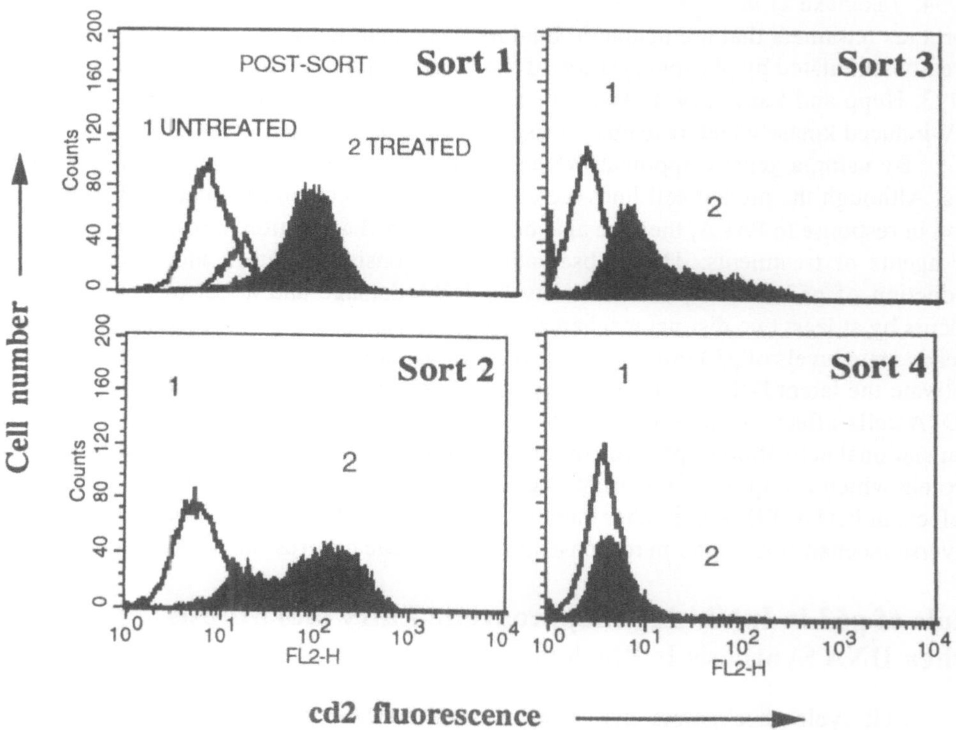

Figure 1. Expression of cell surface cd2 during sequential cell sorting in a mutagenized population of H10 cells. The cells were subjected to four rounds of ICR191 mutagenesis and then treated with 2 mM PALA for 48 hr. Cells were stained with antibody to cd2 and FAC sorted. The least unresponsive 3–5% of cells were isolated and allowed to grow. This population was again treated with PALA for 48 hr and again sorted. Two more cycles of treatment and sorting were carried out. The cd2 distribution of the untreated control (unshaded) and the treated/sorted samples (shaded) are shown for each sequential sort.

Clone PD2A was isolated from a different mutagenized pool. In this clone, the basal levels of p53 mRNA and protein are very low, but the degree of induction of the protein is normal. The quite distinct phenotypes of PD1A and PD2A reveal novel and intriguing features of p53 regulation, confirming and extending the recent observations of Hupp *et al.* (1995). p53 can be induced by PALA (Nelson and Kastan, 1994; Linke *et al.*, 1996), which inhibits *de novo* pyrimidine nucleotide synthesis. Although treatment with PALA may cause DNA damage in cells unable to arrest when conditions are unfavorable for DNA synthesis, p53 induction by PALA occurs without DNA damage in normal fibroblasts (DiLeonardo *et al.*, 1994; reviewed by Chernova *et al.*, 1995). As little as one double strand break in such cells may be sufficient to induce irreversible growth arrest (DiLeonardo *et al.*, 1994). Furthermore, when treated with PALA, p53-null cells fail to arrest, accumulate genomic lesions and die unless they amplify the CAD gene, the target of inhibition by PALA (Chernova *et al.*, 1995). Thus, starvation for pyrimidine nucleotides somehow signals p53 induction, which halts DNA synthesis, under conditions which would otherwise be detrimental. The mechanism of this inhibition has been explored by Linke *et al.* (1996). These observations suggest a signaling network converging on p53, initiated by a variety of stimuli, including lack of DNA precursors, DNA damage, and possibly others.

An important aspect of the stimulation of transcriptional activation in response to p53 involves phosphorylation of the p53 protein itself (Meek *et al.*, 1990; Milne *et al.*, 1994; Takenaka *et al.* 1995). Lane and coworkers have found that p53 exists in a latent form, as tetramers that are unable to bind to DNA, and that DNA binding activity can be greatly stimulated by phosphorylation of the C-terminal region of the protein (Hupp *et al.*, 1992; Hupp and Lane, 1995). Also, UV radiation stimulates phosphorylation of p53 by a UV-induced kinase which resembles JNK1 (Milne *et al.*, 1995).

By using a genetic approach we have isolated two mutants defective in p53 signaling. Although the mutant cell lines were selected for defects in p53-mediated transactivation in response to PALA, they are also defective in p53 activation in response to a variety of agents or treatments. These observations are consistent with a model in which the induction of p53 transcriptional activity by DNA damage and arrest of DNA synthesis occurs by at least two distinct mechanisms (Fig. 2). Activation involves not only increased steady state levels of p53 protein, but also post-translational modification, which serves to activate the latent DNA-binding activity of p53. We speculate that the molecular defect in PD1A cells·affects a protein involved in either steady-state or stimulus-dependent post-translational activation of p53. In contrast, the molecular defect in PD2A cells may be in a protein which is required for basal p53 gene expression. Identification of the nature of the defects in PD1A, PD2A and other mutant cells will provide valuable information about the diverse mechanisms of p53 in response to DNA damage or arrest of DNA synthesis.

Role of p53 in Inhibiting Inappropriate Entry into Mitosis when DNA Synthesis Is Blocked

Cell cycle checkpoints ensure that mitosis does not begin until DNA synthesis is complete, a sequence strictly required for the integrity of the genome and cell viability. Mutant fission yeast carrying deletions of *rad4* alone, or *mik1* and *wee1* together, enter mitosis prematurely when their DNA is not fully replicated, leading to chromatin fragmentation and cell death, a phenotype termed mitotic catastrophe (Nurse, 1994; Saka and Yanagida, *et al.*, 1993; Lundgren *et al.*, 1991). In mammalian cells, the same effect is obtained by overexpressing cdc2/cyclin B or by treating cells with 2-aminopurine or

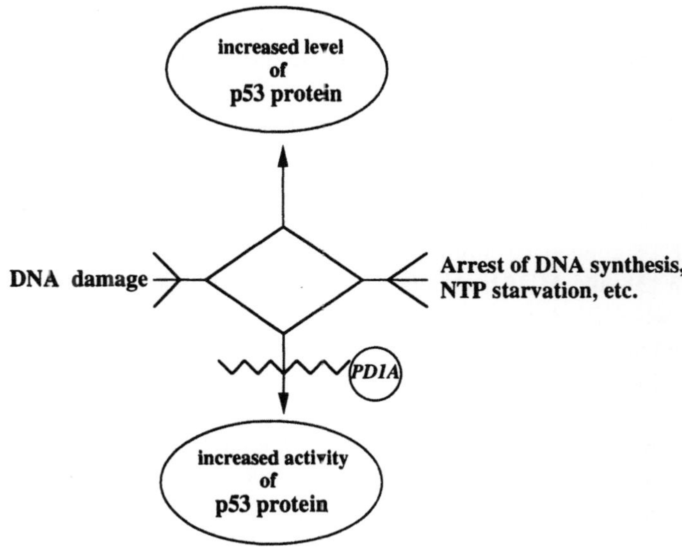

Figure 2. A two-component mechanism of activation of p53 in response to DNA damage or arrest of DNA synthesis. Elevated levels of p53 due to stimulus-dependent stabilization are important for activation of transcriptional activity. However, stabilization of p53 is not sufficient for activation, as revealed by the phenotype PD1A cells which, despite achieving high levels of p53 after treatment, do not activate downstream targets of p53. Full activation of p53-responsive genes may require not only stabilization of the protein but also post-translational activation, for example by phosphorylation.

okadaic acid (Heald *et al.*, 1993; Steinmann *et al.*, 1991). We find that p53 is induced when DNA synthesis is blocked by inhibitors. Cells with p53 are arrested but cells without p53 enter mitosis, as shown by formation of mitotic spindles and extensive phosphorylation of histone H1b. By analyzing the DNA content of individual mitotic cells, we find that a significant fraction of the p53-null cells that enter mitosis in the presence of inhibitors of DNA synthesis contain a DNA complement of less than 4N. These observations reveal that, in addition to its well established roles in G1 and G2 checkpoints, p53 also functions in a checkpoint which maintains the dependence of mitosis on the completion of S phase.

The human fibroblast cell strain WI38, which has wild-type p53 and responds to its induction normally, is stably and reversibly arrested when treated with PALA (Yin *et al.*, 1992). We used incorporation of bromodeoxyuridine (BrdU) to compare the effects of arrest of DNA synthesis in WI38 cells and the p53-null Li-Fraumeni fibroblast cell line MDAH041, using aphidicolin (which inhibits DNA polymerase), hydroxyurea (which inhibits ribonucleotide reductase) or PALA. DNA synthesis in WI38 cells was blocked by treatment with any of these three agents, and aphidicolin or hydroxyurea also abolished DNA synthesis in MDAH041 cells. As previously observed, a large percentage of MDAH041 cells, but not WI38 cells, incorporate BrdU in the presence of PALA (Di Leonardo *et al.*, 1993). p53 was induced in WI38 cells upon treatment with hydroxyurea, PALA or aphidicolin (Linke *et al.*, 1996; Khanna *et al.*, 1993) and we also observed induction of p21/waf1/cip1 in these cells. p53 was also induced in normal mouse embryo fibroblasts by these inhibitors of DNA synthesis. Since p53 can mediate arrest in G2/M as well as in G1/S (Stewart *et al.*, 1995; Agarwal *et al.*, 1995), it seemed possible that induction of p53 by inhibitors of DNA synthesis might also be responsible for blocking premature entry into mitosis. To test this idea, we analyzed the ability of cells lacking p53 to enter mitosis when treated with aphidicolin, hydroxyurea or PALA.

In normal mouse embryo fibroblasts treated with aphidicolin, hydroxyurea or PALA, the phosphorylation of histone H1b was very low and, in every case, was confined to nuclei with interphase morphology (Fig. 3). In addition, no mitotic spindles were observed (Fig. 3). In contrast, many of the p53-null fibroblasts showed evidence of mitosis when treated with any of the three inhibitors. We observed cells with high levels of cytoplasmic phosphorylated histone H1b, condensed chromosomes, and mitotic spindles, and similar results were obtained with W138 and MDAH041 cells (Fig. 3). Many of the p53-null fibroblasts that entered mitosis when DNA synthesis was blocked contained more than two spindle poles. We also observed multipolar spindles in 30% of untreated mitotic p53-null mouse embryo fibroblasts, but in only 13% of untreated mitotic normal cells, in agreement with a recent report (Fukasawa *et al.*, 1996) in which it was shown that 60% of p53-null mouse embryo fibroblasts, but only 5% of normal fibroblasts, enter mitosis with more than two spindle poles. When treated with inhibitors of DNA synthesis, the p53-null mouse embryo fibroblasts that entered mitosis (as revealed by high levels of phosphorylated histone H1b or mitotic spindles) also showed chromatin fragmentation. Fragmented chromatin was not observed in mitotic cells from untreated normal or p53-null mouse embryo fibroblasts.

In the presence of aphidicolin or hydroxyurea, both normal and p53-null cells are tightly arrested with a G1/S complement of DNA, due to the inability of any cell to synthesize DNA in the presence of these potent inhibitors. In this situation, p53 mediates a protective arrest within S phase, ensuring that cells that have not completed DNA synthesis do not enter mitosis prematurely. A similar function has been ascribed to a number of fission yeast genes for which mammalian homologues have not yet been identified (reviewed in Nurse, 1994).

As shown in Fig. 4, p53 functions in regulating several distinct cell cycle processes, including G1 arrest induced by DNA damage, inhibition of endoreduplication, and centrosome homeostasis. Our current results suggest that p53 is also involved in an additional cell cycle checkpoint which prevents premature entry into mitosis when DNA synthesis is blocked (Fig. 4). This control may be important as part of a stress response, for example, perhaps to inhibit premature mitosis when DNA synthesis is blocked transiently by ultraviolet radiation during S phase, or to protect cells if DNA synthesis is blocked transiently by hypoxia when cells grow more quickly than angiogenesis can provide them with oxygen.

Figure 3. Inappropriate entry into mitosis in fibroblasts lacking p53. Cells were treated for 72 hr with Aph: aphidicolin (5 μg/ml), Hu: hydroxyurea (2 mM), or PALA (500 μM). Fixed cells were analyzed by immunofluorescence to detect either mitotic spindles or phosphorylated histone H1B (pH1b). The DNA was counterstained with DAPI. (a) Untreated mitotic WI38 or MDAH041 cells have high levels of pH1b and show discrete spindles. When treated with hydroxyurea or PALA, WI38 cells fail to enter mitosis, have low levels of pH1b, and show only cytoskeletal staining of tubulin. In contrast, MDAH041 cells enter mitosis in the presence of both agents. In the examples shown, the cells were treated with hydroxyurea or PALA. (b) Aphidicolin or PALA prevents p53 +/+ mouse embryo fibroblasts (MEF) from entering mitosis. In contrast, treated p53 -/- MEF do enter mitosis, as shown by high levels of pH1b and formation of discrete spindles. In the examples shown, the cells were treated with aphidicolin or PALA. (c) Chromatin morphology in p53-null fibroblasts that have entered mitosis after S-phase arrest with aphidicolin or PALA. About 39% of the treated mitotic cells contain highly fragmented chromatin. (d) The percentage of normal or p53-null cells entering mitosis in the presence of inhibitors of DNA synthesis, determined by staining for pH1b and spindle formation. Bars on the line represent fewer than one mitotic cell per thousand.

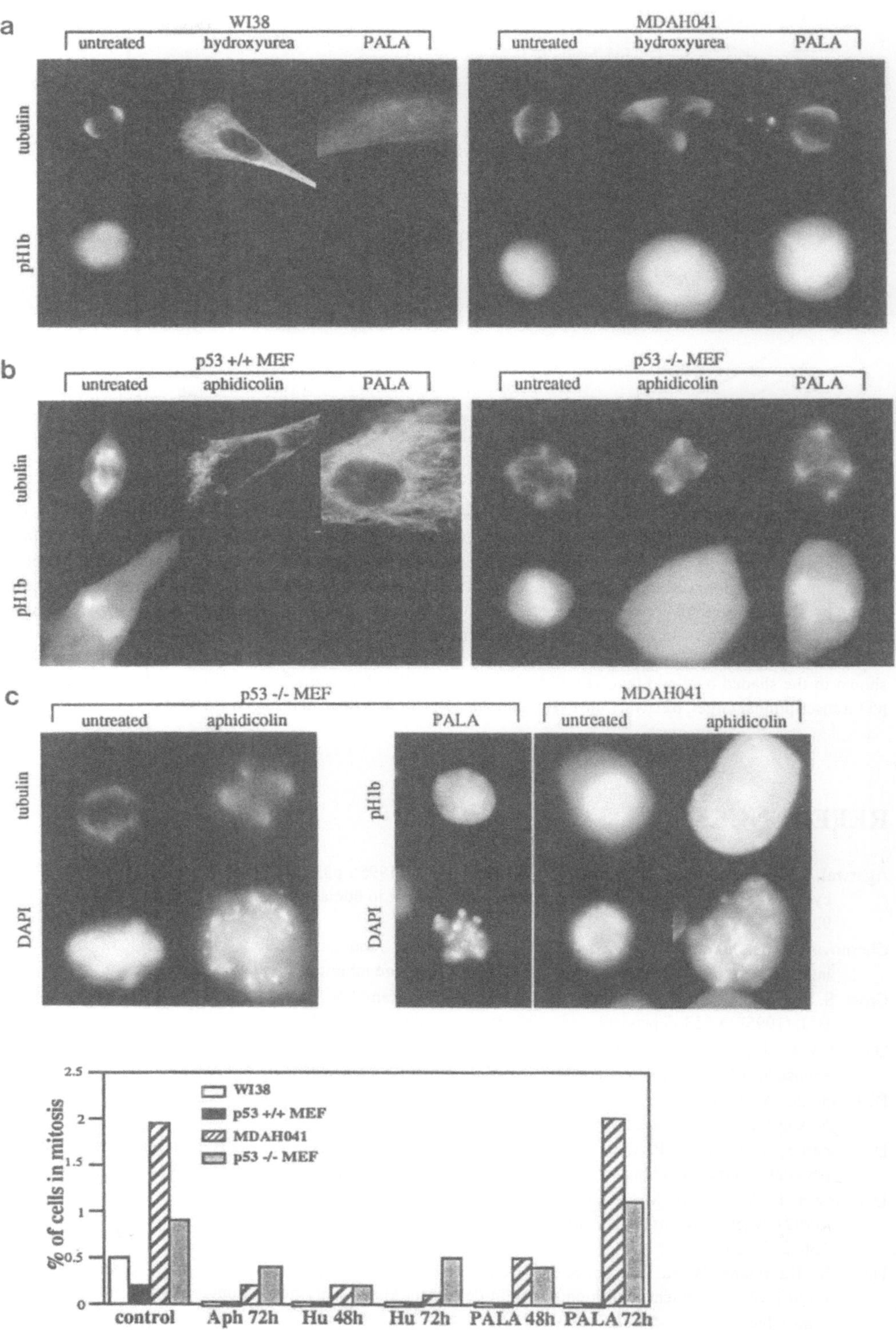

a
WI38
untreated hydroxyurea PALA
MDAH041
untreated hydroxyurea PALA
tubulin
pH1b
b
p53 +/+ MEF
untreated aphidicolin PALA
p53 -/- MEF
untreated aphidicolin PALA
tubulin
pH1b
c
p53 -/- MEF
untreated aphidicolin
PALA
MDAH041
untreated aphidicolin
tubulin
DAPI
pH1b
DAPI
2.5
2
1.5
1
0.5
0
% of cells in mitosis
WI38
p53 +/+ MEF
MDAH041
p53 -/- MEF
control Aph 72h Hu 48h Hu 72h PALA 48h PALA 72h

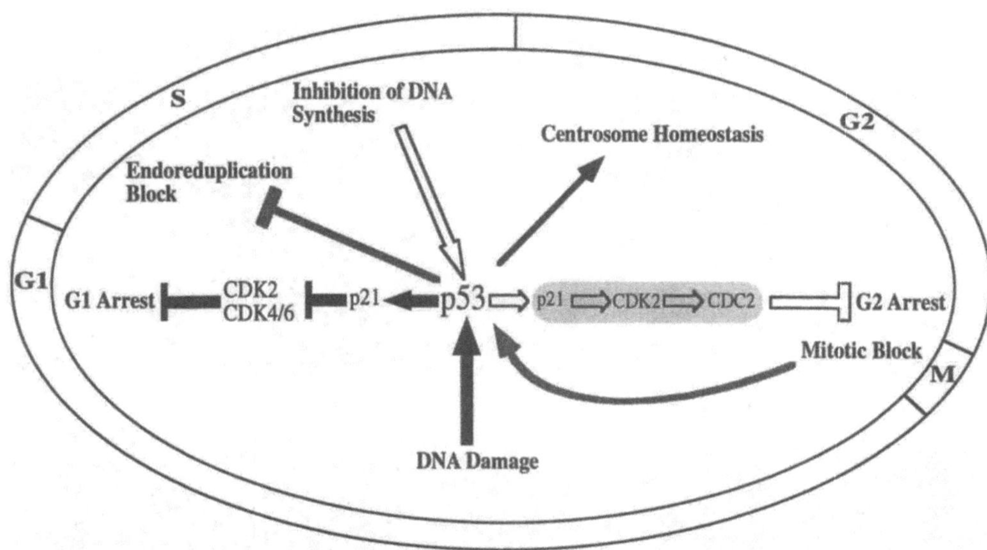

Figure 4. Regulation of cell cycle events by p53. p53 induced by DNA damage causes arrest of cells in G1, through a mechanism dependent on the cdk inhibitor p21/waf1/cip1 (Kastan *et al.*, 1991; Dulic *et al.*, 1994) (thick black arrows). The gray arrow denotes the p53-dependent spindle checkpoint, which ensures that cells do not re-enter S phase when blocked in mitosis (Cross *et al.*, 1995). p53 also regulates centrosome homeostasis, as shown by the presence of more than two spindle poles in p53-null mitotic fibroblasts (Fukasawa *et al.*, 1996) (thin black arrow). The white arrows denote the checkpoint described in the present report, which ensures that cells blocked in S phase do not enter mitosis prematurely. One way in which mitosis may be inhibited under these conditions is shown in the shaded oval and involves a block to cdk2-mediated activation of cdc2, caused by elevation of the p53 transcriptional target, p21/waf1/cip1 (Harper *et al.*, 1995; Guadagno and Newport, 1996).

REFERENCES

Agarwal, M. L., Agarwal, A., Taylor, W. R., and Stark, G. R. (1995). p53 controls both the G2/M and the G1 cell cycle checkpoints and mediates reversible growth arrest in human fibroblasts. *Proc. Natl. Acad. Sci. USA* 92, 8493–8497.

Chernova, O. B., Chernov, M. V., Agarwal, M. L., Taylor, W. R., and Stark, G. R. (1995). The role of p53 in regulating genomic stability when DNA and RNA synthesis are inhibited. *Trends Biochem. Sci.* 20, 431–434.

Cross, S. M., Sanchez, C. A., Morgan, C. A., Schimke, M. K., Ramel, S., Idzerda, R. L., Raskind, W. H., and Reid, B. J. (1995). A p53-dependent mouse spindle checkpoint. *Science* 267, 1353–1356.

Darnell, J. E., Jr., Kerr, I. M., and Stark, G. R. (1994). Jak-STAT pathways and transcriptional activation in response to IFNs and other extracellular signaling proteins. *Science* 264, 1415–1421.

Di Leonardo, A., Linke, S. P., Yin, Y., and Wahl, G. M. (1993). Cell cycle regulation of gene amplication. *Cold Spring Harbor Symposia on Quantitative Biology* 58, 655–667.

Di Leonardo, A., Linke, S. P., Clarkin, K., and Wahl, G. M. (1994). DNA damage triggers a prolonged p53-dependent G₁ arrest and long-term induction of Cip1 in normal human fibroblasts. *Genes Dev.* 8, 2540–2551.

Donehower, L. A., Harvey, M., Slagle, B. L., McArthur, M., Montgomery, L. A., Butel, J. S., and Bradley, A. (1992). Mice deficient for p53 are developmentally normal but susceptable to spontaneous tumours. *Nature* 356, 215–221.

Dulic, V., Kaufmann, W. K., Wilson, S. J., Tlsty, T. D., Lees, E., Harper, J. W., Elledge, S. J., and Reed, S. I. (1994). p53-dependent inhibition of cyclin-dependent kinase activities in human fibroblasts during radiation-induced G1 arrest. *Cell* 76, 1013–1023.

El-Deiry, W. S., Tokino, T., Velculescu, V. E., Levy, D. B., Parsons, R., Trent, J. M., Lin, D., Mercer, E., Kinzler, K. W., and Vogelstein, B. (1993). WAF1, a potential mediator of p53 tumor suppression. *Cell* 75, 817–825.

Farmer, G., Bargonetti, J., Zhu, H., Friedman, P., Prywes, R., and Prives, C. (1992). Wild-type p53 activates transcription *in vitro*. *Nature* 358, 83–86.

Fritsche, M., Haessler, C., and Brandner, G. (1993). Induction of nuclear accumulation of the tumor-suppressor protein p53 by DNA-damaging agents. *Oncogene* 8, 307–318.

Fukasawa, K., Choi, T., Kuriyama, R., Rulong, S., and Vande Woude, G. F. (1996). Abnormal centrosome amplification in the absence of p53. *Science* 271, 1744–1747.

Funk, W. D., Pak, D. T., Karas, R. H., Wright, W. E., and Shay, J. W. (1992). A transcriptionally active DNA-binding site for human p53 protein complexes. *Mol. Cell. Biol.* 12, 2866–2871.

Greenblatt, M. S., Bennett, W. P., Hollstein, M., and Harris, C. C. (1994). Mutations in the p53 tumor suppressor gene:clues to cancer etiology and molecular pathogenesis. *Cancer Res.* 54, 4855–4878.

Guadagno, T. M., and Newport, J. W. (1996). Cdk2 kinase is required for entry into mitosis as a positive regulator of Cdc2-cyclin B kinase activity. *Cell* 84, 73–82.

Harper, J. W., Adami, G. R., Wei, N., Keyomarsi, K., and Elledge, S. J. (1993). The p21 Cdk-interacting protein Cip1 is a potent inhibitor of G1 cyclin-dependent kinases. *Cell* 75, 805–816.

Harper, J. W., Elledge, S. J., Keyomarsi, K., Dynlacht, B., Tsai, L.-H., Zhang, P., Dobrowolski, S., Bai, C., Connell-Crowley, L., Swindell, E., Fox, M. P., and Wei, N. (1995). Inhibition of cyclin-dependent kinases by p21. *Mol. Biol. Cell* 6, 387–400.

Heald, R., McLoughlin, M., and McKeon, F. (1993). Human wee1 maintains mitotic timing by protecting the nucleus from cytoplasmically activated cdc2 kinase. *Cell* 74, 463–474.

Hollstein, M., Sidransky, D., Vogelstein, B., and Harris, C. C. (1991). p53 mutations in human cancer. *Science* 253, 49–53.

Hupp, T. R., Meek, D. W., Midgley, C. A., and Lane, D. P. (1992). Regulation of the specific DNA binding function of p53. *Cell* 71, 875–886.

Hupp, T. R., and Lane, D. P. (1995). Two distinct signalling pathways activate the latent DNA binding function of p53 in a casein kinase II independent manner. *J. Biol. Chem.* 270, 18165–18174.

Hupp, T. R., Sparks, A., Lane, D. P. (1995). Small peptides activate the latent sequence-specific DNA binding function of p53. *Cell* 83, 237–245.

Kastan, M. B., Onyekwere, O., Sidransky, D., Vogelstein, B., and Craig, R. W. (1991). Participation of p53 protein in the cellular response to DNA damage. *Cancer Res.* 51, 6304–6311.

Kern, S. E., Pietenpol, J. A., Thiagalingam, S., Seymour, A., Kinzler, K. W., and Vogelstein, B. (1992). Oncogenic forms of p53 inhibit p53-regulated gene expression. *Science* 256, 827–830.

Khanna, K. K., and Lavin, M. F. (1993). Ionizing radiation and UV induction of p53 protein by different pathways in ataxia-telangiectasia cells. *Oncogene* 8, 3307–3312.

Li, F. P., and Fraumeni, J. F. (1969). Soft tissue sarcomas, breast cancer and other neoplasms. A familial syndrome. *Ann. Intern. Med.* 71, 747–752.

Lin, D., Shields, M. C., Ullrich, S. J., Appella, E., and Mercer, W. E. (1992). Growth arrest induced by wild-type p53 protein blocks cells prior to or near the restriction point in late G1 phase. *Proc. Natl. Acad. Sci. USA* 89, 9210–9214.

Linke, S. P., Clarkin, K. C., DiLeonardo, A., Tsou, A., and Wahl, G. M. (1996). A reversible, p53-dependent G_0/G_1 cell cycle arrest induced by ribonucleotide depletion in the absence of detectable DNA damage. *Genes Dev.* 10, 934–947.

Lowe, S. W., Schmitt, E. M., Smith, S. W., Osborne, B. A., and Jacks, T. (1993). p53 is required for radiation-induced apoptosis in mouse thymocytes. *Nature* 362, 847–849.

Lundgren, K., Walworth, N., Booher, R., Dembski, M., Kirschner, M., and Beach D. (1991). mik1 and wee1 cooperate in the inhibitory tyrosine phosphorylation of cdc2. *Cell* 64, 1111–1122.

Macleod, K. F., Sherry, N., Hannon, G., Beach, D., Tokino, T., Kinzler, K., Vogelstein, B., and Jacks, T. (1995). p53-dependent and independent expression of p21 during cell growth, differentiation, and DNA damage. *Genes Dev.* 9, 935–944.

Malkin, D., Li, F. P., Strong, L. C., Fraumeni, J. F., Nelson, C. E., Kim, D. H., Kassel, J., Gryka, M. A., Bischoff, J. Z., Tainsky, M. A., and Friend, S. H. (1990). Germ line p53 mutations in a familial syndrome of breast cancer, sarcomas, and other neoplasms. *Science* 250, 1233–1238.

Maltzman, W., and Czyzyk, L. (1984). UV irradiation stimulates levels of p53 cellular tumor antigen in nontransformed mouse cells. *Mol. Cell. Biol.* 4, 1689–1694.

Meek, D. W., Simon, S., Kikkawa, U., and Eckhart, W. (1990). The p53 tumour suppressor is phosphorylated at serine 389 by casein kinase II. *EMBO J.* 9, 3253–3260.

Michalovitz, D., Halevy, O., and Oren, M. (1990). Conditional inhibition of transformation and of cell proliferation by a tenperature-sensitive mutant of p53. *Cell* 62, 671–680.

Milne, D. M., Campbell, D. G., Caudwell, F. B., and Meek, D. W. (1994). Phosphorylation of the tumor suppressor protein p53 by mitogen-activated protein kinases. *J. Biol. Chem.* 269, 9253–9260.

Milne, D. M., Campbell, D. G., Caudwell, F. B., and Meek, D. W. (1995). p53 is phosphorylated *in vitro* and *in vivo* by an ultraviolet radiation-induced protein kinase characteristic of the c-jun kinase *JNK1*. *J. Biol. Chem.* 270, 5511–5518.

Mosner, J., Mummenbrauer, T., Bauer, C., Sczakiel, G., Grosse, F., and Deppert, W. (1995). Negative feedback regulation of wild-type p53 biosynthesis. *EMBO J.* 14, 4442–4449.

Nelson, W. G., and Kastan, M. B. (1994). DNA strand breaks: the DNA template alterations that trigger p53-dependent DNA damage response pathways. *Mol. Cell. Biol.* 14, 1815–1823.

Nurse, P. (1994). Ordering S phase and M phase in the cell cycle. *Cell* 79, 547–550.

Pietenpol, J. A., Tokino, T., Thiagalingam, S., El-Deiry, W. S., Kinzler, K. W., and Vogelstein, B. (1994). Sequence-specific transcriptional activation is essential for growth suppression by p53. *Proc. Natl. Acad. Sci. USA* 91, 1988–2002.

Saka, Y., and Yanagida, M. (1993). Fission yeast cut5+, required for S phase onset and M phase restraint, is identical to the radiation-damage repair gene rad4+. *Cell* 74, 383–393.

Steinmann, K. E., Belinsky, G. S., Lee, D., and Schlegel, R. (1991). Chemically induced premature mitosis: Differential response in rodent and human cells and the relationship to cyclin B synthesis and $p34^{cdc2}$/cyclin B complex formation. *Proc. Natl. Acad. Sci. USA* 88, 6843–6847.

Stewart, N., Hicks, G. G., Paraskevas, F., and Mowat, M. (1995). Evidence for a second cell cycle block at G2/M by p53. *Oncogene* 10, 109–115.

Takenaka, I., Morin, F., Scizinger, B. R., and Kley, N. (1995). Regulation of the sequence-specific DNA binding function of p53 by protein kinase C and protein phosphatases. *J. Biol. Chem.* 270, 5405–5411.

Tishler, R. B., Calderwood, S. K., Coleman, C. N., and Price, B. D. (1993). Increases in sequence specific DNA binding by *p53* following treatment with chemotherapeutic and DNA damaging agents. *Cancer Res.* 53, 2212–2216.

Vogelstein, B., and Kinzler, K. W. (1992). p53 function and dysfunction. *Cell* 70, 523–526.

Xiong, Y., Hannon, G. J., Zhang, H., Casso, D., Kobayashi, R., and Beach, D. (1993). p21 is a universal inhibitor of cyclin kinases. *Nature* 366, 701–704.

Yin, Y., Tainsky, M. A., Bischoff, F. Z., Strong, L. C., and Wahl, G. M. (1992). Wild-type p53 restores cell cycle control and inhibits gene amplification in cells with mutant p53 alleles. *Cell* 70, 937–948.

DISCUSSION

Wahl: In the experiment where you mutagenized the HT1080 and then you got cells that are still inducible for p53, but, they are not active with regard to CD2 induction. Did you look at induction of p21 in those cells?

Stark: I should have mentioned it, thanks Geoff. We have looked at several endogenous genes in the mutant cell lines and the messenger RNA induction is defective. It is not completely defective and the degree to which it is not completely defective depends on which gene you look at. So the phenotype is not completely penetrant, and it is gene-dependent. Fortunately, the marker that we have used shows complete loss of induction in these mutants and some genes show one-third of the normal level of induction. There is a clear defect in many genes but it is not the same for every gene. I interpret that to mean that something is wrong with the p53 and the degree to which that is sensed depends on the environment of a particular gene.

Bacchetti: It was not quite clear to me whether you looked at induction or just accumulation of the proteins in response to p53.

Stark: What I showed you on the slide was accumulation of the protein.

Bacchetti: Did you look at induction of transcription?

Stark: Yes, that was, the selection for failure to transcriptionally activate the reporter construct. I guess I should say we have ruled out the trivial possibility that we have mutated the reporter in the cell lines. It is not a cis mutant. And to answer Geoff's question, I have also pointed out that there is a defect on several endogenous p53-dependent genes.

Nasmyth: Going back to Geoff's question, you looked at the induction of endogenous genes, did you look at cell cycle arrest and/or apoptosis?

Stark: It does not work in these cells. Remember these are HT1080 which have an N-RAS mutation, the whole pathway does not work because something is wrong, probably at the end. You can put p53 into these cells under a CMV promoter, express as much as you like and they do not care. So there is not any p53 dependent cell cycle arrest in these cells. Transcriptionally, yes, p53 induces p21 and other genes perfectly normally. So it is a downstream defect, the nature of which we do not know.

Hartwell: The results on the role of p53 is blocking mitosis, when there is an arrest of S; it is very interesting. I just wondered, given all the work that has been done on p53 minus cells, is it just that nobody has done that experiment before?

Stark: As far as I can tell, nobody has done it before.

Mihich: You know, of course, this relatively recent work with truncated p53 and the alternatively spliced forms. There seems to be some evidence that at least the alternatively spliced form can have a preferential effect at the G2M site. Do you have any information on the possibility that different forms of p53 may have differential effect at the different checkpoint?

Stark: We have no information, and an obvious possibility that we have to check is whether there is a defect in alternative splicing in our mutant cell lines. The region that is affected is at the carboxyl terminus which is the tetramerization but also activation domain. It is the region that David Lane thinks is very important in the self-repression of p53 function. So, a very interesting region is affected by the alternative splice. But the physiological significance of the two different forms, I think, is still quite wide open.

Wahl: I would like to raise two points. First, Aldo DiLeonardo looked quite hard to detect uncoupling between replication and mitosis, and I do not recall that he observed it. We certainly will look again.

Stark: You have to look at the mitotic cells and you have to count loss of cells and do it very carefully.

Wahl: I can assure you that Aldo counted the cells carefully.

Stark: I am sorry, yes.

Wahl: The second issue is that in putting cells through cycles of mutagenesis which disrupts DNA replication, I am reminded of numerous stories that describe alterations in DNA methylation. There is a famous story about how MDM2 was first detected as being amplified in cells with normal p53. Nobody knows how it could have been amplified in cells with normal p53. Since MDM2 is a negative regulator of p53 function, you could have normal up-regulation of p53 and defective downstream regulation if in fact your HT1080 had up-regulation of MDM2. Have you looked at that?

Stark: Yes, we have looked at MDM2, among others, as a p53 responsive gene and there is certainly no up-regulation.

Wahl: Have you looked at the protein?

Stark: We have not looked at the protein. We looked at the RNA.

Wahl: I think that may well be a worthy candidate.

Stark: Another very important point that we have not looked at in response to some interesting information that you showed is to see whether there is any defect in cytoplasmic to nuclear translocation of p53. It is an easy experiment, we have just got to do it; as soon as I get home.

Giaretti: By doing immunocytochemistry it is quite rare to find the p53 protein in the cytoplasm if it is this that you referred to, I think it is quite rare.

Stark: Yes, but in these mutant cell lines, I think anything is possible. The other thing, it is very early days in this study and I hope you appreciate that you are receiving this at an early time. We have seven independent mutant isolates, we have not yet done a complementation analysis, so we do not know how many complementation groups we have isolated. There is a lot we have not done in the phenotype. We just have more work to do.

Hartwell: One of the things that potential outcomes of that observation in S-phase would be that p53- cells might be more sensitive to DNA synthesis inhibitors which, to my knowledge, they are not. Do you think it has any therapeutic potential?

Stark: I imagine that would be worth looking at again. Perhaps with some of the same inhibitors that we have used.

Hartwell: In your experiments do the p53- cells die faster than the wild type cells in these inhibitors?

Stark: You know they do not, but that is perhaps something I should comment on independently of your question. We have compared the best pair of cell lines that we can think of, the same pairs that we have studied for S-phase arrest, mouse embryo fibroblasts plus or minus p53 or the human 041 cells with normal p53 restored or not. And we have looked at their sensitivity in a clonogenic assay with respect to a variety of different DNA damaging agents, both chemical and radiation. We can find absolutely no difference in the survival of cells with respect to those of DNA damaging agent when their p53 states are compared. You get the same killing curves for a variety of agents in both pairs. So as far as we can tell, there is, in these two situations at least, no detectable difference in our hands in the survival of cells depending on whether or not they have functional p53.

Wahl: Except for PALA, and other RNTP synthesis inhibitors.

Stark: Of course PALA, as you know Geoff, is completely different. So PALA, as we observed ages ago and you have observed more recently in normal cells, gives long term stable survival. And that is true in the mouse embryo fibroblasts, it is also true in the C11 cells with p53 function restored, and it is true in normal human fibroblasts. So PALA is not behaving like a DNA damaging agent, you will be pleased to know.

Wahl: That is the point I wanted to make, and the other RNTP synthesis inhibitors that we have looked at behave just the same way as PALA. So it is not that PALA is peculiar, it is that class of agents. And that may be worth while looking at with regard to chemotherapeutic potential in $p53^+$ and $p53^-$ tumors.

Stark: I was addressing the issue here of whether DNA damaging agents give a differential effect on cell survival in a p53-dependent manner and what I was saying is that at least in the pairs of cells that we have looked at—one mouse, one human—there is absolutely no difference in survival, with a variety of agents in a dose response experiment with respect to p53 state.

Hartwell: Including the DNA synthesis inhibitors?

Stark: No, I am sorry Lee, I gave you the wrong answer for that. We have done that so far with adriamycin, ionizing radiation, camptothecin and UV radiation. So we have only looked at DNA damaging agents at this point.

Hoeijmakers: One complication concerning this point is that p53 mutations have two opposite effects. It may indeed sensitize cells for damaging agents because cell cycle rest is impaired. But on the other hand, p53 mutations have also an effect on apoptosis so it may compensate for the increased sensitivity.

Stark: There are fibroblasts and they do not apoptose in response to this kind of damage. They just arrest and die. There is no evidence of apoptosis in fibroblasts. That response is characteristic of lymphoid cells, I know, but not fibroblasts which normally respond with arrest rather than apoptosis. Is that correct?

Hoeijmakers: We can clearly induce apoptosis in human fibroblast by injection of p53 cDNAs.

Stark: It may take high levels of expression. And, it is very hard to compare those two experiments because we are not even looking at p53 induction in this experiment, we are just looking at sensitivity to killing by DNA damaging agents and there is no p53 dependence of that.

6

RECOMBINING DNA DAMAGE REPAIR, BASAL TRANSCRIPTION, AND HUMAN SYNDROMES

Jan H. J. Hoeijmakers,[1] Gijsbertus T. J. van der Horst,[1] Geert Weeda,[1]
Wim Vermeulen,[1] G. Sebastiaan Winkler,[1] Jan de Boer,[1] Wouter L. de Laat,[1]
Anneke M. Sijbers,[1] Elizabetta Citterio,[1] Nicolaas G. J. Jaspers,[1]
Jean-Marc Egly,[2] and Dirk Bootsma[1]

[1]Department of Cell Biology and Genetics
Medical Genetics Centre
Erasmus University
P.O. Box 1736
3000DR Rotterdam, The Netherlands
[2]IGMBC, Université Louis Pasteur
1 rue Laurent Fries
BP163, 67404 Illkirch Cédex
C.U. de Strasbourg, France

INTRODUCTION

DNA, the vital carrier of genetic information, is not chemically inert. In the course of time lesions accumulate due to the intrinsic instability of certain chemical bonds. In addition, DNA erodes because of the deleterious effect of numerous exogenous and endogenous genotoxic agents. For instance, the ubiquitous UV component of sunlight induces cyclobutane pyrimidine dimers, 6–4 photoproducts and thymine glycols; X-rays cause various sorts of single strand breaks and the very genotoxic double strand breaks, whereas a wide range of natural and man-made chemicals give rise to many types of DNA adducts, as well as inter- and intra-strand crosslinks (for a review see ref. (5)). Obviously, the corrosion of the double helix interferes with proper functioning, and poses logistic problems for transcription and replication of DNA, which may cause cell death. In addition, DNA lesions frequently give rise to permanent changes in the nucleotide sequence. These mutations can lead to cellular malfunctioning, including carcinogenesis and may contribute to ageing in somatic cells. When occurring in germ cells they may be the cause of inborn genetic defects.

To counteract or prevent these deleterious consequences all living organisms are equipped with a network of DNA repair systems with complementary substrate specificities. These include nucleotide excision repair (for removal of a wide class of damages), base excision repair (for elimination of the class of small alkylation damage and other

Genomic Instability and Immortality in Cancer
edited by Mihich and Hartwell, Plenum Press, New York, 1997

small lesions), and recombination repair (for the correct healing of double strand breaks and removal of inter-strand cross links). These are multi-step pathways, involving the concerted action of a large number of proteins. In addition, single-enzyme systems operate specialized in the repair of one or a narrow range of lesions (for a comprehensive review on different repair systems see ref. (5)). A hallmark of malignant transformation is the frequent functional loss of one or more of these mechanisms. The resulting genetic instability fuels the rapid degeneration of a transformed cell to a more malignant state. Thus, occasional or systematic failure of these security processes constitutes an important factor, that initiates and drives carcinogenesis. Obviously, insight into these mechanisms is of pivotal importance for understanding the origin of cancer.

NUCLEOTIDE EXCISION REPAIR MUTANTS AND HUMAN SYNDROMES

One of the best understood and most versatile repair mechanisms is the nucleotide excision repair (NER) pathway. This process eliminates a remarkably wide spectrum of structurally unrelated lesions, including the major types of UV-induced DNA injury, numerous bulky or helix-distorting chemical adducts as well as intrastrand crosslinks (5). Important insight into the biological and molecular intricacies of NER is derived from a multitude of mutants defective in this process. Typically, they display pronounced hypersensitivity to UV light and numerous chemicals as well as elevated levels of induced mutagenesis. The extensively studied *uvrA, B* and *C* mutants in *Escherichia coli* have provided the basis for the elucidation of the pathway in procaryotes (8, 18, 24). The collection of bakers yeast *Saccharomyces cerevisiae* NER mutants currently includes 16 or more members, comprising the *RAD3* epistasis group. Most of the genes involved have been cloned (8, 9). Similarly, a large number of mammalian NER mutants has been generated in the laboratory using rodent cell lines. Cell fusion has led to the identification of a minimum of 11 complementation groups. By transfection-correction cloning strategies a series of complementing human NER genes could be retrieved. These genes are called *ERCC* (for *e*xcision *r*epair *c*ross *c*omplementing) genes followed by the number of the corrected rodent complementation group. These genes have been instrumental for the present knowledge of the process in eucaryotes (10). Most of these chinese hamster mutants appeared to be affected in the same genes as human patients suffering from a NER deficiency described below.

The biological impact of the NER pathway is most clearly apparent from human inborn mutations in this system. Three distinct, autosomal recessive conditions are associated with compromised NER functions. Strikingly, all are characterized by hypersensitivity of the skin and eyes to sun(UV)light. Cell fusion has also revealed a remarkable genetic heterogeneity underlying the pronounced clinical variability. These conditions are the prototype repair syndrome xeroderma pigmentosum (XP, 7 genetic complementation groups, designated XP-A to -G), Cockayne's syndrome (CS, 2 groups: CS-A and CS-B) and PIBIDS, an acronym for a specific combination of symptoms: photosensitivity, ichthyosis, brittle hair and nails, impaired intelligence, decreased fertility and short stature. It is photosensitive form of a broader syndrome called trichothiodystrophy (TTD, presently 3 NER-deficient groups: TTD-A, one group coinciding with XP-B and one identical to XP-D). Thus, defects in at least 10 genes can give rise to one or more of these conditions (10). The most pertinent symptoms of these diseases are summarized in Table 1. XP shows in addition to extreme sun hypersensitivity other cutaneous and ocular manifestations, including pigmentation abnormalities, a dry parchment-like skin and an over 2000-fold elevated fre-

Table 1. Main clinical symptoms of NER syndromes

Clinical symptoms	XP	XP/CS	CS	TTD
Photosensitivity	++	++	+[1]	+[1]
Abnormal pigmentation	++	+	–	–
Skin cancer	++	+	–	–
Accelerated neurodegeneration	–/+	+	?	?
Neurodysmyelination	–	++	+	+
Wizened facies	–	+	+	+
Growth defect	+/–	+	+	+
Hypogonadism	–	+	+	+
Brittle hair and nails	–	–	–	+
Scaling of skin	–	–	–	+

[1] Also TTD and CS patients occur without photosensitivity and NER defect.

quency of skin cancer in the exposed parts. The disease is often accompanied by progressive neurological degeneration (see (3) for a recent overview). A different type and much more severe and early onset form of neurologic dysfunction is seen in CS. In this disorder it is associated with dysmyelination of neurons. In addition, CS is characterized by impaired physical and sexual development, including cachectic dwarfism, microcephaly, skeletal and retinal abnormalities, and a characteristic 'wizened' appearance (16). Recently, we have identified also a number of patients displaying many of the CS hallmarks except for photo-sensitivity and a NER defect (unpublished observations). Thus, remarkably, this syndrome may also occur in a form without apparent NER involvement. Although XP and CS are very rare (prevalence less than 1 in 10^5) some patients present a combination XP and CS symptoms. They fall exclusively in three XP groups: XP-B, XP-D and XP-G. The third disease PIBIDS manifests many of the CS symptoms and curiously in addition two hallmarks of TTD: ichthyosis and sulphur-deficient brittle hair and nails (reviewed in ref. (13)). The latter is due to a reduced content of a class of cysteine-rich matrix proteins. Intriguingly, CS and PIBIDS individuals do not appear to be cancer-prone. Like in the case of CS there are also TTD patients without photosensitivity and compromised NER function. They are also designated as IBIDS, BIDS, and SIBIDS patients and reflect the wide clinical heterogeneity and the disconnection of some of the symptoms with NER insufficiency. This indicates that CS and TTD are at least in part not necessarily based on a NER deficiency and that except for the photosensitivity many of the other CS and TTD features have a different molecular basis. In summary, a striking clinical heterogeneity is found to be associated with NER impairment which is even more perplexing in view of the notion, that defects in at least two of the NER genes, XPB and XPD can give rise to clinical symptoms that range from XP, to combined XP/CS to even PIBIDS. The extension to repair-proficient TTD (and CS-type) patients suggests, that the syndromes involved in these complementation groups are different manifestations of a much broader clinical entity. The molecular basis for this heterogeneity will be addressed after the reaction mechanism of NER and the components involved have been explained in the following paragraph.

THE REACTION MECHANISM OF NUCLEOTIDE EXCISION REPAIR

The *E.coli* paradigm for NER entails a number of basic steps involving the crucial lesion identification, excision of the damage as part of a small single-stranded oligonucleotide stretch and gap-filling DNA resynthesis (8, 19, 24). Although in outline quite similar, the eukaryotic reaction appears much more complex and on several points different from the *E. coli* model. The cloning of mammalian and yeast NER genes, mainly employing the appropriate mutants, has permitted functional analysis and isolation of the encoded products. Subsequent comparison of the NER gene products of both species has disclosed a striking conservation of the entire pathway within eukaryotic evolution. The principal steps of NER in yeast and in mammals are very similar, and as a rule for each yeast NER protein there is at least one human homolog (for a recent review see (10)). Table 2 summarizes the main properties and other relevant features of the known human NER factors described below as well as proteins not discussed.

The contours of the core of the complex mammalian NER reaction have emerged recently using purified proteins and an *in vitro* reconstitution reaction, in which repair of a damaged DNA substrate can be measured (1). It involves an ordered scenario of steps, schematically outlined in Figure 1 (see also ref. (10) and references therein). First, DNA damage is recognized presumably preceded and followed by chromatin remodelling. This step likely involves at least the XPA product, that exhibits specific affinity for different types of DNA damage. XPA initiates the assembly of the NER machinery and determines its orientation towards the damaged region because it has docking sites for a number of other NER factors, such as RPA, a ss-DNA binding heterotrimer also involved in DNA replication, and the multi-subunit TFIIH complex. The latter is equipped with two DNA helicases, with DNA-unwinding potential in opposite directions: XPB (previously also called ERCC3) unwinds in the 3′→5′ direction and XPD (formerly known as ERCC2) 5′→3′ (for recent references see ref. (11)). It is most logical to assume, that the bidirectional helicase activity of the complex generates a local opening of the double helix around the injury, that is stabilized by RPA. This interpretation is supported by the finding, that the two NER incision factors ERCC1-XPF and XPG both require a ds to ss transition (Y-branch structure) for incision. The ERCC1-XPF hetero dimeric complex has recently been shown to be a structure-specific endonuclease, that is able to make the 5′ incision approximately 22 nucleotides upstream of the lesion in the damaged strand only (21). The XPG protein induces the 3′ incision in the damaged strand 5 bases downstream of the injury (17). The damage-containing 27–29 mer oligonucleotide (12) is released, and the gap is filled by DNA repair synthesis involving a number of proteins borrowed from the replication apparatus: DNA polymerase δ and/or ε, PCNA, RPA and RF-C (1). Finally, the new DNA is ligated to the existing strand by ligase I. In stead of a stepwise assembly of individual factors evidence has been presented in the analogous yeast system for a preassembled 'repairosome' (22). The above tentative model does not incorporate yet the chromatin dynamics that *in vivo* also must play a part. The molecular intricacies of this system are also apparent from the discovery by the group of Hanawalt and collaborators, that in fact two NER subpathways exist: a rapid and efficient removal of transcription-blocking lesions from the transcribed DNA strands (designated here transcription-coupled DNA repair), and a more slow and for some lesions less efficient repair of the remainder of the genome (global genome repair) (for a review and additional references see ref. (6)). The initial damage recognition in the transcription-coupled repair system is probably carried out by the RNA polymerase itself. Most XP complementation groups carry defects affecting both NER subpathways, indicating that they are impaired in a factor common to both processes. XP group C, however, harbours a deficiency that inactivates only global genome repair, whereas both classical CS groups (CS-A and CS-B) are specifically disturbed in the transcription-coupled process (25). This indicates that there are components common for both systems and that also subpathway specific factors

Table 2. Main properties of cloned human NER genes

Gene[1]	Chrom. location	Size protein(aa)[2]	Yeast homolog	Protein properties[3]
XPA	9q34	273	*RAD14*	Zn^{2+}-finger, binds different types of damaged DNA, transient interaction with ERCC1, RPA, and TFIIH(?) complex
XPB/ERCC3	2q21	782	*RAD25*	$3'\rightarrow5'$ DNA helicase, subunit of TFIIH, essential for
XPC	3p25.1[4]	940	*RAD4*	transcription initiation complexed with HHR23B, strong ss-DNA binding, involved in global genome repair only
XPD/ERCC2	19q13.2[4]	760	*RAD3*	$5'\rightarrow3'$ DNA helicase, subunit of TFIIH, essential for
XPE	11	1140	identified[5]	transcription initiation binds UV-damaged DNA, no causative mutations identified in complex with 48 kD, WD-repeat containing protein
XPF/ERCC4	16p13.3	~905	*RAD1*	also identical to ERCC11, complex with ERCC1 makes 5' incision, Y structure-specific endonuclease, dual function in recombination
XPG/ERCC5	13q32-33	1186	*RAD2*	Y structure-specific endonuclease, makes 3'-incision
CSA/ERCC8	5	396	*RAD28*	5 WD-repeats, involved in transcription-coupled repair only
CSB/ERCC6	10q11-21	1493	*RAD26*	DNA-dependent ATP-ase (helicase?), involved in transcription-coupled repair only
ERCC1	19q13.2[4]	297	*RAD10*	partial homology to UvrC and many nucleases, complex with XPF makes 5' incision Y-structure-specific endonuclease, dual function in recombination
HHR23A	19p13.2	363	*RAD23*	ubiquitin-like N-terminus, 2 ubiquitin-associated domains
HHR23B	3p25.14[4]	409	*RAD23*	as HHR23A, fraction of HHR23B complexed with XPC, complex binds ssDNA and is involved in global genome repair only
p62^{TFIIH}	11p14-15.1	548	*TFB1*	subunit of TFIIH, essential for transcription initiation
p52^{TFIIH}	6p21.3	513	*TFB2*	subunit of TFIIH, essential for transcription initiation
p44^{TFIIH}	5q1.3	395	*SSL1*	DNA-binding Zn^{2+}-finger, subunit of TFIIH, essential for transcription initiation
p34^{TFIIH}	12	303	*TFB4*	Zn^{2+}-finger, subunit of TFIIH, essential for transcription initiation

[1] Not included in this table are the genes for proteins involved in the DNA-synthesis step of the NER reaction, such as PCNA, RPA, RF-C, DNA ligase.

[2] aa: amino acids.

[3] Question marks indicate properties inferred but not proven.

[4] The *XPC* and *HHR23B* genes may share a common 650 kb MluI fragment; *ERCC1* and *XPD/ERCC2* are less than 250 kb apart.

[5] A yeast gene encoding a product with clear overall homology to the XPE protein has been identified in the yeast genome data base. In addition, a second human gene with significant similarity to XPE has been discovered (P.J. van der Spek, J.H.J.H., unpublished results).

exist (6). For instance, the XPC protein (in a complex with the HHR23B product) is probably only implicated in global genome repair. This complex has a strong ssDNA affinity (15), but its precise function in the NER reaction is still obscure. The same holds for the CSA and CSB products in the transcription-coupled repair subpathway. CSA is made up of WD-repeats that may provide protein-protein interaction domains (7), whereas CSB is a member of a closely-related family of DNA-dependent ATP-ases, with a presumed (but not yet proven) helicase function (23).

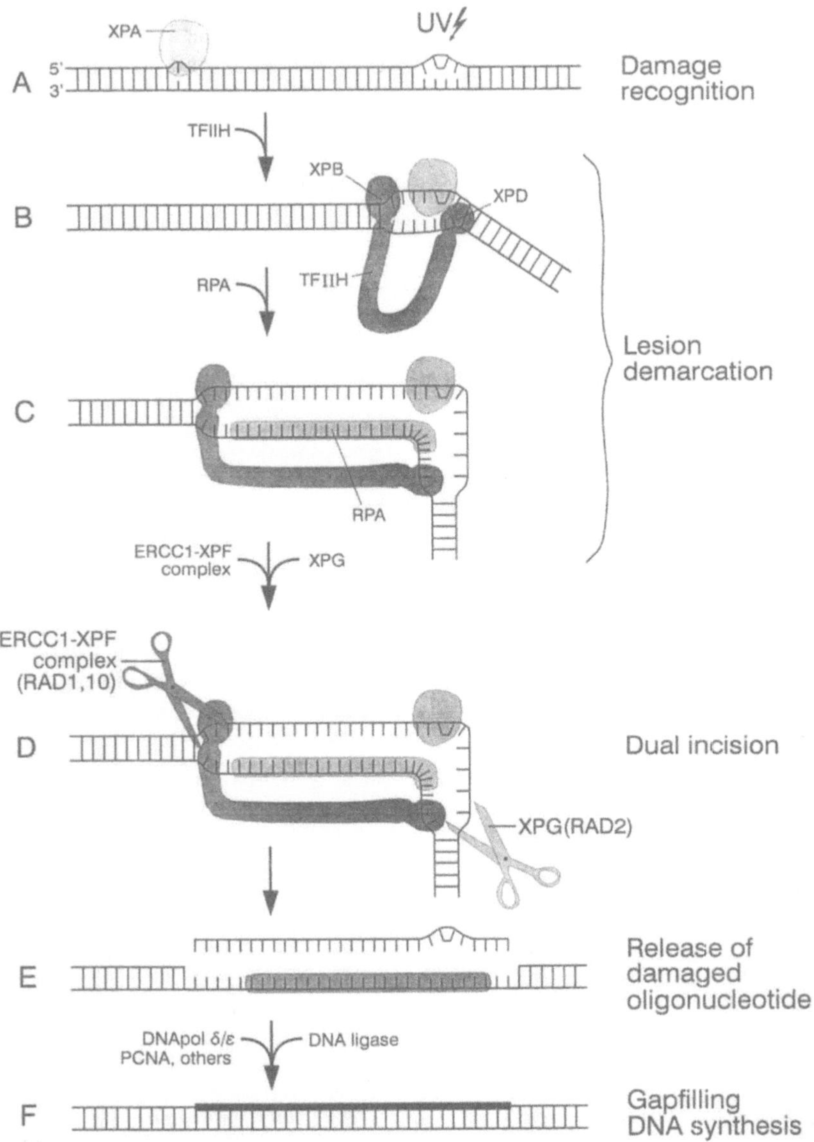

Figure 1. Model for the core of the NER reaction. Lesion recognition presumably involving XPA is followed by the association of ss-DNA binding RPA complex and TFIIH. The 'helix-opener' function accomplished by the XPB and XPD helicases generates a locally melted region. The transition points from duplex to ss DNA constitute substrates for the Y structure-specific endonuclease action of the ERCC1-XPF complex and XPG which are responsible for incising the damaged strand approximately 22 b. upstream and 5 b. downstream of the lesion respectively. Since the helicase function of XPB appears more important for NER and transcription than that of XPD we favour XPB unwinding the largest part of the opened regions in transcription and repair. Note that this model does not accommodate complex chromatin alterations nor the two NER subpathways: transcription-coupled repair and global genome repair. (See text for references and further explanation.)

FUNCTION SHARING BETWEEN NUCLEOTIDE EXCISION REPAIR AND TRANSCRIPTION: CLINICAL CONSEQUENCES

As mentioned above the CS defect already revealed a specific relationship between transcription and NER: the transcription-coupled NER subpathway (25). This process permits rapid clearance of the transcribed strand of active genes from transcription-blocking lesions for which the default repair by the global genome pathway would be too slow. Thus, it enables quick resumption of the vital transcription process. A second unexpected functional link between NER and transcription was discovered in collaboration with the group of J.-M. Egly (Strassbourg) (20). The above mentioned multi-subunit complex, designated TFIIH, was found to have a dual functionality: basal transcription and NER. The presence of the DNA repair helicases XPB and XPD suggests, that one of its functions may be a 'helix-opener' (for references see (11)): in the context of NER as a prerequisite for incision (see Figure 1), within the framework of transcription initiation for loading RNA polymerase II onto the template.

Interestingly, an exclusive relationship was observed between mutations in TFIIH subunits and all three TTD complementation groups (XP-B, XP-D and TTD-A), associated with exceptional clinical variability and symptoms that are difficult to rationalize on the sole basis of a NER defect. This has prompted the concept of 'transcription syndromes': Mutations in TFIIH subunits are envisaged to have multiple effects. Some may inactivate only a NER function, others may also influence the transcription role or both (see also Table 3). A mutation inactivating the NER function can easily account for the XP symptoms. Obviously, total inactivation of the transcription is lethal and this is consistent with the observation that deletion mutants of TFIIH subunits in yeast, Drosophila and mice are inviable (for references see (11), for the mouse mutants: Weeda, de Boer et al., unpublished observations). However, subtle insufficiencies crippling TFIIH transcription functioning may affect the expression of a specific subset of genes whose transcription critically depends on optimal TFIIH activity. For instance, transcription of one or more genes implicated in neuro-development may be affected by such a mutation leading to the neurodysmelination problems of CS and TTD. One can envisage, that strong secondary

Table 3. Model for the relation between TFIIH defects and clinical features[1]

TFIIH-related disorder[2]	TFIIH function[3]			Documented gene mutation
	NER	Transcription	Stability	
XP	−	+	+	*XPD*
XP/CS	−	+/−	+	*XPD, XPB, XPG*[4]
CS (photosensitive)	+/−[5]	+/−	+	*CSA, CSB*[4]
CS (non-photosensitive)[6]	+	+/−	+	?
TTD (photosensitive)	−	+/−	+/−	*XPD, XPB, TTDA*
TTD (non-photosensitive)[6]	+	+/−	+/−	?

[1] For further explanation see text.

[2] For clinical features see Table 1.

[3] Symbol designation: + normal repair or transcription function or stability of TFIIH, +/− partly affected function/stability, − severely impaired function.

[4] TFIIH transcription function may be indirectly affected by mutations in *XPG, CSA,* and *CSB.* In addition XPG affects total NER, CSA, and CSB affect only transcription-coupled repair.

[5] Only the NER subpathway of transcription-coupled repair is disturbed.

[6] Recently several CS patients without a NER defect were identified (A. Raams and N.G.J. Jaspers, unpublished observations). There may be many disorders related to non-photosensitive variants of TTD and CS such as Netherton syndrome, Brain-Hair syndrome, Pollitt syndrome, etc. (see [27]).

structure in a promoter requires the full unwinding capacity i.e. a maximally active complex, to permit efficient transcription. This combined 'repair/transcription syndrome' concept dissects the pleiotropic clinical features into those derived from a NER defect and those explained by a subtle transcriptional deficiency. It provides a rationale for non-photosensitive (NER$^+$) TTD (and CS) patients, in which only the transcription function of TFIIH might be affected. At the same time it accounts for the extreme rarity of these syndromes.

When the CS features in TTD are due to a basal transcriptional problem, this should also apply to the other complementation groups with CS patients, i.e. CS-A, CS-B and XP-G. Although the CSA, CSB and probably XPG proteins are not vital and therefore not essential for transcription, they may influence TFIIH functioning indirectly (Table 3). For instance, CSA and CSB could have an auxiliary function for TFIIH in transcription and in transcription-coupled (but not global genome) repair. Some XP-G (XP/CS) patients are more severe, than XP-A patients ((26) and unpublished results), although the latter are more defective in NER, suggesting that XPG has an additional non-repair function. Consistent with these predictions interactions between the CS proteins and TFIIH components have been claimed (2, 4, 7).

How can we rationalize the differences between CS and TTD, i.e the additional presence of brittle hair and nails in TTD? In view of the intrinsic properties of complexes such as TFIIH it is likely that some mutations also will affect the stability of the complex. In fact, a significant proportion of TFIIH mutations in yeast yields a temperature-sensitive phenotype (28), pointing to complex instability. It is feasible, that such type of mutations also occur among TFIIH patients. Interestingly, an exceptional TTD individual exhibited a reversible sudden, dramatic worsening of the brittleness of the hair formed during an episode of fever (14). This fits with a human, temperature-sensitive TFIIH mutation. Furthermore, it suggests that transcription of the genes for cysteine-rich matrix proteins of hair and nails is affected by TFIIH instability. In other cells the steady state levels of TFIIH may be sufficient, because *de novo* synthesis is high enough. However, the cysteine-rich matrix proteins, are one of the last gene products produced in very large quantities in keratinocytes before they die. Thus, the hallmark of TTD, may be due to a TFIIH stability problem becoming overt in terminally differentiated cells, that are exhausted for TFIIH before completion of their differentiation program. The non-repair CS features can be explained by a malfunctioning of TFIIH, without concomitant complex instability (Table 3). Future research using CS and TTD mouse models should reveal whether these concepts approach reality.

MOUSE MODELS FOR DEFECTS IN NUCLEOTIDE EXCISION REPAIR

To assess the biological consequences of repair defects and to obtain experimental animal models for human repair syndromes for unravelling the complex genotype-phenotype relationships gene targeting in totipotent mouse embryonal stem (ES) cells was utilized to generate NER-deficient mouse mutants. In this manner we have mimicked a mutation of a known CS-B patient in the mouse genome. Analysis of repair parameters in embryonic fibroblasts from the CSB mutant versus wildtype and heterozygous mice showed a specific loss of transcription-coupled repair. In agreement with the human syndrome CSB-deficient mice are photosensitive. Only mild other CS-like clinical features were observed, such as slightly retarded growth and neurological (behavioral and motor coordination) abnormalities. Remarkably, when a global genome repair defect was introduced in CSB-deficient mice a strong augmentation of the CS-phenotype was observed. Such XPC/CSB double mutant mice exhibit a very severe growth retardation of 75%, are unable to walk and die around day 18. These observations suggest that a CS defect becomes exaggerated in the absence of global genome repair, pointing to accumulation of endogenous damage as a contributing factor to the CS symptoms (van der Horst et al., manuscript in preparation).

In striking contrast to human CS, CSB-deficient mice appear clearly prone to skin cancer, when exposed to UV light or to the chemical carcinogen DMBA. Thus, in the mouse intact global genome repair is not sufficient to protect from tumorigenesis. This finding could imply that CS patients may have a hitherto unnoticed cancer predisposition. Alternatively, the species difference in tumorigenesis could be due to the fact that in man compared to rodents the global genome repair pathway is more potent in eliminating cyclobutane pyrimidine dimers, the major UV-induced lesion (van der Horst et al., manuscript in preparation). We expect that the above CSB mice and a number of other NER-deficient mouse models that have been generated or are in progress will be instrumental for unravelling the translation of the molecular defect into the pleiotropic phenotypic consequences, notably with respect to cancer predisposition.

ACKNOWLEDGMENTS

We thank the other members of the DNA repair group of the Medical Genetics Centre for valuable and stimulating discussions. Research in our group is supported by the Netherlands Scientific Organization (NWO), the Dutch Cancer Society, the EEC, Human Frontiers and the Louis Jeantet Foundation.

REFERENCES

1. Aboussekhra, A., M. Biggerstaff, M.K.K. Shivji, J.A. Vilpo, V. Moncollin, V.N. Podust, M. Protic, U. Hubscher, J.-M. Egly and R.D. Wood. 1995. Mammalian DNA nucleotide excision repair reconstituted with purified components. Cell. **80**:859–868.
2. Bardwell, A.J., L. Bardwell, N. Iyer, J.Q. Svejstrup, W.J. Feaver, R.D. Kornberg and E.C. Friedberg. 1994. Yeast nucleotide excision repair proteins rad2 and rad4 interact with RNA polymerase II basal transcription factor b (TFIIH). Mol. Cell. Biol. **14**:3569–3576.
3. Cleaver, J.E. and K.H. Kraemer. 1995. Xeroderma Pigmentosum and Cockayne syndrome, p. *In* Scriver, C.R., A.L. Beaudet, W.S. Sly, D. Valle (ed.), The metabolic basis of inherited disease Seventh edition. McGraw-Hill Book Co., New York.
4. Drapkin, R., J.T. Reardon, A. Ansari, J.C. Huang, L. Zawel, K. Ahn, A. Sancar and D. Reinberg. 1994. Dual role of TFIIH in DNA excision repair and in transcription by RNA polymerase II. Nature. **368**:769–772.
5. Friedberg, E.C., G.C. Walker and W. Siede. 1995. DNA repair and mutagenesis. ASM Press, Washington D.C.
6. Hanawalt, P.C. 1994. Transcription-coupled repair and human disease. Science. **266**:1957–1958.
7. Henning, K.A., L. Li, N. Iyer, L. McDaniel, M.S. Reagan, R. Legerski, R.A. Schultz, M. Stefanini, A.R. lLehmann, L.V. Mayne and E.C. Friedberg. 1995. The Cockayne syndrome group A gene encodes a WD repeat protein that interacts with CSB protein and a subunit of RNA polymerase II TFIIH. Cell. **82**:555–564.
8. Hoeijmakers, J.H.J. 1993. Nucleotide excision repair I: from *E.coli* to yeast. Trends in Genetics. **9**:173–177.
9. Hoeijmakers, J.H.J. 1993. Nucleotide excision repair II: from yeast to mammals. Trends in Genetics. **9**:211–217.
10. Hoeijmakers, J.H.J. 1995. Nucleotide excision repair: molecular and clinical implications. DNA repair mechanisms. in press.

11. Hoeijmakers, J.H.J. 1996. Human nucleotide excision repair syndromes: molecular clues to unexpected intricacies. European J. Cancer. **30A**:1921–1921.

12. Huang, J.C., D.L. Svoboda, J.T. Reardon and A. Sancar. 1992. Human nucleotide excision nuclease removes thymine dimers from DNA by incising the 22nd phosphodiester bond 5' and the 6th phosphodiester bond 3' to the photodimer. Proc. Natl. Acad. Sci. USA. **89**:3664–3668.

13. Itin, P.H. and M.R. Pittelkow. 1990. Trichothiodystrophy: review of sulfur-deficient brittle hair syndromes and association with the ectodermal dysplasias. J. of the American Acadamy of Dermatology. **22**:705–717.

14. Kleijer, W.J., F.A. Beemer and B.W. Boom. 1994. Intermittent hair loss in a child with PIBI(D)S syndrome and trichothiodystrophy with defective DNA repair-xeroderma pigmentosum group D. J. Med. Genet. **52**:227–230.

15. Masutani, C., K. Sugasawa, J. Yanagisawa, T. Sonoyama, M. Ui, T. Enomoto, K. Takio, K. Tanaka, P.J. van der Spek, D. Bootsma, J.H.J. Hoeijmakers and F. Hanaoka. 1994. Purification and cloning of a nucleotide excision repair complex involving the xeroderma pigmentosum group C protein and a human homolog of yeast RAD23. EMBO J. **13**:1831–1843.

16. Nance, M.A. and S.A. Berry. 1992. Cockayne syndrome: Review of 140 cases. American Journal of Medical Genetics. **42**:68–84.

17. O'Donovan, A., A.A. Davies, J.G. Moggs, S.C. West and R.D. Wood. 1994. XPG endonuclease makes the 3' incision in human DNA nucleotide excision repair. Nature. **371**:432–435.

18. Sancar, A. 1994. Mechanisms of DNA excision repair. Science. **266**:1954–1956.

19. Sancar, A. and M.-S. Tang. 1993. Nucleotide excision repair. Photochemistry and Photobiology. **57**:905–921.

20. Schaeffer, L., R. Roy, S. Humbert, V. Moncollin, W. Vermeulen, J.H.J. Hoeijmakers, P. Chambon and J. Egly. 1993. DNA repair helicase: a component of BTF2 (TFIIH) basic transcription factor. Science. **260**:58–63.

21. Sijbers, A.M., W.L. de Laat, R.R. Ariza, M. Biggerstaff, Y.F. Wei, J.G. Moggs, K.C. Carter, B.K. Shell, E. Evans, M.C. de Jong, S. Rademakers, J. de Rooij, N.G.J. Jaspers, J.H.J. Hoeijmakers and R.D. Wood. 1996. Xeroderma pigmentosum group F caused by a defect in a structures-specific DNA repair endonuclease. Cell. **86**:in press.

22. Svejstrup, J.Q., W.J. Feaver, J. LaPointe and R.D. Kornberg. 1994. RNA polymerase transcription factor IIH holoenzyme from yeast. submitted.

23. Troelstra, C., A. van Gool, J. de Wit, W. Vermeulen, D. Bootsma and J.H.J. Hoeijmakers. 1992. *ERCC6*, a member of a subfamily of putative helicases, is involved in Cockayne's syndrome and preferential repair of active genes. Cell. **71**:939–953.

24. van Houten, B. 1990. Nucleotide excision repair in *Escherichia coli*. Microbiol. Rev. **54**:18–51.

25. Venema, J., L.H.F. Mullenders, A.T. Natarajan, A.A. Van Zeeland and L.V. Mayne. 1990. The genetic defect in Cockayne syndrome is associated with a defect in repair of UV-induced DNA damage in transcriptionally active DNA. Proc. Natl. Acad. Sci. USA. **87**:4707–4711.

26. Vermeulen, W., J. Jaeken, N.G.J. Jaspers, D. Bootsma and J.H.J. Hoeijmakers. 1993. Xeroderma pigmentosum complementation group G associated with Cockayne's syndrome. Am. J. Human Genet. **53**:185–192.

27. Vermeulen, W., A.J. van Vuuren, M. Chipoulet, L. Schaeffer, E. Appeldoorn, G. Weeda, N.G.J. Jaspers, A. Priestley, C.F. Arlett, A.R. Lehmann, M. Stefanini, M. Mezzina, A. Sarasin, D. Bootsma, J.-M. Egly and J.H.J. Hoeijmakers. 1994. Three unusual repair deficiencies associated with transcription factor BTF2(TFIIH). Evidence for the existence of a transcription syndrome. Cold Spring Harbor. **59**:317–329.

28. Wang, Z., S. Buratowski, J.Q. Svejstrup, W.J. Feaver, X. Wu, R.D. Kornberg, T.D. Donahue and E.C. Friedberg. 1995. The yeast *TFB1* and *SSL1* genes, which encode subunits of transcription factor IIH, are required for nucleotide excision repair and RNA polymerase II transcription. Molecular and Cellular Biology. **15**:2288–2293.

DISCUSSION

Bacchetti: Do the cells from the knockout mice have a limited life span in culture, and do they have a senescent phenotype?

Hoeijmakers: Yes, that is a very good point. We have the impression that the cells of a number of mutant mice grow more poorly than controls. For instance in the case of the ERCC1 mutants all the cell lines tested go into crisis much earlier than the controls. But more line have to be analyzed to be certain.

Bacchetti: Did you get immortalized cells?

Hoeijmakers: We do get them, yes. Although it is more difficult, we do get immortalized ERCC1 deficient mouse embryonal fibroblasts.

Bacchetti: I have another question. I noticed that your mutations in the TTD and CS or XP although in different amino acids were clustered in a domain or what looked like a domain. What is known about that region?

Hoeijmakers: Actually the XPD mutations have been determined by the laboratories of M. Stefanini, A. Lehmann, and also by Christine Weber in Livermore. In this protein, the helicase motifs occur throughout the protein. Some mutations are in helicase domains indeed. There are no causative mutations as far as we can tell that truncate the protein, except for the very C-terminus. The helicase function itself may be relevant for repair, although we have shown that that is not absolutely indispensable for repair. Still, some repair synthesis occurs with a helicase impairment. But the protein has to be physically present in order to allow the assembly of the complex. If the protein is absent, TFIIH cannot be assembled and you run into a vital problem. Therefore, knockouts of this and other TFIIH genes in yeast, in Drosophila, mouse and in man, are absolutely lethal. There is still another helicase in TFIIH, ERCC3/XPB which can take care for at least some of the DNA helix opening, function probably, and in that way allow partial repair and allow full transcription.

Livingston: Two questions: If you transfect HA tagged ERCC2 or 3, and recapture the transfected protein with anti-HAAB, is it now associated with kinase activity?

Hoeijmakers: These are experiments that we actually would like to do. We are growing these cells in large quantities to be able to get enough pure TFIIH and then we can do functional studies on the TFIIH substrates. What we have done up till now is taking this partially purified complex and showing that it is able to complement the repair defect and repair assay and also microneedle injected into XPB cells which are defective in TFIIH functioning in repair and then this complex completely corrects the repair defects. So from the repair perspective we know that the tagged complex functions, from the perspective of transcription we have not done these assays. These have to be done, they are very relevant to do. It is a very good point.

Livingston: You transfected HA tag ERCC3 and label the cell *in vivo* either with S35 or P32; do you see only the known TFIIH subunits or any other bands?

Hoeijmakers: This is difficult to say because always one can see additional bands. However, we do not know whether they are specific or whether they are non-specific. These experiments should be done under very controlled conditions. So I cannot say for sure there are additional components. One of the intriguing points that I did not have time to discuss with you is the following: In collaboration with J.M. Egly, we are looking for TTD-A, because this is a missing repair factor which must be associated with TFIIH. None of the nine subunits that have been cloned in our assays corrects the TTD-A repair defect. So either in the way we do the experiment we cannot see correction because we have too high expression of that subunit giving rise to other effects that counteract any correction or we are still missing a subunit. Certainly TTD-A is not mutated in the ERCC2 and the ERCC3 subunits. We have done these experiments over and over again with correction with sequencing the gene even, there is not a mutation in these genes.

Hartwell: Do you have knockouts of any of the RAD-50 genes in mice and are they cancer-prone?

Hoeijmakers: It is still early days. In our laboratory Roland Kanaal and colleagues, have cloned in collaboration with Jean-Marie Buenstedde in Basel, the human homologue of RAD-54 (HHR54) and we have inactivated this gene in ES-cells by double targeting HHR54 cells are perfectly viable. They show sensitivity to X-rays, to MMS, to cross-linking agents, but not to UV, as we would expect for a bonafide recombination repair deficient mutant at the cellular level. And also we have preliminary evidence on one of the slides, that was listed, that homologous recombination in these ES cells by targeting is reduced. So from that perspective also a recombination defect may be fairly obvious. We have now germ line transmission of the mice and we are waiting for homozygous mutants to be born within the next couple of weeks. So at this moment I can, unfortunately, not tell you anything about RAD-52 or RAD-54. But we have also other genes of this pathway. Almost all of them have now been cloned in mammals. Not only in our laboratory but also in other laboratories. So soon we will see what is the phenotype of such a mutation, and whether inactivation is viable.

Wahl: You had mentioned association between p53 and XPB and I guess others have reported an association with XPD. I read varying literature on the subject of involvement of p53 and transcription coupled repair. Would you like to say whether you feel that the p53-mediated arrest is required for transcription coupled repair?

Hoeijmakers: That is a very, very difficult point. Because p53 protein has so may different functions that it is difficult to make a distinction between what is fact and what is fiction. But I already indicated our answer to you on one of the slides with the HA tagged TFIIH. One of the questions that was of interest to us is whether under natural conditions, p53 is associated, at least in part, with TFIIH. Geer Weeda, in our laboratory, has performed the experiment a few weeks ago and we can clearly see that a fraction of p53 in the cell line is associated with this HA-tagged TFIIH. If you apply the double tagged TFIIH to a nickel column using the histidine tag, we still see p53 associated. So, at least in these cells under these conditions we see an association but whether it has got functional implications, I do not know. Xin Wang, in the laboratory of Curtis Harris, and in part in our laboratory, did inject a cDNA encoding wild type p53 in normal human fibroblasts: This induces a strong apoptotic reaction within 24 to 48 hours. A very large fraction of the injected cells show all the features of apoptosis. Also p21, for instance, is induced in these cells. What is very remarkable if cells from XP groups B and D, which encode TFIIH subunits, are taken, almost no apoptosis is seen in XPD and in the XPB cells the apoptotic reaction is strongly delayed. This phenomenon can be corrected when the correcting repair gene is co-injected. This points also to a direct or indirect connection between TFIIH mutations and functioning and p53. These are the results. A plausible scenario to how to fit all these findings into a function of p53, directly or indirectly, I cannot provide to you. There may be connection with transcription coupled repair as well, however, we cannot say. In p53 mutant Li-fraumeni cells, Phil Hanawalt and Vilhelm Bohe have shown that there are defects in transcription repair, but I do not know whether they are primary or secondary to the p53 defect.

Wahl: But they are also, at best 2-fold?

Hoeijmakers: Yes, well, but the difference between transcription-coupled repair and global genome repair is also only a factor of two. It is a very difficult assay to do and one has to do many experiments to be certain about a result. But after many experiments, I think the results are quite reliable; even a factor of two difference would be significant.

Egly: Yes, if I can add something. Up to now we were just able to demonstrate that, at least *in vitro*, p53 may inhibit TFIIH—at least the helicase activity. If there are some mutations of p53, we increased this inhibition. But to my knowledge, we were never able to demonstrate that p53 inhibits the transcription reaction *in vitro*. Even if it was demonstrated, we were not able to reproduce experiments where it was also shown that p53 interacts with the DNA binding protein to decrease the transcription reaction *in vitro*. We have results where we show that p53 inhibits helicase activity. And then we got p53 from several laboratories but it was not possible to demonstrate an inhibition of the transcription reaction, at least *in vitro*, in various systems.

Stark: Is that p53 which has been produced, for example, in baculovirus?

Egly: Yes.

Stark: You see, the problem is as David Lane has shown, that p53 requires activation. So it is not functional as DNA binding protein unless you do something.

Bignami: Do you want to make any comments on Isabel Mellon's results on transcription repair coupling defects in mismatch correction mutants in bacteria and human cell lines?

Hoeijmakers: Yes, that is a very good point but a very difficult one as well. What Isabel Mellon has shown is that in *E.coli* and also in some mismatch repair-defective mammalian cell lines there is at the same time a partial or complete defect in transcription coupled repair. Again there, I do not know whether there is a direct or indirect relationship. Also, with mismatch repair defects, a lot of other processes will be preturbent. You may induce a partial stressed state in a cell because of the fact that there are still persisting mismatches in the DNA after replication. Whether under these conditions, for instance, transcription coupled repair behaves abnormally, I do not know. It is certainly not excluded that a direct relationship between mismatch repair and transcription coupled repair exists. A complicated factor yeast, both in the laboratory of Joap Brouwer in the Netherlands and also in the laboratory of Phil Hanawalt in Stanford, in many different mismatch repair deficient yeast mutants no defect in transcription coupled repair was found. It will be very difficult to really sort this out because a primary defect in mismatch repair or primary defect in transcription coupled repair may have all kinds of subtle secondary effects. If that happens, then you may look at some noise that arises from this primary defect. It is a very fundamental problem.

Wahl: Can you use your HA tagged nucleotide excision repair protein to identify or isolate members of other repair pathways in amounts that would not just be expected on the basis of random background association? Have you looked for association with say MSH2?

Hoeijmakers: These things have not been done. We are in the stage of TFIIH purification and biochemical and physical characterization to understand its function. From the phenotype of specific mutant alleles of TFIIH in yeast, one would think that TFIIH has a function in recombination as well. Certain alleles of RAD3, which is the yeast equivalent of XPD are hyper-recombinogenic. But again there, it is difficult to make a distinction between a primary or a secondary effect. A number of mutant alleles of RAD3 will also effect transcription to some extent. If this involves genes which are critical for recombination then one looks at an indirect relationship. It is very difficult to make these direct links.

7

TELOMERE LENGTH REGULATION BY THE Pif1 DNA HELICASE

Ellen K. Monson, Vincent P. Schulz,[*] and Virginia A. Zakian[†]

Department of Molecular Biology
Princeton University
Princeton, New Jersey 08544-1014

Telomeres are specialized protein-DNA structures present at the ends of all linear eukaryotic chromosomes. Telomeres are essential for stable maintenance of chromosomes because they protect chromosome ends from degradation and from fusion with other chromosomes (31, 32, 35). They also allow the cell to distinguish broken chromosome ends, which cause a cell cycle arrest, from intact chromosomes, so that the cell can progress steadily through the cell cycle (40). In addition, telomeres provide a solution to a problem first identified by Watson that replication of linear chromosomes by conventional DNA polymerases would result in their progressive shortening during successive rounds of replication (47). This problem arises because DNA polymerases cannot start synthesis *de novo*. Rather, they use an RNA primer 8–12 nucleotides long to initiate DNA synthesis. When the RNA primer is degraded it leaves 8–12 bases of unreplicated DNA at the terminus. Cells avoid this loss by utilizing a specialized telomeric DNA replication enzyme called telomerase that specifically recognizes telomeric DNA as a substrate. Telomerase is a unique reverse transcriptase that utilizes an RNA template to create a single-stranded 3′ extension at the ends of chromosomes (13). This extended 3′ strand can then presumably be used as substrate by the conventional replication machinery, thus enabling cells to replicate their chromosomes without a net loss of DNA.

Telomeric DNA has been sequenced from a diverse array of organisms. Most of the telomeric DNA repeats that have been characterized are rich in G residues in the strand running 5′ to 3′ towards the end of the chromosome. The repeats can either be regular such as those of humans (C_3TA_2/T_2AG_3) and Tetrahymena (C_4A_2/T_2G_4), or they can be irregular such as those found in the yeast *Saccharomyces cerevisiae* ($C_{1-3}A/TG_{1-3}$) (51). The macronuclear chromosomes in ciliates have a 12–16 bp single-stranded 3′ extension of the G strand ("G-tail") at the ends of their chromosomes (19). *S. cerevisiae* telomeres have a transient G-tail late in S phase (48). This 3′ single-stranded tail may be a common feature of telomeres and may be the specific substrate recognized by telomerase.

---

* Current address: CuraGen Corporation, 322 E. Main St., Branford, Connecticut 06405.
† Corresponding author

Genomic Instability and Immortality in Cancer
edited by Mihich and Hartwell, Plenum Press, New York, 1997

Most telomeres are dynamic structures with the precise number of telomeric repeats at the ends of chromosomes varying widely between organisms, between cell types and even from telomere to telomere within a species. In humans, telomeres vary in length from 10 to 20 kb in germ line and fetal cells to 1–2 kb in some human tumor cell lines (8). In yeast, telomeres are maintained at an average length of 300 ± 75 bp, although this length varies with growth conditions and genetic background. This length heterogeneity suggests that a precise length of telomeric DNA is not required for telomere function (although it is clear that there is a minimum length requirement as discussed below) and that there are competing processes in the cell that can both lengthen and shorten telomeres.

Telomeres are shorter in human somatic cells from older individuals than from younger individuals (17) and many cultured human cell lines show telomere shortening with increasing passages in culture (18). These cell lines lack detectable telomerase activity and have a limited capacity to divide in culture. However, many cancer cells and immortalized cell lines with essentially unlimited replicative capacity possess telomerase activity and show no telomere shortening following multiple rounds of cell division (7). On the basis of these observations it has been proposed that the shortening of telomeres with age is the "molecular clock" by which human cells measure their replicative capacity. This model asserts that once telomeres reach a critically short length, cells stop dividing (16, 17). Cells that have telomerase, such as cancer and immortal cells, can continue to divide indefinitely because their telomere lengths are maintained. Thus, the presence of active telomerase (or an alternative mechanism for telomere maintenance) might be an important factor in determining cell immortality.

Yeast cultures are essentially "immortal" in that a single cell clone can be maintained indefinitely in culture, although the total number of buds that an individual mother cell can give rise to is limited. Yeast also possesses telomerase activity (5, 25, 26, 44) and normally maintains telomere length in the range of 300 ± 75 bp. The distal half of a yeast telomere is subject to a dynamic breakdown and resynthesis process (45). Therefore, an average telomere length must be maintained by a balance of processes that tend to lengthen telomeres (telomerase, recombination) and that tend to shorten telomeres (incomplete replication, nucleases). The fact that this regulation is complex is indicated by the many genes that have been shown to affect telomere length in yeast. These include genes encoding a telomere binding protein, *RAP1* (2, 3), a Rap1p interacting protein, *RIF1* (15), telomerase RNA, *TLC1* (43), a putative protein subunit of telomerase, *EST1* (28), a DNA polymerase subunit, *CDC17* (4), a DNA helicase, *PIF1* (41), components of telomeric and mating type silencing machinery, *SIR3* and *SIR4* (36), an ATM homologue, *TEL1* (12, 30, 34), and an essential gene that lacks homology to known proteins, *TEL2* (30, 39).

The major pathway for maintaining telomeres in yeast is telomerase. The evidence for this is that mutations in either *TLC1*, the RNA component of telomerase or *EST1*, proposed to be a protein component of telomerase, result in gradual shortening of telomeres at a rate consistent with the predicted loss rate resulting from conventional DNA replication (28, 43). Eventually, when telomeres reach a critically short length or "crisis," *tlc1*Δ and *est1*Δ cells senesce and die. Thus, the presence of active telomerase appears to be essential for maintenance of normal telomere length. However, while most *est1*Δ and *tlc1*Δ cells die, there are always cells that manage to survive the critical telomere shortening that occurs. Analysis of these "survivors" shows that they have acquired tandem arrays of Y′ DNA, a subtelomeric repeat, through a *RAD52* dependent recombination pathway (27). Thus, *RAD52* dependent telomere recombination is a second pathway for telomere maintenance that is detectable under conditions that compromise normal telomere length regulation, and this type of recombination may play a constitutive role in maintenance of telomere length in yeast. Interestingly, there are immortal human cell lines that lack telomerase activity, but maintain very long telomeres, presumably through a telomerase-independent mechanism (1). The nature of this alternative pathway is unknown, but could be due to a recombination pathway similar to that observed in *tlc1Δ* and *est1Δ* cells. In addition, a *RAD52* independent recombination pathway may also contribute to telomere maintenance in yeast (37, 46).

Clearly, telomere length regulation in yeast is complex involving many genes and several replication pathways. In this paper, we focus on the interaction of two of these genes, the DNA helicase, *PIF1* and the telomere binding protein, *RAP1*. *RAP1* encodes an essential, multifunctional protein that binds duplex telomeric DNA repeats and is the major protein constituent in *S. cerevisiae* of the specialized telomeric chromatin structures called telosomes (2, 3, 42, 50). In addition, Rap1p activates or represses the transcription of a number of genes and is involved in transcriptional silencing both at telomeres and at the silent mating type loci (10, 21, 23). Deletion of the C-terminus of Rap1p causes telomere lengthening (22). Mutations that affect Rap1p DNA binding cause telomere shortening (29). High level expression of full-length Rap1p or of Rap1p deleted of its DNA binding domain, Rap1ΔBBp, results in telomere lengthening and this lengthening is thought to be due to the titration by the carboxyl terminus of Rap1p of factors that limit telomere length (6, 49). One of the factors that is titrated may be Rif1p, a Rap1p interacting protein whose elimination results in telomere lengthening (14). Similarly, cells carrying extra copies of the yeast telomeric sequence, $C_{1-3}A/TG_{1-3,}$ have longer telomeres (38), again suggesting titration of a length limiting factor or factors. Regulation of telomere length may be controlled by proteins that are targeted to telomeres by specific binding to $C_{1-3}A/TG_{1-3}$ or by protein-protein interactions with Rap1p.

PIF 1 encodes a 5′ to 3′ DNA helicase that was first identified as a gene required for a specialized type of recombination between mitochondrial genomes as well as for mitochondrial DNA repair and maintenance (9, 24). A role for the *PIF1* helicase at telomeres was discovered when *pif1* mutants were identified in a screen for genes that affect the stability of terminal telomeric tracts (41). A marker located just internal to a YAC telomere is lost at a rate several hundred fold higher in a *pif1* mutant than in wild-type cells. Analysis of telomere length revealed that *pif1* mutants have longer and more heterogeneous telomeres compared to wild-type cells. Furthermore, sequence analysis of the YACs that had lost the terminal marker showed that *pif1* mutants allow *de novo* telomere formation to occur at random locations along the YAC, whereas wild-type cells that had lost the terminal marker placed new telomeres almost exclusively at a "seed" of G rich *Oxytricha* telomere sequence on the YAC. Thus, the wild-type *PIF1* helicase must participate in a process that limits telomere length and that influences site selection for *de novo* telomere formation (41). One model that has been proposed is that Pif1p unwinds the telomerase RNA/ telomeric DNA duplex thus acting as an anti-processivity factor that limits the length of telomeric DNA synthesis. Another model is that Pif1p participates in a telomere-telomere recombination process and limits telomere length, by for example, limiting the degree of branch migration in the recombination intermediate. A role of Pif1p in telomere recombination is postulated based on its role in mitochondrial DNA recombination and its homology with another DNA helicase, *RRM3*, that affects recombination in yeast (20) (V. Schulz and V. Zakian, in preparation).

PIF1 appears to affect DNA processes in both mitochondria and in the nucleus because two forms of the protein are produced. *PIF1* has two ATG start sites that result in the production of two different proteins, one of which is localized to the nucleus, and the other which is localized to mitochondria. Mutation of the 1st ATG start site, the *pif1-m1* allele, results in a protein that shows no effect on telomere length or *de novo* telomere formation, but has a mitochondrial defect. Mutation of the 2nd ATG start site, the *pif1-m2* allele, results in the same telomere phenotypes as a *PIF1* deletion, but mitochondrial DNA is unaffected (41). Because it is easier to work with a strain that does not produce a high frequency of petites, the *pif1-m2* allele is used for the experiments in this paper. The longer, mitochondrial version of Pif1p will be referred to as Pif1Lp and the shorter, nuclear form of Pif1p will be referred to as Pif1Sp.

It has previously been shown that *pif1-m2, rif1Δ* cells have longer telomeres than either *pif1-m2* cells or *rif1Δ* cells (41). This result suggests that *PIF1* and *RIF1* affect telomere length maintenance by a different mechanism. Also, *rif1Δ* cells that express high levels of a version of *RAP1* deleted of its DNA binding domain, *RAP1ΔBB* have longer telomeres than *rif1Δ* cells alone, suggesting that telomere lengthening caused by expression of *RAP1ΔBB* is not solely due to titration of Rif1p (49). Here, we continue with this approach by asking what happens to telomere length when *RAPΔBB* is expressed in a *pif1-m2* mutant. We provide evidence that *RAP1* and *PIF1* probably regulate telomere length by different mechanisms. In addition, we provide evidence that the two proteins interact, at least under conditions of Pif1p overexpression.

MATERIALS AND METHODS

The yeast media used are described in Zakian and Scott (52). Yeast strains VPS105 (*MATa ade2 ade3 leu2–3,112 ura3Δ trp1Δ lys2–801 can1*) and VPS105 pif1-m2 (*MATa ade2 ade3 leu2–3,112 ura3Δ trp1Δ lys2–801 can1 pif1-m2*) (41) were used. YEpFAT10 is a 2 μm plasmid containing *TRP1* and a promoter defective *leu2-d* allele as selectable markers (48). Selection on media lacking tryptophan results in a plasmid copy number of 20–50 copies per cell and selection on media lacking leucine results in a plasmid copy number of 100–200 copies per cell. The following YEpFAT10 derivatives were introduced into yeast by transformation: (1) YEpFAT10RAP1ΔBB, which carries a copy of the *RAP1* gene, *RAP1ΔBB*, that is missing amino acids 19 to 497 and (2) YEpFAT10.3 which carries a 276 bp tract of C1–3A/TG1–3 DNA (49). Transformants were initially selected on media lacking tryptophan to select for lower plasmid copy number. The transformants were then restreaked twice on media lacking leucine to select for high plasmid copy number. Cultures were inoculated into 5 mls of liquid media lacking leucine, and genomic DNA was prepared by a glass bead procedure (38). Genomic DNA was digested with XhoI according to the manufacturers instructions (New England Biolabs), and run on a 1% agarose gel. Gels were Southern blotted, and probed with a Y′ subtelomeric probe.

To study the effect of overexpression of *PIF1*, the *PIF1* gene was cloned into plasmid pSH380 (kindly provided by Steve Hahn) such that the expression of the gene is under the control of the galactose inducible promoter, GAL1. pSH380 was cut with SmaI, and ligated to an ApaI *PIF1* restriction fragment from pVS30 (41) that was blunted with T4 DNA polymerase. The *LEU2* gene on this plasmid was replaced with the *URA3* gene by replacing the SalI-SacI *LEU2* fragment with the SalI-SacI *URA3* fragment of pRS316. The resulting plasmid is called pVS52PIF1S. This plasmid (and the control *URA3* plasmid, pVS52) was transformed into yeast strain YPH499 (*MATa ura3–52 lys2–801 trp1-Δ1 his3-Δ200 leu2-Δ1*) that also carries the YEpFAT10RAPΔBB or YEpFAT10 vector plasmid. Cells were streaked on media lacking uracil and tryptophan, and containing glucose or galactose as a carbon source.

RESULTS

Telomere Lengthening in Cells Expressing Large Amounts of Rap1p or Carrying Extra Copies of Internal $C_{1-3}A/TG_{1-3}$ DNA Is Not Caused by Titration of Pif1p

Cells carrying the *pif1-m2* allele have telomeres that are roughly 75 bp longer than wild-type, but they have normal mitochondria (41). Cells that express large amounts of a deletion derivative of Rap1p called Rap1ΔBBp, that consists mainly of the carboxyl third of the protein, have telomeres that are 100 bp longer than wild-type (6). Since Rap1ΔBBp lacks a DNA binding domain, one possible explanation for the lengthened telomeres in cells expressing Rap1ΔBBp is that it titrates Pif1p away from telomeres. To test whether titration of Pif1p could explain the telomere lengthening, we expressed Rap1ΔBBp in a *pif1-m2* mutant and measured telomere length by Southern blotting. As seen in Figure 1, *pif1-m2* cells that express Rap1ΔBBp have even longer telomeres than either *pif1-m2* cells or wild-type cells expressing Rap1ΔBBp. This result shows that the long telomeres seen in cells with large amounts of Rap1ΔBBp are not due solely to the titration of Pif1p and suggests that Pif1p and Rap1p regulate telomere length through different pathways.

Wild-type cells that carry circular plasmids with tracts of $C_{1-3}A/TG_{1-3}$ DNA have longer telomeres than cells carrying the vector alone. Again, this telomere lengthening

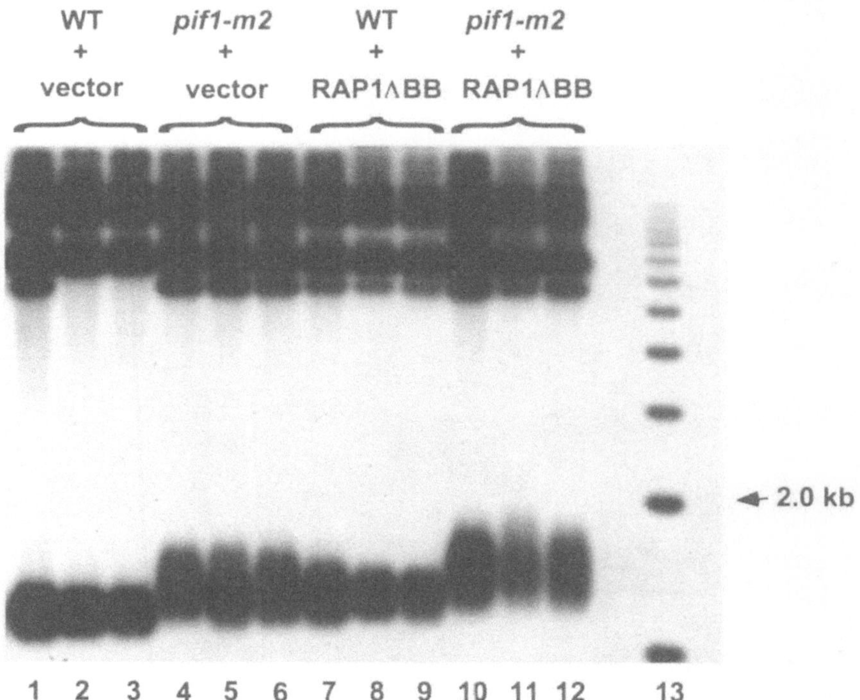

Figure 1. Telomeres get longer in *pif1-m2* cells expressing Rap1ΔBBp. DNA was prepared from three separate cultures from each strain, digested with XhoI, separated on a 1% agarose gel, and Southern blotted. The blot was hybridized to a Y′ probe. Lanes 1–3 contain VPS105/YEpFAT10; lanes 4–6, VPS105pif1-m2/YEpFAT10; lanes 7–9, VPS105/YEpFAT10RAP1ΔBB; lanes 10–12, VPS105pif1-m2/YEpFAT10RAP1ΔBB; lane 13, 1 kb ladder.

could be caused by titration of Pif1p away from the telomere by the plasmid borne $C_{1-3}A/TG_{1-3}$ DNA. To test this possibility, we measured telomere length in *pif1-m2* cells carrying a circular plasmid with a 276 bp tract of $C_{1-3}A/TG_{1-3}$ DNA. *pif1-m2* cells carrying a plasmid with a $C_{1-3}A/TG_{1-3}$ tract have longer telomeres than *pif1-m2* cells carrying the vector alone (Figure 2). This suggests that the telomere lengthening caused by extra copies of internal telomeric repeats is not caused by titration of Pif1p by $C_{1-3}A/TG_{1-3}$ DNA.

Toxicity of *PIF1S* Overexpression Is Suppressed by Expression of Rap1ΔBBp

Overexpression of the full-length, wild-type *PIF1* gene is toxic to cells (24). Similarly, overexpression of the nuclear form of *PIF1*, *PIF1S*, is toxic (Figure 3). The mechanism of *PIF1* toxicity is not understood, but there is evidence that toxicity requires a

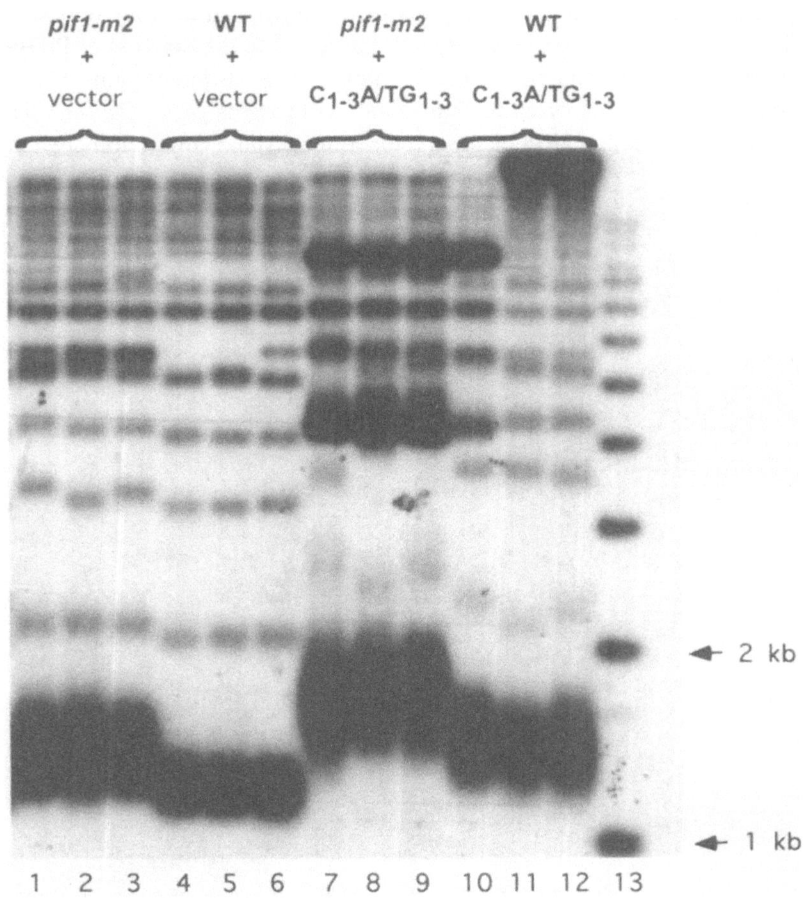

Figure 2. Telomeres get longer in *pif1-m2* cells carrying extra copies of $C_{1-3}A/TG_{1-3}$ DNA. DNA was prepared from three separate cultures from each strain, digested with XhoI, separated on a 1% agarose gel and Southern blotted. The blot was hybridized to a $C_{1-3}A/TG_{1-3}$ probe. Lanes 1–3 contain VPS105pif1-m2/ YEPFAT10; lanes 4–6, VPS105/YEPFAT10; lanes 7–9, VPS105pif1-m2/ YEPFAT10.3; lanes 10–12, VPS105/YEPFAT10.3; lane 13, 1 kb ladder.

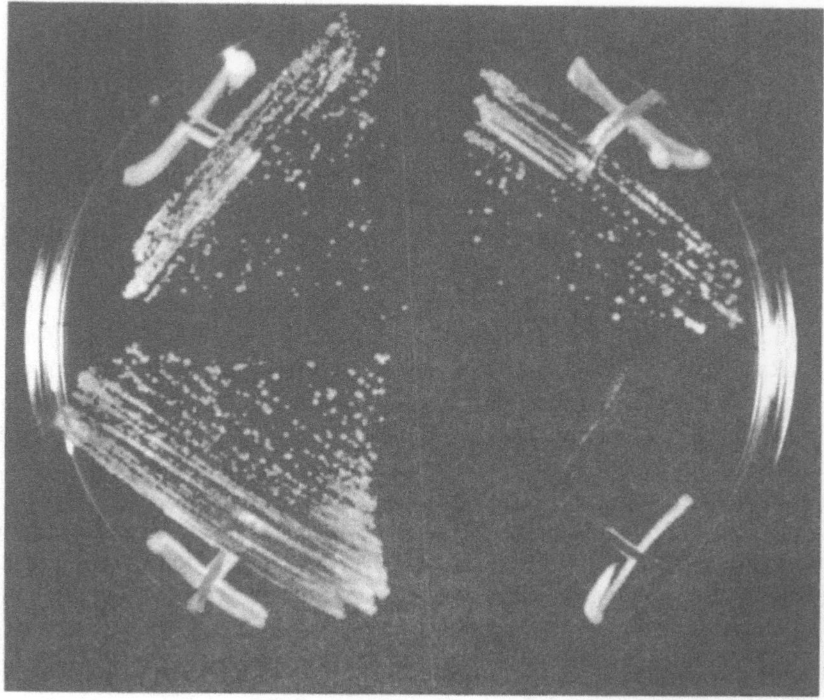

Figure 3. The toxicity of Pif1Sp overexpression is suppressed by expression of Rap1ΔBBp. Cells were streaked onto YC-trp-ura +Gal plates and incubated at 30°C for three days. (A) YPH499/pVS52/YEpFAT10; (B) YPH499/pVS52PIF1S/YEpFat10; (C) YPH499/pVS52/YEpFAT10RAP1ΔBB; (D) YPH499/pVS52PIF1S/YEp-FAT10RAP1ΔBB.

functional Pif1 protein because overexpression of amino acids 43–560, the N-terminal 333 amino acids, or the C-terminal 560 amino acids of Pif1p is not toxic (21, E. Monson and V. Zakian, unpublished). In addition, toxicity is unlikely to be the result of a mitochondrial defect because overexpression of the *pif1-m2* allele, which produces only the mitochondrial form of Pif1p, is not toxic. Furthermore, no increase in the number of rho⁻ cells is observed during GAL induction of the full-length, wild-type *PIF1* gene (24).

As a second approach to testing for interaction between Pif1p and Rap1p, we tested the effects of *RAP1ΔBBp* expression on the toxicity phenotype of high level Pif1Sp expression. We were surprised to find that the growth defect caused by Pif1Sp expression is suppressed by expressing Rap1ΔBBp (Figure 3). One possible explanation is that Pif1p and Rap1p physically interact and that Rap1ΔBBp titrates Pif1p from telomeres or another site where it causes toxicity. If Pif1Sp and Rap1p do interact directly, however, this interaction must be independent of their respective roles in telomere length regulation, based on the results in Figure 1. Cells that overexpress both Pif1Sp and Rap1ΔBBp have a telomere length that is similar to that of wild-type cells expressing Rap1ΔBBp (data not shown), consistent with the conclusion that the telomere lengthening observed in cells with large amounts of Rap1ΔBBp is not due to titration of Pif1Sp. *RAP1ΔBB* expression may not suppress Pif1Sp toxicity simply as a result of telomere lengthening because deletion of *RIF1*, which also causes telomere lengthening, does not suppress Pif1Sp toxicity (data not shown).

DISCUSSION

Since overexpression of *RAP1ΔBB* in a *pif1-m2* mutant results in longer telomeres than those seen in either a *pif-m2* mutant or a *RAP1ΔBB* overexpresser alone (Figure 1), lengthening of telomeres in an otherwise wild-type cell expressing high levels of Rap1ΔBBp cannot be accounted for solely by titration of Pif1Sp. If Pif1Sp titration were the only reason for telomere lengthening in cells with high levels of Rap1ΔBBp, then there should be no difference in telomere length in a *pif-m2* mutant and a *pif-m2* mutant expressing Rap1ΔBBp. This result means either that Pif1Sp and Rap1p do not physically interact, that they interact, but this interaction does not affect telomere length. Thus, *PIF1* and *RAP1* influence telomere length via different "pathways" in the sense that they work through different subsets of telomere length regulatory factors. Whether some of these factors are common to both subsets remains to be seen.

Given that Pif1Sp and Rap1p seem to affect telomere length by different pathways, it was surprising to find that the toxicity of Pif1Sp overexpression is suppressed by expressing Rap1ΔBBp. There are several possible mechanisms for the suppression of Pif1Sp toxicity by Rap1ΔBBp:

1. One possibility is that in a wild-type cell, Pif1Sp interacts with telomere-bound Rap1p. Expression of Rap1ΔBBp titrates Pif1Sp, but can also titrate other telomere-bound factors, such as Rif1p, that limit telomere length (14). We imagine a scenario in which Pif1Sp is only partially titrated by Rap1ΔBBp either because Rap1ΔBBp is limiting or Pif1Sp is in a complex. Removal of Pif1p in the *pif1-m2* mutant results in further telomere lengthening in a Rap1ΔBBp overexpresser because more Rap1ΔBBp is now available to titrate other factors, or because removal of one factor makes other length limiting factors more accessible to titration. Each of these Rap1p interacting factors could have independent roles in, for example, establishing contacts with various components of the telomere replication or telomere degradation machinery. Since Rap1p is known to be a major component of telomeric chromatin, with 10–20 molecules of Rap1p on average per telomere (11, 50), one could imagine that Rap1p is acting as an assembly site for factors affecting telomere length, much as it has been proposed to act as an assembly site for factors affecting transcriptional silencing. A Rap1-like protein has been proposed to be involved in limiting telomere length in *K. lactis*, based on the dramatic telomere lengthening phenotypes of mutants carrying altered telomeric DNA sequences. Telomere lengthening is thought to be due, in part, to altered or lack of binding of this Rap1-like protein to the mutant telomeres (33).
2. A second possibility is that Rap1ΔBBp interacts with and protects the substrate through which Pif1Sp mediates its toxicity. Since Rap1ΔBBp lacks its DNA binding domain, it is unlikely that it is binding to and protecting a telomeric DNA substrate from Pif1Sp. However, Rap1ΔBBp may be interacting with the telomere through protein-protein interactions, thus blocking access of overexpressed Pif1Sp to the telomere.

3. A third possibility is that Pif1Sp does not interact with Rap1p in wild-type cells when the proteins are expressed at normal levels, perhaps because Pif1Sp is sequestered in a replication complex. In this model, suppression of Pif1Sp toxicity by Rap1ΔBB results from an interaction between these proteins that occurs only during Pif1Sp overexpression, and thus may not be relevant to their normal functions in the cell.

This study has shown that *PIF1* and *RAP1* affect telomere length through different mechanisms, although the two proteins may physically interact or interact with common substrates. The precise mechanisms by which Pif1p or Rap1p affect telomere length remain to be seen and may be quite complex. Whether the Pif1 DNA helicase influences, for example, telomerase accessibility to the telomeric DNA substrate, telomerase processivity, the frequency of telomere-telomere recombination, the degree of branch migration of a recombination intermediate, or some other aspect of telomere metabolism is presently unclear. Identification of more components that interact with Pif1p and Rap1p should help our understanding of the complex process of telomere length regulation in yeast.

REFERENCES

1. Bryan, T. M., A. Anglezou, J. Gupta, S. Bacchetti, and R. R. Reddel. 1995. Telomere elongation in immortal human cells without detectable telomerase activity. EMBO J. **14**:4340–4348.
2. Buchman, A. R., W. J. Kimmerly, J. Rine, and R. D. Kornberg. 1988. Two DNA-binding factors recognize specific sequences at silencers, upstream activating sequences, autonomously replicating sequences, and telomeres in *Saccharomyces cerevisiae*. Mol. Cell. Biol. **8**:210–225.
3. Buchman, A. R., N. F. Lue, and R. D. Kornberg. 1988. Connections between transcriptional activators, silencers, and telomeres as revealed by functional analysis of a yeast DNA-binding protein. Mol. Cell. Biol. **8**:5086–5099.
4. Carson, M. J., and L. Hartwell. 1985. *CDC 17*: an essential gene that prevents telomere elongation in yeast. Cell **42**:249–257.
5. Cohn, M., and E. H. Blackburn. 1995. Telomerase in yeast. Science **269**:396–400.
6. Conrad, M. N., J. H. Wright, A. J. Wolf, and V. A. Zakian. 1990. RAP1 protein interacts with yeast telomeres *in vivo*: overproduction alters telomere structure and decreases chromosome stability. Cell **63**:739–750.
7. Counter, C. M., A. A. Avilion, C. E. LeFeuvre, N. G. Stewart, C. W. Greider, C. B. Harley, and S. Bacchetti. 1992. Telomere shortening associated with chromosome instability is arrested in immortal cells which express telomerase activity. EMBO J. **11**:1921–1929.
8. de Lange, T. 1995. Telomere dynamics and genome instability in human cancer, p. 265–293. *In* E. H. Blackburn, and C. W. Greider (ed.), Telomeres. Cold Spring Harbor Laboratory Press, Plainview, NY.
9. Foury, F., and J. Kolodynski. 1983. *pif* mutation blocks recombination between mitochondrial p^+ and p^- genomes having tandemly arrayed repeat units in *Saccharomyces cerevisiae*. Proceedings of the National Academy of Science USA **80**:5345–5349.
10. Giesman, D., L. Best, and K. Tatchell. 1991. The role of RAP1 in the regulation of the MAT*alpha* locus. Mol. Cell. Biol. **11**:1069–1079.
11. Gilson, E., T. Laroche, and S. M. Gasser. 1993. Telomeres and the functional architecture of the nucleus. Trends Cell Biol. **3**:128–134.
12. Greenwell, P. W., S. L. Kronmal, S. E. Porter, J. Gassenhuber, B. Obermaier, and T. D. Petes. 1995. *TEL1*, a gene involved in controlling telomere length in *S. cerevisiae*, is homologous to the human ataxia telangiectasia gene. Cell **82**:823–829.
13. Greider, C. W., and E. H. Blackburn. 1985. Identification of a specific telomere terminal transferase activity in *Tetrahymena* extracts. Cell **43**:405–413.
14. Hardy, C. F. J., D. Balderes, and D. Shore. 1992. Dissection of a carboxy-terminal region of the yeast regulatory protein RAP1 with effects on both transcriptional activation and silencing. Mol. Cell. Biol. **12**:1209–1217.
15. Hardy, C. F. J., L. Sussel, and D. Shore. 1992. A RAP1-interacting protein involved in transcriptional silencing and telomere length regulation. Genes Dev. **6**:801–814.

16. Harley, C. B. 1991. Telomere loss: mitotic clock or genetic time bomb? Mut. Res. **256:**271–282.

17. Harley, C. B., A. B. Futcher, and C. W. Greider. 1990. Telomeres shorten during ageing of human fibroblasts. Nature **345:**458–460.

18. Hastie, N. D., M. Dempster, M. G. Dunlop, A. M. Thompson, D. K. Green, and R. C. Allshire. 1990. Telomere reduction in human colorectal carcinoma and with ageing. Nature **346:**866–868.

19. Henderson, E. R., and E. H. Blackburn. 1989. An overhanging 3' terminus is a conserved feature of telomeres. Mol. Cell. Biol. **9:**345–348.

20. Keil, R. L., and A. D. McWilliams. 1993. A gene with specific and global effects on recombination of sequences from tandemly repeated genes in *Saccharomyces cerevisiae*. Genetics **135:**711–718.

21. Kurtz, S., and D. Shore. 1991. RAP1 protein activates and silences transcription of mating-type genes in yeast. Genes Dev. **5:**616–628.

22. Kyrion, G., K. A. Boakye, and A. J. Lustig. 1992. C-terminal truncation of RAP1 results in the deregulation of telomere size, stability, and function in *Saccharomyces cerevisiae*. Mol. Cell. Biol. **12:**5159–5173.

23. Kyrion, G., K. Liu, C. Liu, and A. J. Lustig. 1993. RAP1 and telomere structure regulate telomere position effects in *Saccharomyces cerevisiae*. Genes Dev. **7:**1146–1159.

24. Lahaye, A., H. Stahl, D. Thines-Sempoux, and F. Foury. 1991. PIF1: a DNA helicase in yeast mitochondria. EMBO J. **10:**997–1007.

25. Lin, J.-J., and V. A. Zakian. 1995. An *in vitro* assay for *Saccharomyces* telomerase requires *EST1*. Cell **81:**1127–1135.

26. Lue, N. F., and J. C. Wang. 1995. ATP-dependent processivity of a telomerase activity from *Saccharomyces cerevisiae*. J. Biol. Chem. **270:**21453–21456.

27. Lundblad, V., and E. H. Blackburn. 1990. RNA-dependent polymerase motifs in EST1: Tentative identification of a protein component of an essential yeast telomerase. Cell **60:**529–530.

28. Lundblad, V., and J. W. Szostak. 1989. A mutant with a defect in telomere elongation leads to senescence in yeast. Cell **57:**633–643.

29. Lustig, A. J., S. Kurtz, and D. Shore. 1990. Involvement of the silencer and UAS binding protein RAP1 in regulation of telomere length. Science **250:**549–553.

30. Lustig, A. J., and T. D. Petes. 1986. Identification of yeast mutants with altered telomere structure. Proc. Natl. Acad. Sci. USA **83:**1398–1402.

31. McClintock, B. 1939. The behavior in successive nuclear divisions of a chromosome broken at meiosis. Proc. Natl. Acad. Sci. USA **25:**405–416.

32. McClintock, B. 1941. The stability of broken ends of chromosomes in *Zea mays*. Genetics **26:**234–282.

33. McEachern, M. J., and E. H. Blackburn. 1995. Runaway telomere elongation caused by telomerase RNA gene mutations. Nature **376:**403–409.

34. Morrow, D. M., D. A. Tagle, Y. Shiloh, F. S. Collins, and P. Hieter. 1995. *TEL1*, an *S. cerevisiae* homolog of the human gene mutated in ataxia telangiectasia, is functionally related to the yeast checkpoint gene *MEC1*. Cell **82:**831–840.

35. Muller, H. J. 1938. The remaking of chromosomes. The Collecting Net **13:**181–195,198.

36. Palladino, F., T. Laroche, E. Gilson, A. Axelrod, L. Pillus, and S. M. Gasser. 1993. SIR3 and SIR4 proteins are required for the positioning and integrity of yeast telomeres. Cell **75:**543–555.

37. Pluta, A. F., and V. A. Zakian. 1989. Recombination occurs during telomere formation in yeast. Nature **337:**429–433.

38. Runge, K. W., and V. A. Zakian. 1989. Introduction of extra telomeric DNA sequences into *Saccharomyces cerevisiae* results in telomere elongation. Mol. Cell. Biol. **9:**1488–1497.

39. Runge, K. W., and V. A. Zakian. 1996. *TEL2*, an essential gene required for telomere length regulation and telomere position effect in *Saccharomyces cerevisiae*. Mol. Cell. Biol. In press.

40. Sandell, L. L., and V. A. Zakian. 1993. Loss of a yeast telomere: Arrest, recovery and chromosome loss. Cell **75:**729–739.

41. Schulz, V. P., and V. A. Zakian. 1994. The Saccharomyces PIF1 DNA helicase inhibits telomere elongation and de novo telomere formation. Cell **76:**145–155.

42. Shore, D., D. J. Stillman, A. H. Brand, and K. A. Nasmyth. 1987. Identification of silencer binding proteins from yeast: possible roles in SIR control and DNA replication. EMBO J. **6:**461–467.

43. Singer, M. S., and D. E. Gottschling. 1994. *TLC1*, the template RNA component of the *Saccharomyces cerevisiae* telomerase. Science **266:**404–409.

44. Steiner, B. R., K. Hidaka, and B. Futcher. 1996. Association of the Est1 protein with telomerase activity in yeast. Pros. Natl. Acad. Sci. USA **93:**2817–2821.

45. Wang, S.-S., and V. A. Zakian. 1990. Sequencing of *Sacchromyces* telomeres cloned using T4 DNA polymerase reveals two domains. Mol. Cell. Biol. **10:**4415–19.

46. Wang, S.-S., and V. A. Zakian. 1990. Telomere-telomere recombination provides an express pathway for telomere acquisition. Nature **345**:456–458.

47. Watson, J. D. 1972. Origin of concatemeric DNA. Nature (New Biology) **239**:197–201.

48. Wellinger, R. J., A. J. Wolf, and V. A. Zakian. 1993. *Saccharomyces* telomeres acquire single-strand TG_{1-3} tails late in S phase. Cell **72**:51–60.

49. Wiley, E., and V. A. Zakian. 1995. Extra telomeres, but not internal tracts of telomeric DNA, reduce transcriptional repression at *Saccharomyces* telomeres. Genetics **139**:67–79.

50. Wright, J. H., D. E. Gottschling, and V. A. Zakian. 1992. *Saccharomyces* telomeres assume a non-nucleosomal chromatin structure. Genes Dev. **6**:197–210.

51. Zakian, V. A. 1995. Telomeres: beginning to understand the end. Science **270**:1601–1607.

52. Zakian, V. A., and J. F. Scott. 1982. Construction, replication and chromatin structure of TRP1 RI circle, a multiple-copy synthetic plasmid derived from *Saccharomyces cerevisiae* chromosomal DNA. Molec. Cell. Biol. **2**:221–232.

DISCUSSION

Cooke: Telomere lengthening in Saccharomyces normally results in a position effect. How do you accommodate that observation with your double mutant findings?

Monson: Well, I think there are cases where telomere length and telomere position effect do indeed correlate. So that if telomeres get longer you see more position effect but we can actually uncouple the telomere length phenotype from telomere position effect. So, it is not always correlated with telomere length.

Nasmyth: RRM3 was involved in suppressing recombination between CUP1 and rDNA repeats, right? And yet, what you are saying is that it is promoting recombination.

Monson: Well it is interesting because with the rDNA repeat assay it appears that RRM3 is repressing recombination. Unpublished data from R. Keil's laboratory shows that PIF1 appears to promote recombination in that assay so it seems to be having the opposite effect in that rDNA repeat assay than it is having at telomeres. So RRM3 and PIF1 are actually acting antagonistically but they are acting antagonistically in opposite ways at rDNA repeats than they are at telomeres.

Hartwell: Two questions: One, is the PIF1 RRM3 G2 delay RAD9 dependent?

Monson: We do not know that yet.

Hartwell: The second question is: Do either of these, or the double mutant, have any effect on the end - end association assay that telomere associations that occur at the end of S-phase in plasmids?

Monson: You mean that CFP assay that Raymund Wellinger had developed? Again, that has not been looked at.

Chapman: I was curious if the pif1 Tel1 double mutant has been constructed to assess if PIFI mutation can abrogate any of the telomere loss that is observed in the absence of telomerase activity?

Monson: No, we have not made that mutant.

Chapman: Because if you can see the enhanced telomere maintenance in the absence of telomerase activity, that would provide evidence that PIF1 is not involved in the telomerase mediated pathway.

Monson: That is one experiment that we are attempting to do at the moment. To look if we see increased healing in the absence of telomerase RNA in a TLC1 mutant. So we do not know whether the healing is dependent on telomerase. It is kind of a technically difficult experiment because when you knock out the telomerase RNA, the cells senesce and die, so you have to catch them in the appropriate window of time. We are trying to do that.

Hoeijmakers: Do you know what is the direction of unwinding of RRM3 gene with respect to Pif1?

Monson: We do not know that and, actually this is also unpublished data, where someone purified the RRM3 protein and they were never able to demonstrate helicase activity for RRM3. So while RRM3 has the helicase motifs conserved, no helicase activity has actually ever been demonstrated for RRM3.

Hoeijmakers: Let us assume that both are helicases, which is the most likely possibility in view of the sequence conservation. Would it be possible that these are types of helicases which are able to shift atop and bottom strand in response to short repeats so that they are still aligned? So if Pif would move the sequences in one direction and RRM3 would do in the opposite direction, one can explain the increased telomere length that you find when Pif is absent.

Monson: So you are talking instead about promoting a branch migration pathway, you are actually talking about misalignment of those sequences. Yes, I have never thought about that. I do not really know—is that really a property of helicases? Can helicases do that? I do not know, but that is an interesting idea.

Bacchetti: Have you looked at the in vitro processivity of the enzyme in this mutant?

Monson: We have not looked at the processivity of Pif1p. We have not done any biochemistry with Pif1p, but Francoise Foury's laboratory purified the protein and it is actually a non-processive or distributive type of helicase. So it falls on and off the substrate quite frequently.

Kolodner: Did you mention what the effect of the RRM3 single mutant is on telomere length?

Monson: Yes and basically we see very little effect. Sometimes when we look at the gels we like to imagine that maybe there is a little bit of a shortening, but it is really not very convincing.

Mihich: Is it known what determines the prevalence of one pathway versus the other?

Monson: You mean in terms of telomerase versus recombination? Well, I think that, clearly in yeast, the telomerase pathway is the major pathway for telomere maintenance and the evidence for that is, if you knock out the telomerase RNA gene in yeast, most of the cells end up dying. But interestingly, there are a certain population of cells in these telomerase minus cells that seem to be able to recover, so they are called survivors. I think we will probably be hearing more about these types of mechanisms in Mike McEachern's talk next. It appears that recombination can act as a kind of backup pathway when the cells do not have any other mechanism to maintain their telomeric DNA.

Mihich: Is there any notion about the signaling that determines the awareness that that pathway is required?

Monson: When yeast cells lack telomerase, the telomeres get shorter and shorter. Presumably it is because of the telomere shortening that you get cellular senescence. So, the idea is that the short telomeres are actually what signals but nothing is known about the mechanism of how that signaling might occur.

Stark: It is a curious property of the recombination mechanism that you can only get a net increase in telomeres through that mechanism if you fail to align completely all the homology. If you lie the shorter telomere down on the longer telomere such that you are making the maximum number of base pairs, you can only make it as long as the longest telomere. With each cycle of replication, the longest one will also get shorter. It is not a general case, it is a special case of recombination that is required. Is there any evidence that that is actually what happens? Through the recombination mechanism can you actually maintain the length through many cell cycles?

Monson: Actually, there is no evidence, as far as I understand it, and Mike may be able to answer this question better than I can, but as far as I understand, there is no evidence that a recombination pathway actually can maintain the telomeres over many cell cycles. I think what is observed in these survivors, when telomerase is not present, there is a recombination event, and it is usually between the Y prime subtelomeric sequences that will allow placement of telomeric seqluence at the end of the chromosome. But then these survivors will be observed to undergo another round of telomere shortening and crisis and senescence. So, it is not as though the cells are stable and happy, it allows them to survive but they have to deal with the problem again as it arises.

Gatti: When you produce chromosomes without telomeres you said that you get cell cycle arrest, then the cells start growing again after twenty four hours. Do you know when in the cell cycle these cells are arrested, in which phase?

Monson: Its RAD9 dependent, so it should be G2, as far as I understand. Maybe Lee Hartwell can answer that better. So it would be a G2 arrest.

Shay: Are there any equivalents of Pif1 human mammalian homologues? Have people actually looked for that?

Monson: Vince Schulz actually tried to do that and he tried to clone a PIF1 homologue by degenerate PCR. He was not able to do that, but he was able to identify, in a database search, a c.elegans homologue and also a candida albicans homologue as well. Those were just cDNA sequences of unknown function in those organisms. So the fact that he sees homologues in c. elegans suggests that it may be present in humans as well.

Shay: What do you think ultimately limits the length of the telomeres in these double Pif1 mutants? They all seem to go from very short to maybe 1.6 or 1.8 kilobases. What eventually limits that? If you get too long telomeres, does that slow down growth rate or do you see any overt changes with the cell carrying extra TRFs?

Monson: Well in terms of what limits telomere length, I think that data from Mike McEachern has shown that telomeres can get really long, really fast and so long, he will probably be showing us that, that they run at the limit mobility of the gel. I think he observes that these cells with these very, very long telomeres grow more slowly. I am not sure what other defects they have. I think Mike has proposed a model where telomere length is limited by the interaction of telomere binding proteins.

8

CONSEQUENCES OF MUTATIONS THAT ALTER TELOMERES IN THE YEAST *K. lactis*

Michael J. McEachern and Elizabeth H. Blackburn

Department of Microbiology and Immunology
Box 0414 University of California
San Francisco, California 94143

1. INTRODUCTION

Telomeres are the structures that comprise the functional ends of eukaryotic chromosomes (For recent reviews, see Blackburn, 1994; Zakian, 1995). In the great majority of eukaryotes, including vertebrates and yeasts, telomeric DNA is composed of tandem arrays of a short repeated sequence. Because the ends of linear DNA molecules cannot be fully replicated by standard DNA polymerases, a specialized means of maintaining chromosome ends is required. In a variety of organisms, these telomeric repeat arrays are known to be maintained by telomerase, a ribonucleoprotein reverse transcriptase that synthesizes single strand copies of the unit telomeric repeat onto the 3' end of the chromosomal terminus by using part of its RNA subunit as a template (Blackburn, 1994; Zakian, 1995).

In addition to preventing gradual sequence loss from chromosome ends, telomeres are also essential because of the role they play in "capping" chromosome ends, thereby distinguishing them from DNA double strand breaks (DSBs). This can be crucial to cells because DSBs often elicit drastic responses such as stopping the cell cycle (Busse et al., 1978; Weinert and Hartwell, 1988). Presumably, this capping function involves proteins that specifically interact with telomeres. Telomeric DNA from a number of organisms appears to be packaged in non-nucleosomal form, and instead is organized into some other form of nuclease-resistant structure (Blackburn and Chiou, 1981; Edwards and Firtel, 1984; Gottschling and Cech, 1984; Lucchini et al., 1987; Price, 1990).

In this paper, a genetically tractable yeast model system for studying telomeres and telomerase is described. As described at the end of this paper, what is learned about the consequences of altered telomeres in simple organisms like the yeast *K. lactis* should prove helpful in understanding telomere function in healthy and diseased human cells.

2. VARIABILITY OF THE TELOMERIC REPEAT UNIT AMONG YEAST SPECIES

Representatives from a very wide phylogenetic range of eukaryotic species, including plants, vertebrate and invertebrate animals, fungi, and protozoans have been found to have telomeric DNA composed of 5–8 bp G-rich repeats (Blackburn, 1994; Zakian, 1995). The human repeat unit, TTAGGG, is a typical example and is quite similar to most other unit repeats that have been identified. These data indicate that telomeres composed of very short tandem repeats is the norm among eukaryotes. In recent years however, it has emerged that not all telomeres are quite so simple in their DNA sequence composition. In *Drosophila*, for example, telomeric sequences are very different, being composed of sizable retroelements that occasionally transpose to the chromosome ends, thereby maintaining their length (Biessmann et al., 1990; Levis, 1993). A more modest example of telomere sequence variability is seen among some ascomycetous yeasts and their asexual kin in which telomeric DNA is composed of tandem repeats of up to 26 bp in length which are often not G-rich (see Fig. 1) (McEachern and Hicks, 1993; McEachern and Blackburn, 1994; Cohn and Blackburn, 1995). However, as discussed below, telomeres from these yeast species appear to function and be maintained like more typical telomeres.

The first long telomeric repeat sequence unit was identified in the pathogenic yeast *Candida albicans* where it was initially found as part of a clone identified in a screen for middle-repetitive sequences (Sadhu et al., 1991; McEachern and Hicks, 1993). When used as a hybridization probe in Southern blots, the clone produced a pattern of diffuse bands, resembling the telomeric fragment patterns seen in other systems. Sequencing of the clone revealed the presence of a tandem array of 23 bp repeats directly abutting vector sequences (Sadhu et al., 1991). These repeats were confirmed to reside very near to *C. albicans* telomeres by their relative sensitivity to exonuclease digestion, a characteristic property of telomeric DNA. Strong support that these repeats were in fact the functional, terminal telomeric DNA and not merely sequences near telomeres came from the demonstration that the 23 bp repeats were: 1) present at the terminus of every cloned *C. albicans* telomere; 2) present on every *C. albicans* chromosome; 3) present on a newly formed "healed" telomeric end at a previously internal chromosomal position, and 4) increased in number when environmental conditions produced telomere lengthening.

t/aRRTGYaYRGRt	No. of repeats	Species
ttgattaggtatg T G G T G T A C G G A T	1	*K. lactis*
g T G G T G T G T G G G T	1*	*S. cerevisiae*
gtct G G G T G T C T G G G T	2*	*S. castellii*
gctg T G G G G T C T G G G T	1	*C. glabrata*
gtctaacttct T G G T G T A C G G A T	1	*C. albicans*
gtac T G G T G T A C T G G T	2	*C. guillermondii*
gtcacgatcat T G G T G T A A G G A T	1	*C. tropicalis*
gcagactcgct T G G T G T A C G G A T	1	*C. maltosa*

Figure 1. Binding site for Rap1 protein is present in yeast telomeric sequences. Each of eight related sexual and asexual yeast species contain a potential Rap1p binding site (consensus derived from *S. cerevisiae* and *K. lactis* [Larson et al., 1994]) within either one long telomeric repeat or two short telomeric repeats. Asterisks indicate species with heterogeneous telomeric sequences. The *S. cerevisiae* sequence shown is the complement to the template region of the telomerase RNA gene *TLC1* (Singer and Gottchling, 1994). The *S. castellii* sequence shown is two repeats of the most common form of telomeric repeat in that species (Cohn and Blackburn, 1995).

By using telomeric sequences from either *C. albicans* or *S. cerevisiae* (which was previously identified) as hybridization probes under reduced stringency conditions, telomeric sequences were identified from a variety of other related yeast species (McEachern and Blackburn, 1994; Cohn and Blackburn, 1995). This work demonstrated that long telomeric repeats (16–26 bp) were present in a number of species but that short telomeric repeats (8 bp) were present in others. The sequences of most of these telomeric repeat units are shown in Figure 1. Like the *C. albicans* telomeric repeats, many of the other long telomeric repeats are not very G-rich.

This point is relevant because many G-rich telomeric sequences have been shown to be capable of forming a variety of two, three, and four stranded structures involving hydrogen bonding between two to four G residues (reviewed in Rhodes and Giraldo, 1995) and it has been suggested that these unusual structures may exist *in vivo*. In *S. cerevisiae*, telomeric ends have been shown to develop 3' single stranded tails of >30 bp during S phase (Wellinger, 1993; Wellinger et al., 1996). These single stranded tails have been shown to be capable of interacting with one another *in vitro* through non-covalent bonds that are dependent upon G-rich telomeric sequences. A fraction of the molecules of a linear plasmid isolated from *S. cerevisiae* cells were shown to exist in a circular form stabilized by interactions between the single stranded tails (Wellinger et al., 1996). Short 3' overhangs are widely known to occur at the telomeres of ciliated protozoans (Klobutcher et al., 1981; Pluta, 1982; Henderson and Blackburn, 1989). It is not known whether similar single strand overhangs are present at telomeres in other yeast species. However, the lack of conservation of overall G-richness among yeast telomeric sequences suggests that unusual DNA structures dependent upon guanine-guanine interactions might not be a universal feature of telomeres.

Our telomere cloning data hints that yeast telomeres might often exist as blunt ends within cells. We have been able to readily clone telomeres from most of the yeast species shown in Fig. 1 by direct ligation of yeast DNA isolated from stationary phase yeast cells to a blunt ended site of a plasmid vector without prior enzymatic treatment of the telomeres that would ensure that they would be blunt ended (McEachern and Hicks, 1993; McEachern and Blackburn, 1994). Telomeric clones obtained always appeared to be full length and, in the case of *C. albicans* telomeres, calculations suggested that telomeric fragments were present at anticipated frequencies in the specialized genomic library used (McEachern and Hicks, 1993).

Although there is considerable sequence variability among yeast telomeric sequences, each of the ones we have studied contains (either in one long repeat or in two tandem short repeats) a close match to the *S. cerevisiae* consensus binding site for the Rap1 protein (Fig. 1). This suggests that the capacity to interact with a Rap1 homologue is crucial to telomere function in each of these yeast species.

The Rap1 protein (Rap1p) has been extensively studied in *S. cerevisiae*, in which it has been shown to be a DNA binding protein essential for cell viability (reviewed in Shore, 1994). Rap1p functions as a transcriptional activator of many genes, is involved in transcriptional silencing at mating type loci and at telomeres, and is involved in telomere length regulation. Within the arrays of irregular repeats that comprise *S. cerevisiae* telomeres, it has been estimated that Rap1p binding sites are present, on average, every 18 bp (Gilson, 1993). The *RAP1* gene has also been identified from the yeast *K. lactis* (Larson et al., 1994).

In the sequenced *C. albicans* telomeric clones a pronounced bias was found in the position of the last base pair of the telomeric repeat array within the 23 bp telomeric repeat. In 19 of 21 telomeric clones examined, the telomere end point fell between two to ten bp distal to the TGGTGTACGGAT core of the Rap1 binding site (McEachern and Hicks, 1993). This tendency for telomeric sequences to terminate at positions just past the Rap1p binding site (at least in stationary phase cells) has been observed in the other yeast species we have sequenced, though fewer telomeric clones have been examined from them (unpublished data). One possibility to explain this would be that Rap1 protein is an essential part of the "cap" that protects telomeric ends from degradation or ligation to other ends, at least during some part of the cell cycle. The properties of mutant telomeric repeats in *K. lactis* deficient in their ability to bind Rap1 protein are consistent with this possibility (see below).

3. IDENTIFICATION OF THE *K. lactis* TELOMERASE RNA GENE

An immediate question that was raised by the highly variable telomeric repeats in yeasts is whether or not they are maintained by telomerase or through some other mechanism such as recombination. Recent work has demonstrated that both the long, uniform telomeric repeats of *S. lactis* and the highly irregular telomeric repeats of *S. cerevisiae* are in fact the product of telomerase [Singer, 1994 #360; Cohn and Blackburn, 1995; McEachern and Blackburn, 1995). It is therefore highly likely that telomerase will be responsible for telomere maintenance in all the yeasts we have studied.

For both *K. lactis* and *S. cerevisiae*, identification of telomerase initially depended upon identification of their respective telomerase RNA genes and proving that part of the RNAs served as the template for synthesis of telomeric repeats (Singer and Gottschling, 1994; McEachern and Blackburn, 1995). Until recently, telomerase RNA genes had previously been known only from ciliated protozoans (Greider and Blackburn, 1989; Shippen-Lentz and Blackburn, 1990; Romero and Blackburn, 1991; Lingner et al., 1994). Though there is a well conserved secondary structure in the small (< 200 nt) RNAs encoded by related ciliates (Romero and Blackburn, 1991; Lingner et al., 1994), their primary sequences have diverged widely. Hence, it has not been possible to use the ciliate genes to identify telomerase RNA genes from other taxa. Other approaches have therefore been necessary to identify homologues from yeast.

The *K. lactis* telomerase RNA gene (named *TER1*) was identified by taking advantage of the long telomeric repeats of this species. Any telomerase RNA gene is expected to contain at least one full length permutation of the telomeric repeat unit. By using different permutations of the 25 bp *K. lactis* telomeric repeat as hybridization probes it was possible to identify all of the DNA fragments containing telomere homology within the genome, a strategy that is feasible only with organisms with long telomeric repeats. Most of the *K. lactis* fragments observed were the telomeres themselves, identifiable by their intense hybridization (because they each consist of multiple repeats) and their preferential sensitivity to Bal31 exonuclease digestion. The single fragment exhibiting hybridization to a telomeric probe at high stringency was cloned and was found to contain 30 bp of exact homology to *K. lactis* telomeres (Fig. 2). This sequence was demonstrated to function as the template for telomere synthesis by showing that two base changes made within a unique region of it (outside the 5 bp terminal redundancies) that created a *Bgl*II restriction site (*TER1-Bgl*, Fig. 2) led to telomeres whose most terminal repeats contained *Bgl*II sites (McEachern and Blackburn, 1995). This template region was found to be transcribed as part of a 1.3 kb non-polyadenylated RNA within *K. lactis* cells.

The *S. cerevisiae* telomerase RNA gene (*TLC1*) was initially found through a genetic screen for cDNA clones that interfered with the transcriptional silencing that can occur at

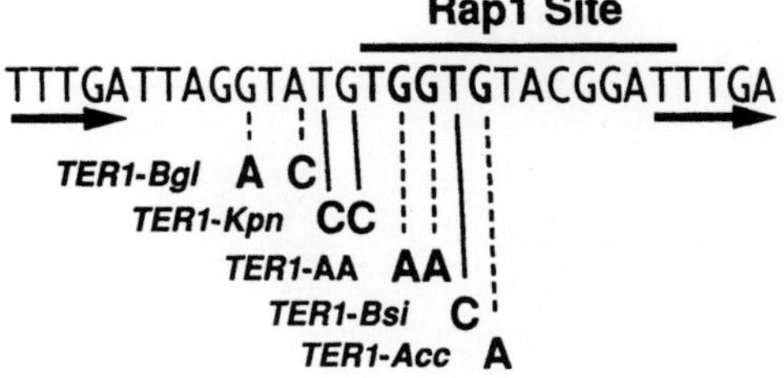

Figure 2. Template region within the *K. lactis* telomerase RNA gene *TER1*. The strand shown is the opposite to that which appears in the RNA molecule. The Rap1p binding site and the 5 bp terminal redundancies are indicated by the bar and the arrows, respectively. The sequence alterations of five *TER1* mutations are shown.

S. cerevisiae telomeres (Singer and Gottschling, 1994). The clone did not contain an extensive open reading frame but did contain a short stretch of telomere homology, which, when mutated, led to telomeres containing the expected alterations. Like the *K. lactis TER1* gene, *TLC1* encodes an RNA of ~1.3 kb.

Normal maintenance of *K. lactis* telomeres, as expected, was found to be dependent upon the presence of a functional *TER1* gene. Initially, this was determined by studying mutations that deleted ~300 bp within the *TER1* gene including the template region. Subsequently, it has been shown that essentially the same results are obtained when the entire *TER1* gene is deleted (J. Roy and E. Blackburn, unpublished results). Elimination of *TER1* function resulted in cells with telomeres that progressively shortened at a rate of ~5 bp per cell division (McEachern and Blackburn, 1995). This gradual loss of telomeric sequences was accompanied by a gradual loss of cell viability. Though *TER1* deletion mutants produced colonies indistinguishable from wild type on the transformation plates immediately after their creation, upon successive restreaks on solid medium, the deletion mutants showed colonies that progressively became smaller and rougher in appearance. Within 100 cell divisions of growth on solid medium, the great majority of cells were unable to grow. At the point of maximal senescence, telomeres were found to have shortened to below 100 bp in length. *S. cerevisiae* mutants defective in telomere maintenance (*est1* and *tlc1*) were observed to have rates of telomere shortening and viability loss very similar to those observed with *K. lactis TER1* deletion mutants (Lundblad and Szostak, 1989; Singer and Gottschling, 1994).

The abnormal colony morphologies of *TER1* deletion mutants appear to be caused by abnormal cellular phenotypes (McEachern et al., 1996). Highly senescent cultures of *TER1* deletion mutants have cells that are enlarged relative to wild type *K. lactis*. They also exhibit cell division defects as suggested by the presence of frequent groups of cells attached together in short, often branching strings (Fig. 3).

The growth problems of *K. lactis TER1* mutants appear to be solely a function of shortening telomeres, rather than the absence of telomerase activity per se. This was demonstrated by studying *TER1* deletion mutants with elongated telomeres. *K. lactis* cells that had telomeres two to three times normal length were created by using the *TER1* template mutation *TER1*-AA (McEachern et al., 1995, McEachern et al., 1996, and see below). Subsequent deletion of the *TER1*-AA allele resulted in cells lacking telomerase function but initially having elongated telomeres. These cells exhibited the same rate of loss of telomeric sequences (~5 bp per cell division) as *ter1* deletion mutants that began with normal length telomeres. However, unlike strains starting with normal length telomeres (which invariably showed some growth defects with their initial streaking on YPD medium), in most (six of eight tested) *ter1* deletion strains that started with elongated telomeres, growth defects were not observed at either the colony or the cellular level until after two to five restreakings on YPD medium (McEachern et al., 1996). Two of the

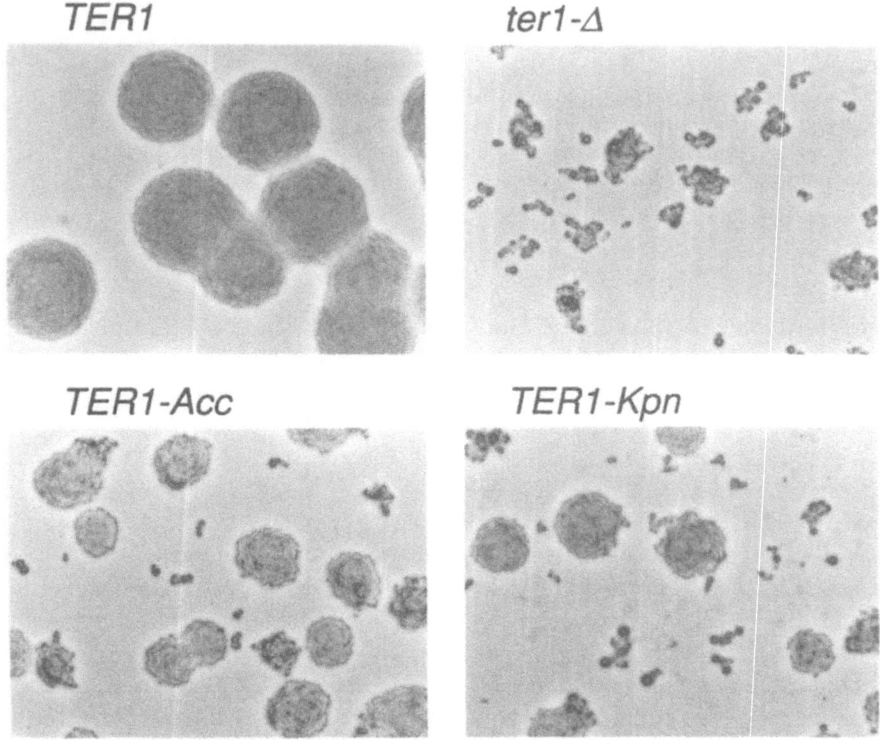

Figure 3. Abnormal cells present in *TER1* mutants. Microcolonies of wild type *K. lactis* and three *TER1* mutant strains are shown after overnight growth. The length of time cells were grown is not exactly the same for all samples.

eight strains showed growth defects earlier, and were found to have single telomeres that were shorter than normal. These results imply that while *K. lactis* telomerase activity is required to maintain telomeres at their proper length, it does not produce any specific telomeric structure, such as a 3′ single strand overhang, that is required or important for normal mitotic cell divisions. Additionally, our results suggest that a single short telomere may be enough to cause cell growth defects in *K. lactis* cells.

4. MAINTENANCE OF TELOMERES IN THE ABSENCE OF TELOMERASE

Deletion of the *TER1* gene from *K. lactis* cells did not result in 100% cell death. Some cells survived beyond the initial senescence and produced lineages that appeared capable of indefinite growth. These "post-senescence survivors" have been extensively studied (McEachern et al., 1996) and reveal a number of interesting features of cells lacking telomerase. First, the growth of post-senescence survivors is unstable. When assayed by serial restreaks on solid medium, the growth of individual survivor lineages fluctuated erratically, ranging from completely normal to highly senescent. It should be noted that the growth rate instability of post-senescence survivors is apparent in these experiments only because colonies are derived from single cells with each restreak. Cultures of post-senescence survivors passaged by dilution in liquid medium would not exhibit pronounced changes in mean population growth rates because of rapidly-generated telomere length heterogeneity in these cells (see below).

Second, post-senescence survivors still require the more or less continuous presence of at least some telomeric repeats at all chromosome ends. This is suggested by our inability to ever detect chromosome ends without some detectable telomeric repeats. In no instance did we see fragments which hybridized in Southern blotting experiments to a subtelomeric probe (which normally detects 11 of 12 *K. lactis* telomeres) but not to a telomeric probe. This is similar to observations in *S. cerevisiae*, where loss of a single telomere can arrest growth and greatly destabilize the affected linear plasmid or chromosome (Bennett et al., 1993; Sandell, 1993).

Third, the emergence of post-senescence survivors was invariably associated with not only telomere maintenance, but with elongation of telomeric arrays relative to those found in their senescent predecessors. This telomere elongation was highly irregular, producing cells that often had telomeres of many different lengths, ranging from several times larger than wild type length to very short (<100 bp) (McEachern et al., 1996). Elongated telomeres in post-senescence survivors were found to progressively shorten at the same rate as during the initial senescence, indicating that the defect in *TER1* function had not been corrected.

Fourth, suppressor mutations do not appear to be responsible for the telomere elongation, based on the following lines of evidence: a) post-senescence survivors arise at frequencies well above normal mutation rates. Additionally, when secondary instances of severe senescence (i.e., after the initial senescence and recovery) were observed, faster growing survivors emerged at frequencies comparable to those seen after the initial senescence; b) when post-senescence survivors were transformed with a wild type *TER1* gene, thereby restabilizing telomeres, then the *TER1* gene was again deleted, the ensuing senescence and occasional growth recovery was indistinguishable from the initial senescence event (J. Roy and E. Blackburn, unpublished data).

To test whether telomere maintenance in cells lacking *TER1* function was due to some form of homologous recombination, we combined a *ter1* deletion mutation with a mutation in the *K. lactis RAD52* gene (McEachern et al., 1996). *RAD52* has been extensively studied in *S. cerevisiae*, where its function is known to be required for most homologous recombination (reviewed in Petes et al., 1991). Double mutant *K. lactis* strains (*ter1 rad52*) produced post-senescence survivors at frequencies that were reduced by two or three orders of magnitude, clearly implicating homologous recombination in some form as being responsible for telomere maintenance in *ter1* deletion strains. Because this telomere elongation through recombination appears to be induced by the loss of telomere capping (see below), we have termed it telomere cap-prevented recombination (telomere CPR). The few post-senescence survivors that emerged from the double mutant strains were similar to *ter1 RAD52* post-senescence survivors in that they had undergone telomere elongation and were still subject to gradual shortening. However, the degree of telomere elongation appeared to be less, on average, than that observed in *ter1 RAD52* strains. Conceivably, the post-senescence survivors in the *rad52* background may arise through a relatively inefficient homologous recombination pathway independent of *RAD52* function. The diagram in Fig. 4 summarizes the behavior of telomeres in *ter1* mutants.

Questions remain as to how telomere CPR is able to maintain telomeric repeat arrays at the twelve chromosome ends of *K. lactis*. As normal mitotic recombination rates are too low to permit incidental recombination to maintain telomeres, even in the presence

wild type *K. lactis*

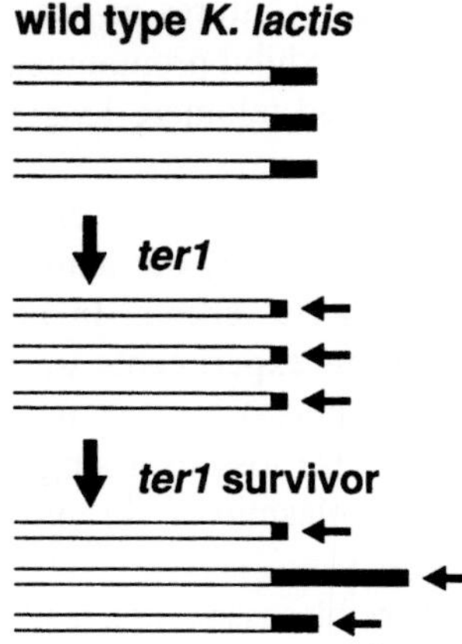

Figure 4. Schematic of telomere structure in *ter1* deletion mutants. Telomeric repeat arrays (shown as black) of wild type (*TER1*) *K. lactis* are uniform in size range, varying between ~250–500 bp. Deletion of *TER1* leads to all telomeres gradually shortening in relative unison. Loss of telomere capping function causes growth arrest and induces telomeric recombination. Telomere cap-prevented recombination (telomere CPR) leads to telomeres that are, on average, elongated relative to those of senescent cells, but highly variable in length. These elongated telomeres in post-senescence survivors are still subject to gradual telomere shortening.

of very strong selective pressure, it is necessary to propose that telomeric recombination is somehow induced as telomeres become critically shortened. An important, and perhaps the only, means by which telomeric recombination is induced appears to involve a partial or complete loss of the capping function of telomeres, thereby leading shortened telomeres to be treated as DNA double strand breaks (DSBs). In *Saccharomyces cerevisiae*, it has been shown that DSBs, are normally repaired through gene conversion events (Petes et al., 1991; Haber, 1995). Recision of the 5' end occurs by an exonuclease, leaving a long single stranded 3' tail that then invades a homologous DNA duplex and initiates a gene conversion event (Haber, 1995). A similar process occurring at shortened telomeres, and involving the staggered alignment of the two telomeric arrays could create a telomere longer than either original as shown in Fig. 5.

This gene conversion model to explain telomere CPR is supported by a number of different observations. First, senescent *TER1 K. lactis* cells arrest with highly enlarged cells that are often attached together in small groups suggestive of cell division problems. Similar phenotypes have been seen with *S. cerevisiae* cells arrested as a result of the presence of DSBs (Weinert and Hartwell, 1988). Second, both cells that completely lack telomerase function (*ter1* deletions) and cells that have telomeres that are stably maintained at shorter than normal lengths (as a result of other *TER1* mutations) display high

Telomere Lengthening through Gene Conversion

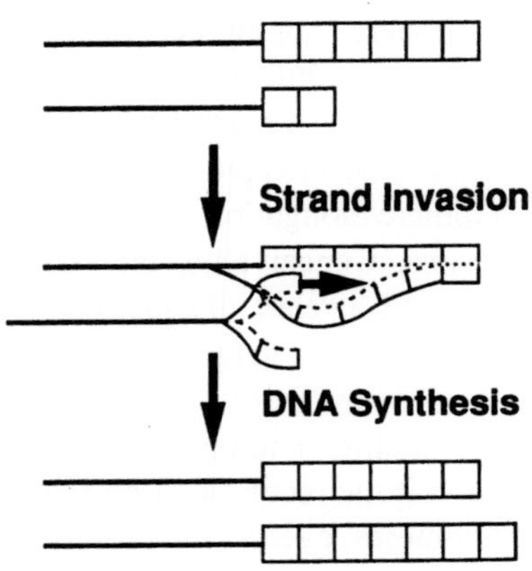

Figure 5. A possible mechanism for telomere CPR. Loss of telomere capping from telomere shortening, followed by strand invasion of a 3' end of one telomere into another telomere would initiate gene conversion. Staggered alignment of residual repeat arrays could maximize extent of the elongation brought about by the gene conversion event.

rates of gene conversion near telomeres (unpublished data). This is manifested as changes in subtelomeric sequences (which share homology in 11 of 12 telomeres in the *K. lactis* strain with which we work) or in increased loss rates of a marker gene placed near a telomere (unpublished data). Third, even the telomere that lacks subtelomeric homology with the other *K. lactis* telomeres can be maintained in the absence of telomerase activity (indicating that telomeric homology alone is sufficient to permit elongation).

5. TELOMERIC LENGTH REGULATION AND REPEAT TURNOVER

K. lactis telomeres are normally maintained within a defined length range of approximately 250–500 bp (10–20 repeats). These lengths are thought to be maintained by a balance of the action of two opposing processes: sequence loss through the inability of DNA polymerase to replicate DNA ends completely and sequence addition by telomerase. When the latter process is eliminated (in *ter1* deletion mutants), only the former process occurs and a constant rate of gradual sequence loss is seen. It is presumed that the extent of sequence loss at a DNA end from incomplete replication is simply a consequence of the length of RNA primers made by DNA primase. If so, the rate of sequence loss will be probably be constant under many different conditions for a given organism. Regulation of telomere length therefore is likely to center upon regulation of the ability of telomerase to add telomeric repeats onto chromosome ends.

The heterogeneity of telomere length observed within a clonal population of cells suggests that telomerase does not add a fixed number of telomeric repeats to each chromosome end at each cell division. In fact, for the majority of telomeres within a *K. lactis* cell, telomerase may not add any DNA in a given cell cycle. With a 25 bp telomeric repeat unit and a loss rate of ~5 bp per cell division in the absence of telomerase, *K. lactis* can maintain its telomeres with only a single telomeric repeat added onto a given telomere, on average, once every five cell divisions.

The dynamics of telomeric repeat turnover can be studied by replacing the wild type *TER1* gene with a *TER1* gene containing a mutation within the template region. By utilizing mutations that both create restriction sites and produce no detectable changes in telomere phenotype it is readily possible to follow the progress of replacement of wild type repeats by mutant repeats. Using this approach we have demonstrated that cells carrying a *TER1* template mutation contain 1–5 mutant repeats per telomere after approximately 30–50 cell divisions from the time of the introduction of the mutation (McEachern and Blackburn, 1995 and unpublished data). This result illustrates both the relatively slow rate of telomeric repeat turnover and the initial variability of turnover between different ends within the same cell. Following such clonal lines over much longer periods of continued restreakings shows that mutant telomeric repeats continue to encroach further inward at progressively slower rates, so that even after thousands of cell divisions the most internal repeats remain wild type (unpublished data). These results clearly demonstrate that the more internal telomeric repeats within *K. lactis* telomeric arrays are faithfully replicated by standard DNA polymerases and rarely, if ever, turned over by the action of telomerase. Other data supporting the relative stability of the more internal regions of telomeric repeat arrays has been reported in other systems (Wang and Zakian, 1990; Ponzi et al., 1992).

The initial evidence that Rap1p plays some role in telomere length regulation in yeast came from the identification of *RAP1* mutations in *S. cerevisiae* that result in telomeres ranging from about twenty percent longer than normal to up to ten times normal

length, depending upon the allele (Kyrion et al., 1992; Liu, 1994). Some of these mutations affect a small C-terminal region of the protein. A similar small C-terminal truncation
of the *K. lactis* Rap1 protein also produces mild telomere elongation in (~100 bp increase)
in *K. lactis* (Krauskopf and Blackburn, 1996).

6. DISRUPTION OF TELOMERE LENGTH REGULATION

Our knowledge about telomere length regulation in *K. lactis* has mostly come from
analyzing *TER1* template mutations that disrupt that regulation, resulting in telomeres of
abnormal length. Most informative so far are mutations that lead to telomere lengthening.
Several such mutations were made, and these can be loosely divided into two general categories: those causing telomeres to elongate immediately after replacement of the wild type
TER1 gene, and those causing lengthening only after a delay of hundreds or thousands of
cell divisions. Two single base substitutions at adjacent positions (*TER1-Acc* and *TER1-
Bsi*) within the conserved Rap1 protein binding site (Fig. 2) resulted in immediate
telomere lengthening (McEachern and Blackburn, 1995). The telomeric fragments of
TER1-Acc mutants largely migrate near limit mobility in Southern blotting experiments
and hybridize much more intensely to telomeric probes, indicative of extremely elongated
telomeric repeat arrays (Fig. 5). This massive elongation has occurred as soon as *TER1-
Acc* mutant cells can be analyzed (~30 cell divisions after their creation). The *TER1-Bsi*
allele produced less extreme telomere elongation which also occurs more gradually,
requiring at least several hundred cell divisions to be complete. Nonetheless, this elongation is also readily apparent as soon as cells can be analyzed.

The telomere elongation caused by the *TER1-Acc* and *TER1-Bsi* mutations is largely
prevented if a wild type *TER1* gene is also present (Fig. 6 and McEachern and Blackburn,
1995). Hence, the *TER1-Acc* and *TER1-Bsi* mutations are recessive, showing that they do
not act by producing hyper-active telomerases, which would be expected to be genetically
dominant. Instead, the data suggest that these mutations act through the sequence alterations in the telomeric repeats that are synthesized by the mutant telomerases. Furthermore,
because the elongation in these mutants occurs on telomeric arrays that still contain nearly
a full complement of internal wild type telomeric repeats (McEachern and Blackburn,
1995), the telomeric target disrupted by the mutations must only be the very terminus of
the telomere and not the entire telomeric array. This contrasts with the delayed lengthening mutants discussed below.

The *TER1-Acc* and *TER1-Bsi* mutations produce sequence alterations at particularly
well conserved positions (as judged by known binding sites in *S. cerevisiae*) within the
Rap1p binding site that is present in *K. lactis* telomeric repeats (Fig. 1 and Fig. 2). It was
therefore of great interest to determine if binding of Rap1p was affected by these changes.
By using gel shift experiments it was determined that both the *Acc* and *Bsi* mutations
produced severe reductions in the ability of *K. lactis* Rap1p to bind to a double-stranded
telomeric repeat. Binding of Rap1 was estimated to be reduced ~30 fold by the *Bsi* mutation and ~100 fold by the *Acc* mutation (Krauskopf and Blackburn, 1996). Thus, the severity of the telomere lengthening phenotype correlated with severity of disruption of Rap1p
binding. In contrast, binding of *K. lactis* Rap1p to the telomeric repeat sequences specified
by the delayed lengthening mutants *TER1-Bgl* and *TER1-Kpn* (see below), was unaffected
or showed more modest changes (Krauskopf and Blackburn, 1996). The *Bgl* mutant
telomeric repeat bound normally, the *Kpn* mutant repeat bound twice as well as normal,
and the AA mutant bound ten times less well than normal.

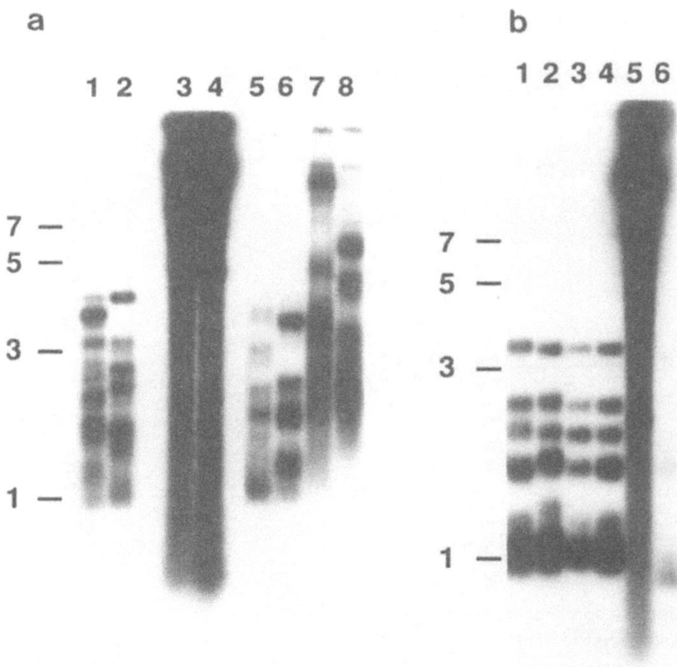

Figure 6. Elongated telomeres as a result of *TER1* template mutations. Shown are Southern blots of *Eco*RI digested *K. lactis* genomic DNA from each of several mutants run on a 0.8% agarose gel and probed with a telomeric oligonucleotide probe. Panel A. Lanes 1 and 2 show strains carrying one copy of wild type *TER1* and one copy of *TER1-Acc*. Lanes 3 and 4 show two clones carrying only *TER1-Acc*; lanes 5 and 6 show two clones carrying one copy of wild type *TER1* and one copy of *TER1-Bsi*. Lanes 7 and 8 show clones carrying only *TER1-Bsi*. Samples from Panel A had been grown for ten passages (200–250 cell divisions) prior to isolation of DNA. Panel B. Lanes 1 and 2 show clones of *TER1*-AA and lanes 3 and 4 show two clones of a *TER1-Bcl*, two mutants that are exhibiting wild type telomere length. Lanes 5 and 6 show two clones of *TER1-Kpn*, one after runaway telomere elongation, and the other prior to runaway telomere elongation. All samples from Panel B had been grown for fifteen passages prior to isolation of DNA. The *TER1-Kpn* clone in lane six was found to have undergone runaway telomere elongation by passage 20. Duplicate clones, in both panels, were followed independently for the duration of growth.

Our results therefore suggest a role for Rap1p in regulating telomere length by acting at or very near the telomeric terminus, at the point where telomerase adds new telomeric repeats. Though the *S. cerevisiae* Rap1p has been shown to have a very weak ability to bind to single strand telomeric sequences (Giraldo et al., 1994), it would seem more likely that its double strand DNA binding activity would be involved in regulating telomere function. Conceivably, Rap1p binding to the most terminal double-stranded telomeric repeat(s) in a telomeric array would interfere with the ability of telomerase to bind to the terminus and begin adding new telomeric repeats as illustrated in the model shown in Fig. 7.

TER1-Acc cells immediately display slowed growth and abnormal colony and cell morphologies (Fig. 3) while *TER1-Bsi* cells initially grow normally, eventually developing mild growth problems when their telomeres approach their maximal lengths. The delayed runaway mutants *TER1-Bgl* and *TER1-Kpn*, discussed below, also cause similar abnormal

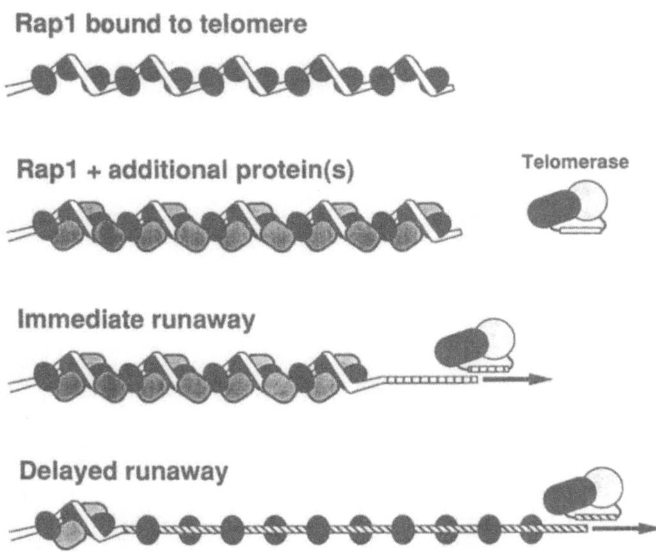

Figure 7. Model for telomere length regulation in *K. lactis*. Telomeric DNA is depicted as bound by molecules of Rap1p (black ovals) and producing DNA bends upon binding wild type telomeric repeats. Rap1 protein recruits one or more additional proteins (gray ovals) to the telomere to form a complex that regulates access of telomerase to the end of the telomere and perhaps that also serves as the telomere "cap." The ability to block the access of telomerase to the terminus is postulated to increase as the length of the telomere is increased, possibly because of cooperativity during assembly of the complex, which would create the negative feedback loop needed to maintain constant telomere lengths. The runaway elongation mutants that act immediately disrupt Rap1p binding and might therefore allow unfettered access of telomerase to the terminus. The runaway elongation mutants that act in a delayed fashion do not disrupt Rap1p binding but presumably act by eventually destabilizing the telomeric complex, once enough of the initially wild type internal repeats have been replaced by mutant repeats. Conceivably, these mutations might act by affecting the manner in which Rap1p bends DNA.

cell growth phenotype (Fig 3). Interestingly, the colony and cellular abnormalities produced by elongated telomeres are reminiscent of those of strains deleted for *TER1* function (Fig 3). This suggests that cells may respond similarly to shortening telomeres of wild type sequence and lengthened telomeres of abnormal sequence. Both situations could conceivably be due to defects in telomere capping, thereby eliciting recombinational repair and other responses to what the cell may perceive as unprotected DNA ends.

Other mutations within the *TER1* template region, *TER1-Kpn*, *TER1-Bgl* and *TER1-AA* (Fig. 2) also produce telomere elongation, but only after an extended latent period lasting hundreds or even thousands of cell divisions (McEachern and Blackburn, 1995). Telomeres in these strains are initially stable at either wild type length (*TER1-AA*) or at shorter than normal lengths (*TER1-Kpn* and *TER1-Bgl*). The elongation that eventually occurs in *TER1-Kpn* and *TER1-Bgl* mutant strains is substantial, with most telomeric signal observed on Southern blots being in fragments running at limit mobility (Fig. 6). Quantitation of the radioactive signal suggests that in some *TER1-Bgl* strains the telomeres average 100 times longer than normal (i.e., several tens of kilobases).

As described above, replacement of the wild type telomerase RNA gene with a *TER1* gene carrying a template mutation initially leads to telomeres containing the altered repeats only at the most terminal few repeats, with substantial replacement of the more internal wild type repeats in the telomeric arrays taking much longer. The latent period prior to telomere elongation in the *TER1-Kpn* and *TER1-Bgl* mutants appears to be due to a requirement for replacing a sufficient number of wild type telomeric repeats with mutant telomeric repeats. We tested whether this normally slow replacement of the pre-existing wild type repeats is responsible for the delayed runaway elongation phenotypes by showing that telomere elongation in *TER1-Kpn* strains can be markedly accelerated by transforming the *TER1-Kpn* gene directly into *K. lactis* cells with wild type repeat arrays that were already short (McEachern and Blackburn, 1995). These were produced by using cells deleted for *TER1* function as the recipients of the *TER1-Kpn* allele. The resulting *TER1-Kpn* cells lost most wild type telomeric repeats essentially immediately. Similar experiments with *TER1-Bgl* did not produce a marked acceleration of telomere elongation. However, *TER1-Bgl* strains normally take much longer to begin runaway telomere elongation than *TER1-Kpn* strains (and not all *TER1-Bgl* strains showed elongation even after > 2000 cell divisions), consistent with requiring a greater extent of repeat turnover. Also, unlike *TER1-Kpn* strains, the onset of runaway elongation in *TER1-Bgl* strains was correlated with the appearance of telomeric fragments lacking detectable wild type repeats (McEachern and Blackburn, 1995), suggesting that complete replacement with mutant repeats on at least one telomere was needed to trigger elongation in *TER1-Bgl* strains. Therefore, the extent of telomeric repeat turnover required to produce runaway elongation in *TER1-Bgl* strains is likely to be greater than that required for *TER1-Kpn* strains. Such a high degree of replacement of wild type repeats might not be easily attained by transformation into *TER1* deletion strains. The considerable extent of wild type repeat turnover required for runaway elongation in *TER1-Bgl* strains may also require telomeric recombination events. As described below, shorter-than-normal telomeres can greatly increase recombination near telomeres. Thus, ironically, the delayed runaway elongation phenomenon of *TER1-Bgl* cells may be dependent upon its initial short telomere phenotype.

Another mutation that can produce telomere elongation is the *TER1-AA* allele, which produces G to A transitions at a pair of neighboring positions in the Rap1p binding site within the *K. lactis* telomeric repeat, close to the positions altered by the *TER1-Bsi* and *TER1-Acc* mutations (Fig. 2). If the *TER1-AA* allele is introduced into cells containing a wild type *TER1* gene, and then the wild type *TER1* gene is deleted by plasmid "loop out", telomere lengths in the resulting *TER1-AA* cells remain essentially normal for at least 2000 cell divisions (McEachern and Blackburn, 1995, and unpublished data). Occasionally, one or more telomere was observed to lengthen up to ~200 bp longer than normal length; however, these modestly elongated telomeres always shortened back to near normal sizes quite quickly. Even after 2000 cell divisions, the more internal parts of the telomeric repeat arrays were still composed of wild type telomeric repeats (unpublished data). That disruption of telomere length regulation by the *TER1-AA* allele can occur is more apparent when the *TER1-AA* gene is transformed into senescent cells carrying a deletion of *TER1* (as described above for *TER1-Kpn*), thereby leading to an essentially instantaneous replacement of most wild type telomeric repeats. The *TER1-AA* clones derived from the same senescent *TER1* deletion parent displayed variable telomere lengthening rates. For example, two *TER1-AA* clones recovered from such an experiment, at the time of their isolation, showed telomeres that were heterogeneous in length with a mean size two to three times normal length. However, the telomeres in these strains had not fully stabilized in length, as they continued to drift rather erratically upward in length, at an average rate of ~1 bp per cell division, for hundreds of cell divisions. After this period, telomeric fragment patterns ceased to exhibit net elongation but remained heterogeneous, with telomere lengths in the range of 3–10 kb in size.

The requirement that some fraction of the wild type telomeric repeats located internally from the very telomeric terminus be replaced to produce delayed elongation indicates that the disrupted telomere length regulation process normally acts on the internal repeats of a telomeric array, and not simply the most terminal repeat or two. These results implicate the internal repeats within a telomeric array as being crucial to the process of telomere length regulation. This appears logical if length regulation of telomeres requires sensing the total number of repeats present in telomeric arrays. The properties of the two classes of elongation mutations, immediate and delayed, suggest that telomere length regulation can be disrupted by either disrupting the very terminus of the telomere or by disrupting the more internal repeats of the telomere.

Because the internal repeats within the telomeric arrays are a significant component of the target affected in the delayed lengthening mutants, it seems likely that these mutations acts by affecting a protein complex bound to this double-strand telomeric DNA. If the *TER1*-AA mutation acts through its moderate disruption in Rap1p binding to telomeric sequences, such moderate binding problems might then only be manifested when much of the telomeric array is replaced by mutant AA repeats, especially if assembly of the telomeric protein complex is cooperative. Though the *TER1-Bgl* and *TER1-Kpn* mutations produce telomeric repeats that bind Rap1p as well or better than wild type repeats *in vitro*, they still could produce their phenotypic consequences by affecting the Rap1p-containing telomeric complex. The *S. cerevisiae* Rap1p is known to produce a large bend upon binding to DNA and this bend that has been shown to require more than just the minimal DNA-binding domain of the protein (Müller et al., 1994; König et al., 1996). As suggested in the model in Fig. 7, the *Bgl* and *Kpn* mutations, which lie adjacent to the Rap1p binding site, may act by influencing the ability of *K. lactis* Rap1p to bend telomeric DNA. Altered bending of individual telomeric repeats might then disrupt the geometry of the protein complex assembled at telomeres, which in turn could prevent proper regulation of telomere length.

Consistent with this model, in *K. lactis*, combining the C-terminal Rap1p truncation, which alone causes slight telomere lengthening, with the *TER1*-AA or *TER1-Bsi* mutations, markedly exacerbated the telomere lengthening effects of these mild mutations (Krauskopf and Blackburn, 1996).

7. GENOMIC INSTABILITY FROM ALTERED TELOMERES IN YEAST

Altered telomeres in *K. lactis* and *S. cerevisiae* can lead to a variety of forms of genomic changes and instability. The most obvious, of course, are changes in the sequence and size of the telomeres themselves. In *K. lactis*, these size changes can be large enough to cause telomeres to be an appreciable part of total genomic DNA. For example, the telomere length increases observed with some of the mutants described here (e.g., *TER-Bgl*) lead to telomeric sequences increasing from ~0.03% of total genomic DNA in wild type *K. lactis* to up to ~5% of total DNA in some clonal cell populations. Also, telomere lengths in some *TER1* mutants are often unstable, as for example in strains deleted of *TER1* function. Another prominent example of this is the telomere lengthening that starts only after turnover of a sufficient number of the more internal wild type repeats in the delayed runaway mutants. Even once elongated, the telomeres of *TER1-Bgl* strains are still unstable, and occasionally are seen to abruptly shorten appreciably (McEachern and Blackburn, 1995).

TER1 mutations also can cause extreme instability in subtelomeric sequences. This appears to stem from defects in telomere capping and the ensuing attempts at recombinational repair by the cell. By measuring loss rates of a marker gene placed near a telomere, we estimate gene conversion rates near telomeres of *TER1* mutants with stably shortened telomeres to be increased by up to 1000-fold over the wild type rate (unpublished data). Similar experiments done with senescing *TER1* deletion strains show an even higher gene conversion rate. This greatly enhanced recombination near telomeres helps rationalize the ability of *ter1* deletion strains to rebuild telomeres to many times the lengths seen at the peak of senescence. It is also consistent with models such as the one described in Fig. 5, in suggesting that increased recombination as a result of telomere capping deficiencies is largely or entirely localized to near telomeres. High rates of recombination within several tens of kilobases of telomeres, but not elsewhere in the genome, has been described to occur in *S. cerevisiae cdc13* mutants (Garvik et al., 1995). Which have been suggested to be defective in a component of the telomere cap.

In addition, shortened telomeres in senescing *S. cerevisiae est1* strains are known to increase the rates of chromosome loss (Lundblad and Szostak, 1989). Very high rates of chromosome loss are observed if one of its telomeres is completely removed by cleavage with HO endonuclease *in vivo* (Sandell, 1993). The mechanism behind such chromosome loss is unclear. Perhaps the recombinational repair machinery of cells is not freely diffusable in the nucleus, thereby preventing normal segregation of chromosomes stuck in unproductive repair intermediates.

Another consequence of altered telomeric sequences in *K. lactis* appears to be telomere-telomere fusions. Certain mutations that disrupt the Rap1p binding site have been shown to result in sharply defined bands in agarose gels that hybridize to telomeric probes, but unlike telomeres, are not preferentially sensitive to exonuclease digestion (unpublished data). This and other data suggest that these bands are the result of fusions between two mutant telomeres. Such strains, which may have all their telomeres fused much of the time, are capable of growing surprisingly well. This may suggest that they have circularized their chromosomes rather than having dicentric chromosomes. Investigations into the structure of the chromosomes in these strains is currently under way.

8. TELOMERE FUNCTION AND HUMAN CANCER

The importance of telomeres and telomerase for the biology of human cells, while yet to be fully defined, has attracted considerable interest in recent years. Telomerase activity appears to be absent in many human somatic cell types and some evidence suggests that telomeres in these cells are gradually shortening over time (reviewed in de Lange, 1995; Harley, 1995). In contrast, in most human cancers that have been examined, telomerase activity is present (Kim et al., 1994).

It may be that telomeres and telomerase are connected to human cancer progression in a number of ways (de Lange, 1995; Wright and Shay, 1995). Somatic cells may generally be indifferent to gradual telomere loss occurring during the limited number of cell divisions that occur in an organism's lifetime. However, the eventual erosion of telomeres might serve as a barrier to the many additional cell divisions expected to occur in the clonal development of a cancer. Loss of telomere function might in turn be the source of many forms of genetic changes that occur in cancer cells. Obvious mutations that could be selected for during cancer progression include those that restore a means to maintain telomeres, such as restoring telomerase activity, or conceivably, activating an alternative telomere maintenance pathway such as recombination. Other mutations that could conceivably be selected for by loss of telomere function include checkpoint genes involved in blocking cell division when DNA double strand breaks are present (uncapped telomeres could mimic DSBs). The p53 gene, which is known to be mutated in a large fraction of human tumors (Hollstein et al., 1991; Livingstone et al., 1992; Yin et al., 1992), would be a candidate for such a gene. Finally, loss of telomeres can lead to end-to-end fusions (McClintock, 1939; McClintock, 1941). The dicentric chromosomes that would often be formed from such fusions could in turn produce a plethora of chromosomal aberrations including aneuploidies, deletions, amplifications, and translocations (for review, see de Lange, 1995). While the potential for telomere function in human cells to be connected to the development of cancer is clearly large, the exact extent of those connections awaits further research.

REFERENCES

Bennett, C. B., Lewis, A. L., Baldwin, K. K. and Resnick, M. A. (1993). Lethality induced by a single site-specific double-strand break in a dispensable yeast plasmid. Proc. Natl. Acad. Sci. USA *90*, 5613–5617.

Biessmann, H., Carter, S. B. and Mason, J. M. (1990). Chromosome ends in Drosophila without telomeric DNA sequences. Proc. Natl. Acad. Sci. USA *87*, 1758–1761.

Blackburn, E. H. (1994). Telomeres- No end in in sight. Cell *77*, 621–623.

Blackburn, E. H. and Chiou, S.-S. (1981). Non-nucleosomal packaging of a tandemly repeated DNA sequence at termini of extrachromnosomal DNA coding for rRNA in *Tetrahymena*. Proc. Natl. Acad. Sci USA *78*, 2263–2267.

Busse, P. M., Bose, S. K., Jones, R. W. and Tolmach, L. J. (1978). The action of caffeine on X-irradiated HeLa cells. III. Enhancement of X-ray-induced killing during G2 arrest. Radiat. Res. *76*, 292–307.

Cohn, M. and Blackburn, E. H. (1995). Telomerase in yeast. Science *269*, 396–400.

Edwards, C. A. and Firtel, R. A. (1984). Site-specific phasing in the chromatin of the rDNA in *Dictyostelium discoideum*. J. Mol. Biol. *180*, 73–90.

Garvik, B., Carson, M. and Hartwell, L. (1995). Single-stranded DNA arising at telomeres in *cdc13* mutants may constitute a specific signal for the *RAD9* checkpoint. Mol. and Cell. Biol. *15*, 6128–6138.

Gilson, E., Roberge,M. , Giraldo, R. , Rhodes, D. , and S. Gasser. (1993). Distortion of the DNA double helix by RAP1 at silencers and multiple telomeric binding sites. J. Mol. Biol. *231*, 293–310.

Giraldo, R., Suzuki, M., Chapman, L. and Rhodes, D. (1994). Promotion of parallel DNA quadruplexes by a yeast telomere binding protein- a circular dichroism study. Proc. Natl. Acad. Sci. USA *91*, 7658–7662.

Gottschling, D. E. and Cech, T. R. (1984). Chromatin structure of the molecular ends of *Oxytricha* macronuclear DNA: phased nucleosomes and a telomeric complex. Cell *38*, 501–510.

Greider, C. W. and Blackburn, E. H. (1989). A telomeric sequence in the RNA of *Tetrahymena* telomerase required for telomere repeat synthesis. Nature *337*, 331–337.

Haber, J. E. (1995). *In vivo* biochemistry: physical monitoring of recombination induced by site-specific endonucleases. BioEssays *17*, 609–620.

Henderson, E. R. and Blackburn, E. H. (1989). An overhanging 3' terminus is a conserved feature of telomeres. Mol. Cell. Biol. *9*, 345–348.

Hollstein, M., Sidransky, D., Vogelstein, B. and Harris, C. C. (1991). p53 mutations in human cancers. Science *253*, 49–53.

Kim, N. W., Piatyszek, M. A., Prowse, K. R., Harley, C. B., West, M. D., Ho, P. L. C., Coviello, G. M., Wright, W. E., Weinrich, S. L. and Shay, J. W. (1994). Specific association of human telomerase activity with immortal cells and cancer. Science *266*, 2011–2015.

Klobutcher, L. A., Swanton, M. T., Donini, P. and Prescott, D. M. (1981). All gene-sized DNA molecules in four species of hypotrichs have the same terminal sequence and an unusual 3' terminus. Proc. Natl. Acad. Sci. USA *78*, 3015–3019.

König, P., Giraldo, R., Chapman, L. and Rhodes, D. (1996). The crystal structure of the DNA-binding domain of yeast RAP1 in complex with telomeric DNA. Cell *85*, 125–136.

Krauskopf, A. and Blackburn, E. H. (submitted).

Kyrion, G., Boakye, K. A. and Lustig, A. J. (1992). C-terminal truncation of RAP1 results in the degregulation of telomere size, stability, and function in *Saccharomyces cerevisiae*. Mol. Cell Biol. *12*, 5159–5173.

Larson, G. P., Castanotto, D., Rossi, J. J. and Malafa, M. P. (1994). Isolation and functional analysis of a *Kluyveromyces lactis* RAP1 homologue. Gene *150*, 35–41.

Levis, R. W., Ganesan, R., Houtchens, K., Tolar, L. A., and Sheen, F. (1993). Transposons in place of telomeric repeats at a *Drosophila* telomere. Cell *75*, 1083–1093.

Lingner, J., Hendrick, L. L. and Cech, T. R. (1994). Telomerase RNAs of different ciliates have a common secondary structure and a permuted template. Genes and Dev. *8*, 1984–1998.

Liu, C., Mao, X. , and A. Lustig (1994). Mutational analysis defines a C-terminal tail domain of RAP1 essential for telomeric silencing in *Saccharomyces cerevisiae*. Genetics *138*, 1025–1040.

Livingstone, L. R., White, A., Sprouse, J., Livanos, E., Jacks, T. and Tlsty, T. D. (1992). Altered cell cycle arrest and gene amplification potential accompany loss of wild-type p53. Cell *70*, 923–935.

Lucchini, R., Pauli, U., Braun, R., Koller, T. and Sogo, J. M. (1987). Structure of the extrachromosomal ribosomal RNA chromatin of *Physarum polycephalum*. J. Mol. Biol. *196*, 829–843.

Lundblad, V. and Szostak, J. W. (1989). A mutant with a defect in telomere elongation leads to senescence in yeast. Cell *57*, 633–643.

McClintock, B. (1939). The behavior of successive nuclear divisions of a chromosome broken at meiosis. Proc. Natl. Acad. Sci. *25*, 405–416.

McClintock, B. (1941). The stability of broken ends of chromosomes in *Zea mays*. Genetics *26*, 234–282.

McEachern, M. J. and Blackburn, E. H. (1994). A conserved sequence motif within the exceptionally diverse telomeric sequenced of budding yeasts. Proc. Natl. Acad. Sci., USA *91*, 3453–3457.

McEachern, M. J. and Blackburn, E. H. (1995). Runaway telomere elongation in telomerase RNA gene mutants of *Kluyveromyces lactis*. Nature *376*, 403–409.

McEachern, M. J., and Blackburn, E. H. (1996). Recombination between terminal telomeric repeat arrays maintains telomeres in *K. lactis* lacking telomerase. Genes & Dev. (in press).

McEachern, M. J. and Hicks, J. B. (1993). Unusually large telomeric repeats in the yeast *Candida albicans*. Mol. Cell Biol. *13*, 551–560.

Müller, T., Gilson, E., Schmidt, R., Giraldo, R., Sogo, J. M., Gross, H. and Gasser, S. M. (1994). Imaging the asymmetric DNA bend induced by repressor activator protein 1 with scanning tunneling microscopy. J. Struct. Biol. *133*, 1–12.

Pluta, A. F., Kaine, B. P., and Spear, B. B. (1982). The terminal organization of macronuclear DNA in *Oxytricha fallax*. Nuc. Acids Res. *10*, 429–433.

Ponzi, M., Pace, T., Dore, E., Picci, L., Pizzi, E. and Frontali, C. (1992). Extensive turnover of telomeric DNA at a *Plasmodium berghei* chromosomal extremity marked by a rare recombinational event. Nucleic Acids Res. *20*, 4491–4497.

Price, C. M. (1990). Telomere structure in *Euplotes crssus*: Characterization of DNA-protein interactions and isolation of a telomere-binding protein. Mol. Cell. Biol. *10*, 3421–3431.

Rhodes, D. and Giraldo, R. (1995). Telomere structure and function. Curr. Op. Struc. Biol. *5*, 311–322.

Romero, D. and Blackburn, E. H. (1991). A conserved secondary structure for telomerase RNA. Cell *67*, 343–53.

Sadhu, C., McEachern, M. J., Rustchenko-Bulgac, E. P., Schmid, J., Soll, D. R. and Hicks, J. B. (1991). Telomeric and dispersed repeat sequences in *Candida* yeasts and their use in strain identification. J. Bacteriol. *173*, 842–850.

Sandell, L. L., and Zakian, V. A. (1993). Loss of a yeast telomere- arrest, recovery, and chromosome loss. Cell *75*, 729–739.

Shippen-Lentz, D. and Blackburn, E. H. (1990). Functional evidence for an RNA template in telomerase. Science *247*, 546–552.

Shore, D. (1994). RAP1: a protein regulator in yeast. Trends Genet. *10*, 408–412.

Singer, M. S. and Gottschling, D. E. (1994). *TLC1*- template RNA component of Saccharomyces cerevisiae telomerase. Science *266*, 404–409.

Wang, S. S. and Zakian, V. A. (1990). Sequencing of *Saccharomyces* telomeres cloned using T4 DNA polymerase reveals two domains. Mol. Cell. Biol. *10*, 4415–4419.

Weinert, T. A. and Hartwell, L. H. (1988). The *RAD9* gene controls the cell cycle response to DNA damage in *Saccharomyces cerevisiae*. Science *241*, 317–322.

Wellinger, R. J., Wolf, A. J., and Zakian, V. A. (1993). *Saccharomyces* telomeres acquire single-strand $TG_{(1–3)}$ tails late in S-phase. Cell *72*, 51–60.

Wellinger, R. J., Ethier, K., Labrecque, P. and Zakian, V. A. (1996). Evidence for a new steop in telomere maintenance. Cell *85*, 423–433.

Wright, W. E. and Shay, J. W. (1995). Time, telomeres and tumours: is cellular senescence more than an anticancer mechanism? Trends Cell Biol. *5*, 293–297.

Yin, Y., Tainsky, M. A., Bischoff, F. Z., Strong, L. C. and Wahl, G. M. (1992). Wild-type p53 restores cell cycle control and inhibits gene amplification in cells with mutant p53 alleles. Cell *70*, 937–948.

Zakian, V. A. (1995). Telomeres: Beginning to understand the end. Science *270*, 1601–1607.

DISCUSSION

Hartwell: The telomere-telomere fusion is a very interesting phenomenon. I do not know if you can tell us anymore about that. In the experiment you showed us, on Bal31 there was a little bit of shortening then, all of a sudden, just dropped off, everything went away.

McEachern: You mean in terms of the wild type? That is very characteristic because the telomeres were all the same length. So what happens is, once you chew the telomeres off, the fragment is still there, getting shorter, but it does not have any telomeric sequences on it. If I probee with the subtelomeric sequences then you would see the shortening continue to go down. The fragments have not disappeared, they have just lost the telomeres.

Hartwell: But you would still expect the shortening to be much greater as they disappear. It seemed more like there was some sort of stoichiometric effect where there was an inhibitor of the Bal31.

McEachern: No, if you look with a subtelomeric probe, or if you look at just the ethidium pattern, you see a gradual shortening that is occurring in these things. The abruptness is just because the telomeres are all the same length and once you lose them, all that signal simply goes away.

Hartwell: Is there any other evidence for this telomere-telomere fusion besides the restriction pattern?

McEachern: One thing that I want to do is pulsed-field gels. I have not done that yet.

Hartwell: It should also have very strange meiotic behavior.

McEachern: Right. I have not looked at any of the mutations that I have made in meiosis, but that is obviously another very interesting area.

Wahl: Do you think that the fused telomeres are covalently joined?

McEachern: That is right.

Wahl: So, do you heat the DNA beforehand in order to ask whether the structures can become sensitive to Bal31 upon denaturation?

McEachern: No.

Wahl: So it could be a non-covalent association. Is that possible?

McEachern: Well, it would have to be an awfully stable one because these things are going through DNA extraction procedures and so on.

Wahl: I agree.

McEachern: I have not cloned them so there is still work to be done on really nailing the nature of the fusions down, that is true. One thing that I forgot to mention that I think was pretty obvious from the slide in the initial fusion is that they are quite unstable over time. They are constantly changing over the fairly long period of time that I was looking at them. In Saccharomyces when Andrew Murray and Jack Szostak, years ago, had looked at an artificial telomere-telomere fusion and transforming that into Saccharomyces cells on a plasmid, that turned out to be very unstable being converted back to a linear molecule at a rate of about one to a hundred. So that would be sort of consistent with what I see here, an appreciable rate of conversion back to a linear molecule. Probably, in my case, because the only telomerase that is present is the mutant one that is still making bad repeats, the telomeres are still going to be prone to fusion again which would not have occurred in the initial Saccharomyces experiment.

Wahl: How big are the chromosomes in this?

McEachern: They should be probably, on average, two to three times the size of the Saccharomyces chromosomes. K.lactis has fewer chromosomes than Saccharomyces with about the same size genomes.

Wahl: Can you see them under the appropriate microscope? Because you might look for rings directly.

McEachern: That I do not know. That is something I actually hope to do with a pulsed-field gel. The idea would be that big circles really do not enter into these gels so you would not be able to tell by that gel itself. But the way circles are identified in these pulsed-field gels is by then hitting them with ionizing radiation which makes them linear and then looking at the size of the linear molecules. So what I would expect to see in the pulsed-field gels is that there may initially be very little sign of linear molecules, but after being hit with radiation and they should, if they were circular, they should now be visible as linear molecules of the original size.

Wahl: So that is why I was suggesting that it may be more direct if you could see the chromosomes to actually look to see if there are rings versus dicentrics, which would also be very large and hard to analyze by a pulsed-field method.

Hoeijmakers: In the Ter1 deletion mutant you would predict the following scenario: If an organism has more chromosomes there are more telomeres and thus more sequences close to the ends which become endangered. So, if such cells go through several rounds of crisis and recover, you would expect that the number of chromosomes would reduce because that would solve the problem in part. In fact, you showed an example where this problem occurred over and over again and you found fusion of the telomeres. Did you do a long lasting experiment where you could see a reduction in the number of chromosomes or when an organism is used which has many chromosomes, does it have more problems to regain long chromosome ends by recombination?

McEachern: No I have not done any experiments in situations where there would be altered numbers of chromosomes. That would be good to do it in a diploid but K. lactis is not generally very stable as a diploid, so it is problematic to do.

Bacchetti: What happens to Rap1 synthesis when telomeres elongate?

McEachern: That we have not looked at.

Bacchetti: Do you think that Rap 1 may be rate limiting when you elongate telomeres?

McEachern: It is a very interesting question, whether or not that becomes a problem. In these situations where the telomeres become much longer, they are up to a hundred or more times longer than what they started with. So you can easily imagine that it is titrating all the Rap1 that is present in the cell. It is a good question.

Bacchetti: Can you look at that?

McEachern: Well it is something that should be looked at. What is interesting to note is that the phenotype of the mutant with these enormously elongated telomeres is quite reminiscent of senescent Ter1 deletion strains with the same sorts of cellular abnormalities. So it does suggest that they could be relatively uncapped in the way they are behaving consistent with that idea.

Shay: What is the phenotype of a Rap1 deletion? Is it non-viable?

McEachern: It is non-viable. You can get telomere phenotypes, this was originally done by Art Lustig with Saccharomyces. If you chop off chunks of the C terminus you can get modest telomere lengthening for the most part. Anat Krauskoft has done that sort of thing with K. lactis Rap1 gene and seen the same thing. If you lop off a little bit of the C terminus you get some telomere lengthening.

Shay: In another area you talked about these RAD52 mutants where you could not get survivors in the Ter1 knockouts. However, you hinted that sometimes there was something that survived and I was curious if there was a RAD52 independent pathway.

McEachern: There apparently is. The RAD52 knock-out decreases the occurrence of survivors two to three order of magnitude, so they are very appreciable. But you do see some. These guys turn out, at least, the handful that I have looked at, to also show telomere elongation that is unstable, in other words, they are still prone to gradual shortening. So they do not appear to be bypasses of the requirement for telomeres, but instead, they may use a RAD52 independent form of recombination.

Shay: So they did not circularize?

McEachern: That is right. If they are different in any way, it might be that their telomeres might not get as long as they do in the RAD52 plus situation.

Shay: You sort of wonder why we ever evolved linear chromosomes. I think many viruses, bacteria, are circular, so they solved it in some of these mutants. It seems as if you are reforming circular chromosomes. What happens to these things in mitosis? Does that create a problem separating?

McEachern: Where circular chromosomes in Sacchromyces have been looked at, they do not seem to really cause any problems in mitosis. There would be expected to be some degree of problems in meiosis because if you have a crossover event between linear molecules, it will just create a crossover. In a circle, it will create a co-integrant that is going to be a dicentric and that is a real problem. So I would say probably why we have linear chromosomes is to avoid that sort of problem.

Gatti: Many ring chromosomes occur in Drosophila, some of them are quite unstable because with the sister chromatid exchange you can lose them and become dicentric chromosomes or sister chromatid exchanges. Nobody understands why some rings are very stable and other rings which are not stable.

McEachern: One thing that I did not mention about these parent fusion strains in K. lactis is that they grow quite well. They are not as quickly growing as wild type cells but they grow remarkably well.

Bacchetti: If you co-express a mutant and the wild type telomerase RNA, do you get both sequences in the telomere, and what happens to Rap1 binding and telomere lengthening?

McEachern: I have not looked at that a great deal since I primarily concentrated on just the pure mutant situations. But they generally appear to be. That certainly is the expectation and what data I have suggest that that is true. In the case of the lengthening mutants, I think I can say in all mutations I have looked at, the telomere phenotypes are recessive to wild type, or largely so. So in these extreme elongation mutants, if there is a wild type present, which is actually shown on one of the slides though I did not talk about it. When you have a heteroallelic situation (wild type and mutant) you do not get anywhere near that degree of elongation.

Bacchetti: Does that tell you how much of the wild type sequence you need to have at the end in order for the telomere to behave as wild type? What is the minimum?

McEachern: That I do not know.

9

REGULATION OF TELOMERE LENGTH IN MAMMALIAN CELLS

Brenda R. Grimes,[1] David Kipling,[1] Niolette I. McGill,[1] Claudia Teschke,[1] Sally H. Cross,[2] Patricia Malloy,[1] Helen E. Wilson,[1] Christine J. Farr,[3] and Howard J. Cooke[1*]

[1]MRC Human Genetics Unit
Western General Hospital
Crewe Road, Edinburgh EH4 2XU
Scotland, United Kingdom
[2]Institute of Cell and Molecular Biology
University of Edinburgh
King's Buildings, Mayfield Road
Edinburgh EH9 3JR, Scotland
[3]Department of Genetics
University of Cambridge
Downing Street, Cambridge CB2 3EH, England

1. INTRODUCTION

The ends of linear mammalian chromosomes consist of a simple repeated sequence (T_2AG_3)n [reviewed in 1,2]. These terminal repeats play a role in prevention of end to end fusion and prot7ection from exonucleolytic degradation [3]. A ribonucleoprotein complex called telomerase, which was first isolated in *Tetrahymena* and shown to be a specialised reverse transcriptase, facilitates *de novo* synthesis of terminal repeats to compensate for sequence loss caused by incomplete replication at chromosome ends [4]. Telomerase uses an RNA molecule as a template for the addition of terminal repeat monomers. This enables the cell to overcome the problems associated with replication of terminal sequences encountered by other DNA polymerases. These can only synthesise DNA in a 5' to 3' direction using a primer. At chromosome ends, this would result in a 3' overhang which could not be replicated and over many cell divisions would lead to a progressive loss of DNA. Mammalian telomerase activity was first identified in HeLa cell extracts which can add over sixty T_2AG_3 repeats onto a T_2AG_3 oligonucleotide *in vitro* [5]. Mouse

---

* To whom correspondence and reprint requests should be addressed. Telephone: (031) 332 2471 ext 2205; fax: (031) 343 2620; email: howard@hgu.mrc.ac.uk

Genomic Instability and Immortality in Cancer
edited by Mihich and Hartwell, Plenum Press, New York, 1997

cell extracts were shown to add at most one or two T_2AG_3 repeats onto a T_2AG_3 oligonu-cleotide *in vitro* suggesting a non processive behaviour which is in contrast to human telomerase [6]. Mammalian telomerase activity is thought to be restricted to germline cells *in vivo* and immortalised cells *in vitro* [7–11].

The amount of human terminal repeat sequence varies, with average tract sizes rang-ing from approximately 5–15 kb in lymphocytes and 15–25 kb in sperm [7,9,12–15]. Telomere repeat arrays have been shown to shorten with age in some somatic tissues [7,16–18, reviewed in 19]. This phenomenon is also observed in primary fibroblast cultures where telomeres shorten with increasing number of cell divisions until the cells reach a cri-sis, when they senesce and stop dividing [8,17,20]. After crisis, spontaneous transformation can occur in a few cells giving rise to rare foci which can be established as immortalised clonal populations. Cells can also become immortalised following infection with DNA tumour viruses [reviewed in 21]. Telomerase activation and telomere length stabilisation have been observed in immortalised cell lines transformed with DNA tumour viruses [10,22] and in human tumour cells [23]. In one example it has been shown that telomere length increases following epithelial cell transformation with papilloma virus [11].

The observation that telomere length in germline cells and immortalised cells is maintained within a defined size range suggests that there has to be a balance between the rate of telomere loss, as a result of incomplete replication, and telomere repeat addition. A small alteration in the rate of telomere repeat addition or loss would result in either con-tinuous telomere elongation or shortening so it seems necessary that some control mecha-nism is operative. Although the nature of this homeostatic mechanism is unknown, a number of gene products with possible roles in telomere maintenance have been identified in lower eukaryotes (see Discussion).

We have taken two approaches to investigate telomere length regulation in mammal-ian cells. Firstly, we show an increase in the length of a human X chromosome pseudoautosomal telomere (Xpter) following transfer to a mouse cell background. In our second approach we show that cloned telomere repeats which have healed a broken ham-ster chromosome in a Chinese hamster hybrid cell line increase in length initially and then stabilise to a new average telomere length.

2. RESULTS

2.1. Behaviour of the Human Xpter Pseudoautosomal Telomere in a Human X Chromosome–Mouse Hybrid Cell Line

Somatic cell fusion between a human male lymphoblastoid cell line (PES) and a mouse hprt⁻ cell line X63 / NS1 was used to generate the hybrid cell line called XMGU-1, which contains a human X chromosome as the sole human component. The fate of the Xpter pseudoautosomal telomere on the human X chromosome was analysed following transfer to the mouse cell background. The pseudoautosomal region is involved in X and Y chromosome pairing during meiosis and the terminal *Bam* HI fragment can be detected as a 17 kb smear in the PES cell line using the probe 29C1 [28]. We have refined the re-striction map of the Xpter pseudoautosomal region in PES cells [26] as shown in Fig. 1A. We established that the *Hind* III, *Sfi* I and *Eco* RI fragments at Xpter were the same size in both the parental PES cell line and XMGU-1 using 29C1 [22,28]. This shows there have been no gross rearrangements at Xpter within the terminal 15.9 kb, up to the final *Hind* III site. However, digestion with *Bam* HI and Southern analysis using 29C1 reveals that the

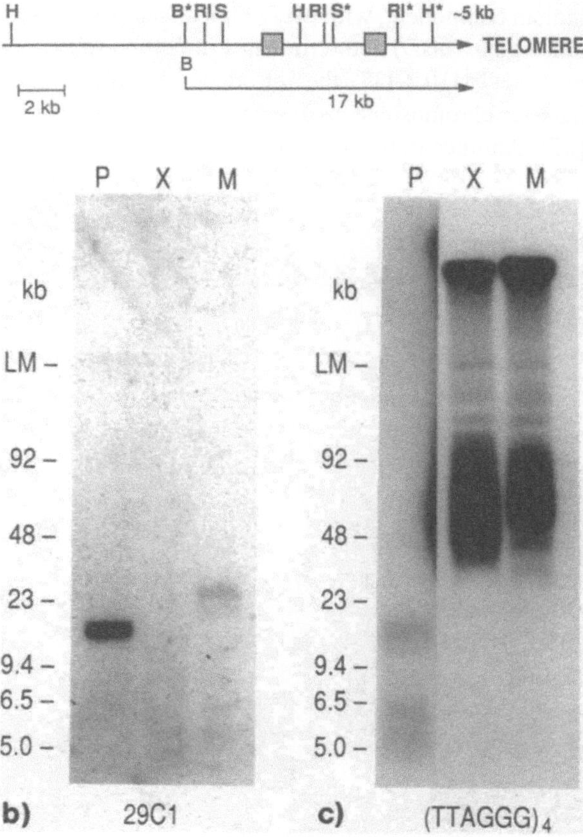

Figure 1. Length of human Xpter in a human X chromosome–mouse hybrid cell line. (a) Restriction map of a human X chromosome pseudoautosomal telomere (Xpter) in PES cells. Regions of 29C1 homology are marked by boxes. Restriction sites: B= *Bam* HI, RI= *Eco* RI, S= *Sfi* I, H= *Hind* III. * indicates last site for this enzyme before the telomere. DNA was digested with Bam HI, and separated in a 1% agarose gel using pulsed field gel electrophoresis. Lane 1= PES, parental human cell line; lane 2= X63/NS1, parental mouse cell line; lane 3= XMGU-1, human X chromosome–mouse hybrid cell line. Terminal 17 kb Bam HI fragment recognised by 29C1 is indicated. Size markers derived from phage lambda DNA, phage lambda DNA digested with *Hind* III and *Hind* III plus *Eco* RI, and *S.cerevisiae* chromosomes (YPH148) are indicated. Following Southern transfer the filters were probed with 29C1 (b) and then subsequently with (TTAGGG)$_4$ (c). Fig. 1C lane 1 shows a longer exposure of PES cells to allow visualisation of the (TTAGGG)n hybridisation signal.

Bam HI terminal restriction fragment is 10 kb larger in XMGU-1 (Fig. 1B). According to the restriction map this indicates that there has been an addition of 10 kb of terminal sequence beyond the most distal *Hind* III site.

The Xpter array of (T_2AG_3)n is estimated to have an upper size limit of 5 kb in PES cells (Fig. 1A) whereas the mouse telomeres of the XMGU-1 cell line are 40–100 kb in length (Fig. 1C). The 10 kb increase in length of Xpter was measured after XMGU-1 cells had undergone approximately 20 divisions. If our interpretation is correct then this would mean an average increase in Xp telomere length in XMGU-1 of approximately 500 bp / cell generation if $(T_2AG_3)_n$ repeats are added continuously. However since this is only one observation and we could not rule out the possibility that some other process, such as a recombination event, has generated the longer terminal Xpter fragment, we decided to examine the fate of a defined telomere at regular intervals, over 300 generations.

2.2. Regulation of Telomere Length in an X Chromosome–Chinese Hamster Hybrid Cell Line

Several groups have now shown that linear plasmids terminating in (T_2AG_3)n repeats can heal broken chromosome ends when introduced into mammalian cells, thereby forming a functional telomere [25,29–31]. Farr *et al.*, [27] transformed a human X chromosome–Chinese hamster hybrid cell line, containing a human X chromosome as the only human component, with pHTM linearised with *Nde* I. This plasmid carries a histidinol resistance gene (*hisD*) and terminates in 500 bp of $(T_2AG_3)n$ repeats (Fig. 2A). In one transformant, HTM10 Cl37, the introduced telomere repeats have healed the end of a broken hamster chromosome, as shown by chromosome *in situ* hybridisation and BAL-31 digestion [27]. Another transformed cell line, HTM18 TC8, contains the pHTM plasmid integrated

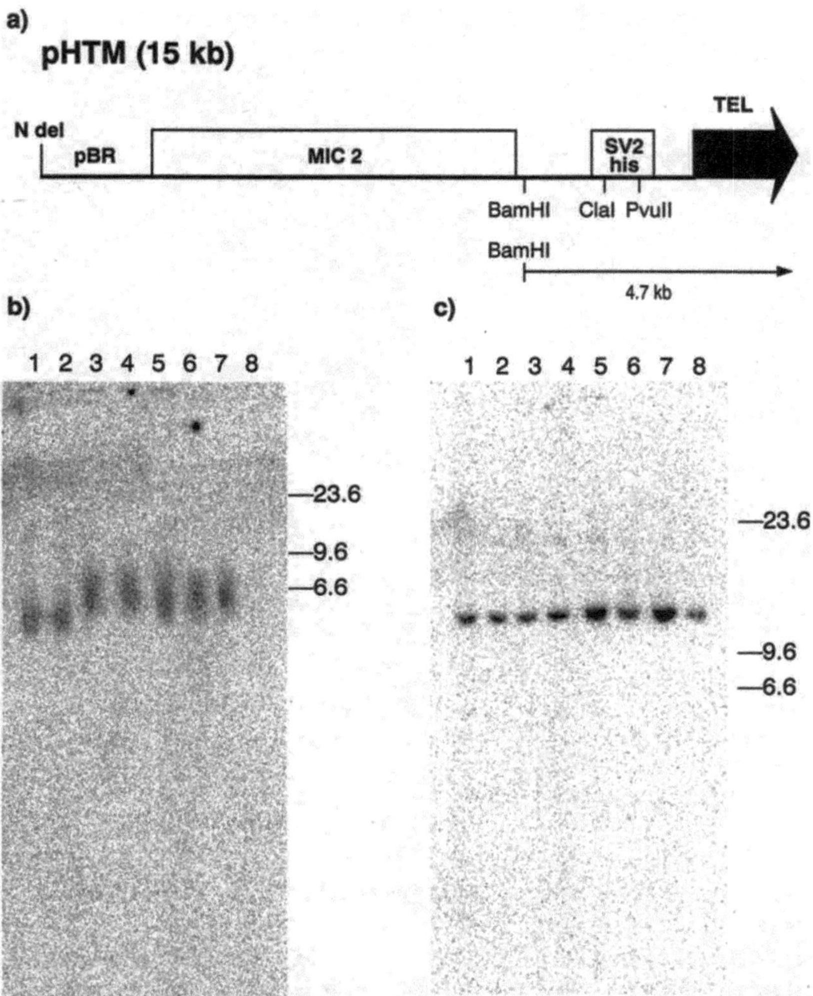

Figure 2. Length regulation of a telomere generated by chromosome fragmentation in an immortalised cell line. (a) Structure of pHTM used to introduce cloned telomeres into a human X chromosome–CH hybrid cell line. (b) Telomere growth and stabilisation following healing of a broken hamster chromosome by cloned telomere repeats in HTM10 Cl37. (c) In the control cell line HTM18 TC8, the construct has integrated at an interstitial site and was detected as a discrete fragment. Each cell line was selected initially in histidinol-containing medium for approximately 20 generations. DNA was extracted from each cell line after a further 10, 100, 200, and 300 generations grown either in the presence (lanes 1, 3, 5, and 7 respectively) or absence of histidinol selection (lanes 2, 4, 6, and 8 respectively). The DNA samples were digested with *Bam* HI, separated using conventional gel electrophoresis, Southern blotted and probed with *hisD*. Size markers (Lambda DNA digested with *Hind* III) are indicated on the left-hand side of each figure.

within the MIC2 locus, an interstitial site on the human X chromosome [27,29]. Each cell line was selected initially on histidinol—containing medium and grown for approximately 20 generations following plasmid introduction before this set of experiments was initiated. From this point each cell line was grown either in the presence or absence of histidinol. In order to examine the fate of the introduced construct, DNA was extracted after approximately a further 10, 100, 200 or 300 generations. Digestion of genomic DNA from each sample with *Bam* HI was followed by Southern analysis using a *hisD* probe. This probe recognises a 4.7 kb fragment in pHTM digested with *Nde* I and *Bam* HI (Fig. 2A). Fig. 2B shows that the *his D* probe recognises a terminal restriction fragment smear which is 5.2–6.6 kb in length in HTM10 Cl37, after approximately 30 generations following the introduction of pHTM, which would be consistent with an addition of between 0.5 and 1.9 kb of $(T_2AG_3)n$ repeats. After a further 100 generations the terminal restriction fragment smear ranges in size from 5.5–7.5 kb (Fig. 2B, lane 3), which would be consistent with a net addition of a further 0.3–0.9 kb of terminal $(T_2AG_3)n$ repeats.

It has been suggested that there may be a correlation between telomere length and repression of an adjacent gene (reviewed in [32]). Consequently, it was considered important to demonstrate that our telomere growth results were not distorted by a selective bias towards shorter telomeres resulting from the telomeric location of the *hisD* gene. We therefore conducted a similar series of measurements on cell line HTM10 C137 grown in the absence of histidinol. As shown in Fig. 2B, the results obtained confirm that the telomere length increase and stabilisation are independent of selection for *hisD* expression up to generation 200. Unfortunately we were unable to determine telomere length in the absence of selection at generation 300 since the terminal *Bam* HI fragment carrying the *hisD* gene was evidently lost in the intervening period, probably as a result of chromosome loss. This was confirmed by chromosome *in situ* hybridisation (data not shown) and failure of the cell line to grow when transferred to histidinol containing medium.

In HTM18 TC8, the control cell line containing pHTM at an interstitial location, the *hisD* probe recognises a discrete 10.5 kb *Bam* HI fragment at all generations analysed either in the presence or absence of histidinol selection (Fig. 2C).

3. DISCUSSION

We have examined the fate of telomeric sequences following their introduction into a foreign cell environment and find that the response of the host cell machinery to the exogenous telomere points to the existence of a control mechanism governing telomere length. In our initial experiment it was demonstrated that the most distal 5 kb sequences of a human X chromosome pseudoautosomal telomere (Xpter) increase to a new length of 15 kb following transfer of a human X chromosome to a mouse cell line using somatic cell fusion. Although we cannot rule out the possibility that a rearrangement involving the most distal 5 kb of Xpter has generated this increase, no gross rearrangements at Xpter were detected using restriction analysis. Consequently, it seems a reasonable interpretation of the findings that the mouse cell possesses a system for monitoring telomere length which recognises that the Xpter telomere is below the threshold length for this cell type, resulting in the addition of a further 10 kb of terminal $(T_2AG_3)n$ repeats. The length of Xpter was analysed after approximately 20 divisions following the establishment of the hybrid XMGU-1. If the addition of terminal repeats is a continuous process, the observed 10 kb increase suggests a rate of addition of approximately 500 bp of $(T_2AG_3)n$ per generation. We speculate that continued growth of this cell line would be accompanied by a further increase in the Xpter telomere length to a new stabilised value, within the range (20–100 kb) of the other telomeres in XMGU-1.

Our telomere mediated chromosome breakage constructs enabled us to define the terminal restriction fragment with greater resolution than was possible for Xpter. Cell lines generated in this way were therefore analysed in greater detail. In cell line HTM10 Cl37, the cloned 500 bp of (T_2AG_3)n repeats which formed a new telomere, were subject to a net increase in length of 0.5–1.9 kb after 30 generations following introduction into the Chinese hamster (CH) hybrid cell line. Between generation 30 and 100 the introduced telomere increased by a further 0.3–0.9 kb, bringing the total net increase to between 0.8–2.8 kb. In accordance with our interpretation of the net increase at Xpter in a mouse cell background, we propose that in HTM10 C137 there has been an addition of terminal (T_2AG_3)n repeats to the introduced telomere, an argument strengthened in this case by the demonstration that the length increase was a gradual process. Since no further net change in telomere length was detected during subsequent growth of HTM10 C137 for 200 generations, it appeared that a new stable length had been reached by generation 100. This finding gives support to the hypothesis that the steady state telomere length seen in immortalised cells represents an equilibrium between processes of telomere repeat synthesis and degradation. Although the nature of this regulatory mechanism is unknown, it is interesting to speculate that a comparatively short telomere introduced into a new host cell environment is a preferred substrate for telomerase, and after an initial period of terminal repeat addition, the telomere is then stabilised at a new length.

Telomere length varies widely between different cell types in culture. A fundamental question therefore, concerns what determines the equilibrium telomere length in a particular cell type and whether it is a property of the chromosomes themselves or their cellular environment. In this regard, it would clearly be useful to ascertain if the exogenous telomeres in the present experiments were subject to the same regulation as their native counterparts, a question which remains unanswered. Whilst the initial observation of a net increase in length at Xpter in a mouse cell background was at least consistent with this possibility, in the second set of experiments the length of the endogenous hamster telomeres could not be determined due to the presence of abundant internal (T_2AG_3)n tracts in the CH genome [33]. However, we have other examples of CH transformant cell lines generated by telomere mediated chromosome fragmentation, and after 30 generations the introduced telomeres have been subject to a net increase in length similar to that observed in HTM 10 C137 (data not shown). These similar responses to several independent breakage events suggest that cellular environment is indeed the critical determinant of telomere length. Future experiments should include testing for the predicted reduction in length of an introduced telomere longer than the telomeres of the host cell.

Elsewhere it has been shown that the amount of (T_2AG_3)n repeats added to a newly formed telomere following telomere mediated chromosome breakage and healing does depend on the host cell type. Barnett *et al.*, [31] used cloned telomeric DNA to seed the formation of new telomeres in three immortalised human cell lines and a mouse embryonic stem (ES) cell line. The length of (T_2AG_3)n repeats added to the introduced telomeric repeats was measured following establishment of stable transfectants. The added sequence of between 1 and 20 kb in the human cell lines contrasted with the addition of 40–75 kb in a mouse embryonic stem cell line.

We have presented evidence for length regulation of a defined telomere in a Chinese hamster hybrid cell line and for widely divergent telomere lengths in mouse cells in culture and between two mouse species. However, we still do not understand how telomere length is monitored and kept within upper and lower limits in immortalised cell lines and in germline cells. To date, components of this regulatory mechanism, such as mammalian telomerase and other *trans* acting factors, have not been cloned. Nevertheless, since telomere structure is highly conserved it is possible that some of the loci involved in telomere maintenance in lower eukaryotes will have homologs in higher eukaryotes. For example in *S.cerevisiae*, cells carrying a mutation in the *est 1* gene have shortened telomeres and show a senesence phenotype [34]. The *est 1* gene encodes a protein with weak similarities to RNA dependent polymerases and may be a component of yeast telomerase [35]. RAP 1, a yeast protein which is involved in transcriptional activation and repression, also binds to yeast double stranded telomere repeats [36, 37]. A temperature sensitive mutation of RAP 1 causes the yeast telomeric repeats to shrink to a new steady state level over 100 generations [38, 39]. Genetic analysis has facilitated the identification of RIF 1 which interacts with RAP 1 [40]. Strains carrying RIF 1 mutations show increased telomere length. Both RIF 1 and RAP 1 may be involved in regulating access of telomerase to the chromosome ends, thereby exerting control over telomere length.

There is some evidence in ciliates that a protein binds to the G rich single stranded overhangs found at telomeres which may be preventing either degradation or telomerase access. The *Euplotes* protein binds with very high affinity [41] and is thought to "cap" telomeres since the protein can be purified roughly in 1:1 proportions with the number of chromosome ends [42]. The *Oxytricha* protein behaves in a similar fashion but forms a heterodimer [43, 44]. A few proteins have been identified in mammalian cells which bind to single stranded [45] and double stranded [46] (T_2AG_3)n repeats *in vitro*. The gene encoding this last protein has been cloned [47]. However, there is no evidence at present for their potential role *in vivo*.

Expression of antisense RNA to the RNA component of telomerase in HeLa cells results in the shortening of their telomeres and in some cases a senescent phenotype suggesting that the abundance of this RNA and or protein is likely to be a significant regulator of telomere length in cell populations expressing the enzyme.

On the basis of known telomeric function, the maintenance of telomeric sequences is a requisite of chromosomal stability. However, it is difficult to explain the observed diversity in telomere length, both between species and even between chromosomes within the same cell. Conceivably, these differences may be of little biological significance or simply reflect the minimum length required to accommodate sequence loss at the ends of different chromosomes. However, it remains possible that as our knowledge grows of how telomere length is regulated in different cell types, so it may become apparent that telomere length diversity reflects a range of functional constraints on different telomeres and a greater complexity of telomeric function than hitherto suspected.

CONCLUDING REMARKS

We have taken two approaches to investigate the regulation of telomere length in mammalian cells. In our initial experiment we examined the length of the Xpter pseudoautosomal telomere of a human X chromosome in both the original human lymphoblastoid cell line and after transfer into a mouse cell line. Although analysis revealed no gross rearrangements at Xpter following transfer to the mouse cell line, the terminal 5 kb fragment at Xpter increased in length. We speculate that the observed increase is due to addition of further terminal (T_2AG_3)n repeats at Xpter as the result of a response from the mouse cell environment which regulates telomere length. It has been shown that cloned human (T_2AG_3)n repeats can function as telomeres when introduced into mammalian cells since they seed the formation of a new telomere at broken chromosome ends. In our second approach, we determined the rate of growth of (T_2AG_3)n repeats which have healed a broken hamster chromosome in a human X chromosome-Chinese hamster hybrid cell line for 300 generations following the introduction of the cloned telomeres. The introduced telomere increased in length by approximately 0.8–2.8 kb, during the first 100 generations with no further detectable telomere growth during generations 100–300. We propose that the introduced telomere repeats have been subject to length regulation.

ACKNOWLEDGMENTS

We wish to thank Dominique Broccoli, Rosemary Bayne, Roger Slee and Prof. Nick Hastie for useful comments on the manuscript, and Eric Thomson for technical assistance. This work was supported by the UK Medical Research Council, the EC Bridge T Project (BRG) and the HGMP (CJF, DK, HEW). DK is a Beit Memorial Fellow. SHC was supported by a SERC studentship.

REFERENCES

1. Kipling D., Cooke H.J. (1992) Beginning or end? Telomere structure, genetics and biology. *Human Molecular Genetics* **1**, 3–6

2. Greider C.W. (1994) Mammalian telomere dynamics: healing, fragmentation, shortening and stabilization. *Curr. Opinion Genet. Dev.* **4**, 203–211

3. Blackburn E.H. (1991) Structure and function of telomeres. *Nature* **350**, 569–573

4. Blackburn E.H. (1992) Telomerases. *Ann. Rev. Biochem.* **61**, 113–129

5. Morin G.B. (1989) The human telomere terminal transferase enzyme is a ribonucleoprotein that synthesises TTAGGG repeats. *Cell* **59**, 521–529

6. Prowse K.R., Avilion A.A., Greider C.W. (1993) Identification of a non processive telomerase activity in mouse cells. *Proc. Natl. Acad. Sci. USA* **90**, 1493–1497

7. Hastie N.D., Dempster M., Dunlop M.G., Thompson A.M., Green D.K., Allshire R.C. (1990) Telomere reduction in human colorectal carcinoma with ageing. *Nature* **346**, 866–868

8. Harley C.B., Futcher A.B., Greider C.W. (1990) Telomeres shorten during ageing of human fibroblasts. *Nature* **345**, 458–460

9. de Lange T., Shiue L., Myers R.M., Cox D.R., Naylor S.L., Killery A.M., Varmus H.E. (1990) Structure and variability of human chromosome ends. *Mol. Cell. Biol.* **10**, 518–527

10. Counter C.M., Avilion A.A., Le Feuvre C.E., Stewart N.G., Greider C.W., Harley C.B., Bacchetti S. (1992) Telomere shortening associated with chromosome instablity is arrested in immortal cells which express telomerase activity. *EMBO J.* **11**, 1921–1929

11. Klingelhutz A.J., Barber S.A., Smith P.P., Dyer K., Mc Dougall J.K. (1994) Restoration of telomeres in human papilloma virus- immortalised human anogenital epithelial cells. *Mol. Cell. Biol.* **14**, 961–969

12. Cooke H.J., Smith B.A. (1986) Variability at the telomeres of the human X/Y pseudoautosomal region. *Cold Spring Harbor Symp. Quant. Biol.* **51**, 213–219

13. Allshire R., Gosden J.R., Cross S.H., Cranston G., Rout D., Sugawara N., Szostak J.W. (1988) Telomeric repeats from *T. thermophila* cross hybridise with human telomeres. *Nature* **332**, 656–659

14. Allshire R., Dempster M., Hastie N.D. (1989) Human telomeres contain at least three types of G-rich repeat distributed non-randomly. *Nucleic Acids Res.* **17**, 4611–4627

15. Cross S.H., Allshire R.C., McKay S.J., McGill N.I., Cooke H.J. (1989) Cloning of human telomeres by complementation in yeast. *Nature* **338**, 771–774

16. Lindsey J., McGill N.I., Lindsey L.A., Green D.K., Cooke H.J. (1991) In vivo loss of telomeric repeats with age in humans. *Mutation Research* **256**, 45–48

17. Allsopp R.C., Vaziri H., Patterson C., Goldstein S., Younglai E.V., Futcher A.B., Greider C.W., Harley C.B. (1992) Telomere length predicts replicative capacity of human fibroblasts. *Proc. Natl. Acad. Sci. USA* **89**, 10114–10118

18. Vaziri H., Schächter F., Uchida I., Wei L., Zhu X., Effros R., Cohen D., Harley C.B. (1993) Loss of telomeric DNA during aging of normal and trisomy 21 human lymphocytes. *Am. J. Hum.Genet.* **52,** 661–667

19. Broccoli D., Cooke H. (1993) Ageing, healing and the metabolism of telomeres. *Am. J. Hum. Genet.* **52,** 57–660

20. Levy M.Z., Allsopp R.C., Futcher A.B., Greider C.W., Harley C.B. (1992) Telomere end-replication problem and cell aging. *J. Mol. Biol.* **225,** 951–960

21. Shay J.W., Wright W.E., Werbin H. (1991) Defining the molecular mechanisms of human cell immortalisation. *Biochem. Biophys. Acta.* **1072,** 1–7

22. Counter C.M., Botelho F.M., Wang P., Harley C.B., Bachetti S. (1994a) Stabilization of short telomeres and telomerase activity accompany immortalisation of Epstein-Barr virus - transformed human B lymphocytes. *J. Virol.* **68,** 3410–3414

23. Counter C.M., Hirte H.W., Bachetti S., Harley C.B. (1994b) Telomerase activity in human ovarian carcinoma. *Proc. Natl. Acad. Sci. USA* **91,** 2900–2904

24. Kipling D., Cooke H.J. (1990) Hypervariable ultra long telomeres in mice. *Nature* **347,** 400–402

25. Starling J.A., Maule J., Hastie N.D., Allshire R.C. (1990) Extensive telomere repeat arrays in mouse are hypervariable. *Nucl. Acids. Res.* **18,** 6881–6888

26. Cross S.H. (1989) Isolation and characterisation of human telomeres. Ph.D. thesis University of Edinburgh, Scotland, U.K.

27. Farr C.J., Fantes J., Goodfellow P., Cooke H.J. (1991) Functional reintroduction of human telomeres into mammalian cells. *Proc. Natl. Acad. Sci. USA* **88,** 7006–7010

28. Cooke H.J., Brown W.R.A., Rappold G.A. (1985) Hypervariable telomeric sequences from the human sex chromosomes are pseudoautosomal. *Nature* **317,** 687–692

29. Farr C.J., Stevanic M., Thomson E.J., Goodfellow P.N., Cooke H.J. (1992) Telomere associated chromosome fragmentation: applications in genome manipulation and analysis. *Nature Genetics* **2,** 275–282

30. Itzhaki J.E., Barnett M.A., MacCarthy A.B., Buckle V.B., Brown W.R.A., Porter A.C.G. (1992) Targetted breakage of a human chromosome mediated by cloned human telomeric DNA. *Nature Genetics* **2,** 283–287

31. Barnett M.A., Buckle V.J., Evans E.P., Porter A.C.G., Rout D., Smith A.G., Brown W.R.A. (1993) Telomere directed fragmentation of mammalian chromosomes. *Nucleic Acids Res.* **21,** 27–36

32. Sandell L.L., Zakian V.A. (1992) Telomeric position effects in yeast. *Trends Cell Biol.* **2,** 10–14

33. Meyne J., Baker R.J., Hobart H.H., Hsu T.C., Ryder O.A.,Ward O.G., Wiley J.E., Wurster-Hill D.H., Yates T.L., Moyzis R.K. (1990) Distribution of non-telomeric sites of the (TTAGGG)n telomeric sequences in vertebrate chromosomes. *Chromosoma* **99,** 3–10

34. Lundblad V., Szostak J.W. (1989) A mutant with a defect in telomere elongation leads to senescence in yeast. *Science* **57,** 630–643

35. Lundblad V., Blackburn E.H. (1990) RNA-dependent polymerase motifs in EST 1:tentative identification of a protein component of an essential yeast telomerase. *Cell* **60,** 529–530

36. Buchman A.R., Lue N.F., Kornberg R.D. (1988a) Connections between transcriptional activators, silencers, and telomeres as revealed by functional analysis of a yeast DNA-binding protein. *Mol. Cell. Biol.* **8,** 5086–5099

37. Buchman A.R., Kimmerly W.J., Rine J., Kornberg R.D. (1988b) Two DNA binding factors recognize specific sequences at silencers, upstream activating sequences, autonomously replicating sequences, and telomeres in *Saccharomyces cerevisiae. Mol. Cell. Biol.* **8,** 210–225

38. Lustig A.J., Kurtz S., Shore D. (1990) Involvement of the silencer and UAS binding protein RAP 1 in regulation of telomere length. *Science* **250,** 549–553

39. Conrad M.N., Wright J.H., Wolf A.J., Zakian V.A. (1990) RAP 1 protein interacts with yeast telomeres in vivo: overproduction alters telomere structure and decreases chromosome stability. *Cell* **63,** 739–750

40. Hardy C.F.J., Sussel L., Shore D. (1992) A RAP1-interacting protein involved in transcriptional silencing and telomere length regulation. *Genes Dev.* **6,** 801–814

41. Price C.M. (1990) Telomere structure in *Euplotes crassus*: characterization of DNA-protein interactions and isolation of a telomere-binding protein. *Mol. Cell. Biol.* **10,** 3421–3431

42. Price C.M., Skopp R., Krueger J., Williams D. (1992) DNA recognition and binding by the *Euplotes* telomere protein. *Biochemistry* **31,** 10835–10843

43. Hicke B.J., Celander D.W., MacDonald G.H., Price C.M., Cech T.R. (1990) Two versions of the gene encoding the 41-kilodalton subunit of the telomere binding protein of *Oxytricha nova. Proc.Natl. Acad. Sci. USA* **87,** 1481–1485

44. Gray J.T., Celander D.W., Price C.M., Cech T.R. (1991) Cloning and expression of genes for the Oxytricha telomere-binding protein: specific subunit interactions in the telomeric complex. *Cell* **67,** 807–814

45. McKay S.J., Cooke H. (1992) hnRNP A2/B1 binds specifically to single stranded vertebrate telomere repeat TTAGGGn. *Nucleic Acids Res.* **20,** 6461–6464
46. Zhong Z., Shiue L., Kaplan S., de Lange T. (1992) A mammalian factor that binds telomeric TTAGGG repeats in vitro. *Mol. Cell. Biol.* **12,** 4834–4843

DISCUSSION

Gatti: Did you find any difference between individual telomeres within the same cell in mouse?

Cooke: There are differences between different internal restriction fragments in mouse but that is not entirely due to the internal repeats themselves. If you are asking if we have measured the internal repeat length on different chromosomes, then the answer is, no we have not.

Gatti: Is that doable by *in situ* hybridization because they are quite big?

Cooke: I think it is now doable by *in situ* hybridization in human cells as well where they are not so big.

Shay: I think Peter Landsdorf in Vancouver has recently demonstrated a quite sophisticated FISH analysis of individual telomeres in human cells; whereas, I believe the results are that on an individual chromosome, the ends are about the same, there is at least a six fold variation in telomere length size if you look at all the chromosomes within a given cell. So there is a bit of heterogeneity intracellularly, I do not know what that means.

Cooke: I am not sure if he has looked at homologues within the same cell either. I think that is an interesting question.

Wahl: The examples that you showed, if I understood correctly, all involved increasing length starting from certain size. Could you find any examples of decreasing length, going from something long into a species that had shorter average telomere size?

Cooke: Well, there are technical problems there. It is actually very difficult to maintain a length of telomeric repeat much over about a kb in *coli* to actually do the experiment with. All those experiments have been done with a range of repeats which have been cloned and which are about a kb in length. It would be rather tricky to start with a defined substrate to get something that was larger than a cell line which was immortal and had short telomeres, because even they would have telomeres in the order of 2 to 5 kb in length. In principle, yes, it is an interesting question but I think technically it is not actually possible to study it at the moment.

Shay: Maybe if I am understanding your question, we know that if you take a human cell which does not have detectable telomerase, fuse that to a telomerase expressing cell, that you transdominantly inhibit telomerase. So there is something in the normal cell that will turn off telomerase. Then what you see is that the cells will grow for a while and eventually, the immortal cells which generally have shorter telomeres get critically short telomeres and that is what limits the life span of normal cross tumors cell hybrids. At least that is what we think. We have known that for decades, back from Henry Harris in the 60's where he made mouse-mouse fusions or human-human fusions. So the question that I was really curious about, is when you cross a mouse cell to a human cell, it has been known for several decades in fact, that human chromosomes tend to be segregated out. I was wondering if that might have something to do with the survival of those mouse chromosomes in that type of a cross.

Cooke: People have speculated about that, they have also speculated that there may be an impairment of centromere function in transferring a centromere from one species to another. That would lead to loss of human chromosomes. These are certainly possibilities but deciding which of the many possibilities is right, I think, has not been done yet.

Hartwell: I have a question about these hybrids. Is it the case that when you make the hybrid there is an immediate cessation of telomerase activity? It is not just slowly diluted out?

Shay: As soon as you can test, I mean again, all of these are made by some sort of genetic selection using dominants selectable genes, and so there is an obvious finite period before you can actually test these. But it seems to be immediate. It is not a diffusable inhibitor. So, if you take a normal cell and take an extract from a normal cell and mix it with an extract from a telomerase expressing cell, there is nothing in that normal cell extract that can inhibit telomerase in an *in vitro* assay. So I think it is something that is obviously being regulated at some other level that we do not really completely understand at this point in time.

Stark: If it is true that short telomeres are finally sensed as ends, is the cell arrest mechanism that responds to that due to p53? Is the evidence for or against that at the moment?

Shay: We do not really know. The thought collective is that young cells with long telomeres are capped or protected and that telomere erosion occurs with continued proliferation and perhaps eventually a few telomeres have relatively short telomeres whereas the bulk of the ends of the chromosomes still have relatively long telomeres. So there is no evidence that you are going to simultaneously shrink all of your telomeres on all ends of all ninety-two chromosomes. So, the idea is that perhaps finally one chromosome which is no longer capped or protected is now recognized as damaged DNA and that sets off some p53 mediated, p21 checkpoint arrest. That is one possibility and that is what I think the thought collective in the field is that that would be the mechanism. Woody Wright, a colleague of mine, and I actually speculated on a positional effect alternate mechanism, which there is also no direct experimental evidence for, except by homology to yeast. The basic idea is perhaps there is a gene that is silent that is somewhere in the subtelomeric region. The gene is silent in young cells and as telomeres erode perhaps heterochromatin domains change. So as cells get old, there might be a turning-on of a gene that was previously silenced and the expression of this gene may participate in the senescence blockade. The whole concept that telomeres in fact are the direct counting mechanism that regulate when cells senesce is still a pretty tenuous hypothesis. The most direct evidence was alluded to by Howard Cooke in an experiment which we just recently published. We were able to treat immortalized cells with short oligonucleotides that were similar to telomeres. We fed these cells small 12-mer oligonucleotides that were of the sequence T2AG3 with the idea that these might compete for telomere binding proteins or telomerase and therefore, inhibit telomerase. Therefore what we expected to see was telomere shortening. We were actually very surprised when we looked at those cells and the telomeres had actually elongated by 2 to 3 kb pairs. So, we repeated this and we did this with four or five different tumor cell lines of varying lengths of telomeres. Some started with 2 or 3 kb pairs, some with 4 or 5. In each case, by feeding these small oligonucleotides, we showed that the telomeres actually elongated. But this only occurred in cells that expressed telomerase. What we did next was to utilize that observation to make cell hybrids. Essentially, we took a normal human fibroblast with very long telomeres and fused it to an immortal cell with very short telomeres that had telomerase, and then made another cross between that normal fibroblast and an experimentally elongated immortal cell. So we had two sets of cell hybrids, one with long and short telomeres, and one with long and experimentally elongated telomeres. The idea was to determine which hybrids would grow for the longest. We found that the ones that had the 2 or 3 kb pairs of increased telomeres actually grew for about twenty or thirty population doublings longer than those that had the very short telomere. There were two important conclusions about this: first, when you make a hybrid between a telomerase negative and a telomerase positive cell, you turn off telomerase, and, second, that the longevity of that cell hybrid seems to be dependent on the length of the telomeres of the shortest parent cell that is in that fusion. Obviously that is an initial observation and much more work needs to be done to really prove that. But I think it also gives us a first evidence towards a cause and effect relationship on telomeres and senescence. I should point out Howard Cooke's group was probably one of the earliest to show that if you look at telomere length *in vivo* with age in skin or in almost any tissue, that the telomeres are shorter in older individuals than in younger individuals. So the phenomenon of telomere shortening is very real and supports the hypothesis that the telomeres are some sort of timing mechanism.

Cooke: There are a couple of points that are worth making about that. One is that almost any oligo has that effect, is that right?

Shay: Not any oligo, G rich oligos. We used complementary oligos, the C rich, which would be the complement of the T2AG3, and those did not work. We are continuing to explore this finding, and a number of other groups are actually looking at that as well right now.

Cooke: The other point, I guess, is that the anti-sense to the human telomerase RNA has been expressed in HeLa-cells and shows a switch in phenotype from immortality to mortality with concomitant telomere shortening.

Shay: As far as I know nobody has replicated that either. But that was an expected result similar to what you see in yeast.

Stark: I would like to try to get back to the question that I asked, again, and ask a question to Geoff Wahl who showed us yesterday that if you micro-inject in the limit a single copy of a double strand break into a p53 competent cell, that you can trigger cell cycle arrest in that way. I wonder, Geoff, have you looked at the sequence dependence of that? What happens if you micro-inject a TTAGGG oligomer that is linear?

Wahl: If I had had more time yesterday, I would have shown that there is a length dependence and that short oligos do not work. You need to get beyond a certain size. What we are doing now is the direct experiment. We are making hairpin substrates that have telomeric extensions and we have also made substrates that we can put inside chromosomes and break them so we can look at bona fide chromatin, rather than in the micro-injected DNA. So, in the next few months we should know the answer to that experiment.

Evan: My understanding is that the levels of CDKs inhibitors, that is, p15, p16, as well as p21, and for all I know p27 as well, all seem to increase in certainly fibroblast cultures as you grow them and as they senesce. Now, the question is, is it that all cells gradually increase the levels of these inhibitors in the culture, or is it that certain cells suddenly go positive for that on account of that perhaps they have run out of telomere size and they are responding to some sort of DNA damage? Does anybody know? It seems to me that most of these assays are done at the level of cell cultures where you just boil up all the cells and then you see a gradual increase. Now what I want to know is, is it gradual at the level of individual cells?

Bacchetti: Actually I do not know that anybody has looked at individual cells. What comes to mind is work done with other senescent markers like the beta-galactosidase that Judith Campisi has looked at. I think in that case she could see it in all of the cells in the population. For that marker, therefore individual cells were looked at.

Evan: If these markers gradually go up in all the cells in the population as the culture ages, I do not actually understand how that maps on to the idea that at an individual cell, the trigger for cessation of growth is that you run out of telomeres.

Bacchetti: We have no indication that there is a single telomere that triggers growth arrest. I favor the hypothesis that it is an overall shortening of all the telomeres that somehow is sensed at senescence rather than a critically short telomere. The reason why I favor that is because, if you lengthen the lifespan of normal cells by transformation you see that telomeres remain functional for quite a number of cell divisions. But, again, we do not have any proof of that.

Evan: That would be a fundamentally different mechanism.

Bacchetti: Right.

Shay: One other thing that I think you need to be aware of, is that the *in vitro* models of aging, are based on a population kinetics that is going on. During the last ten or fifteen population divisions before cells senesce, there still are some young cells in the population. So it would be very easy, at least theoretically, to imagine that a few cells in the population that are senescing due to a telomere checkpoint or aging mechanism might be upregulating proteins like p21, whereas a large percentage of the cells still have replicative ability and have not reached that point. It is also important in these experiments to distinguish between senescent cells and cells that are quiescent—e.g., the young cells that are quiescent. I think that we have to be very careful to distinguish those components. Recently, Judith Campisi has reported a new marker of aging. In an old mouse she does not see a hundred percent of the cells that are betagalactocydase positive, only maybe ten or fifteen percent. So we really do not know how many of these cells have to be senescent to give you an aging or senescence phenotype.

Livingston: In respect to Gerard Evan's question, if you inject a linear vs the same DNA with increasing telomere lenghts, does one reach a point at which the cell stops replicating.

Bacchetti: Actually, you cannot introduce telomeres that way in cells that have no telomerase is my understanding. So I do not think the experiment has actually been done.

Cooke: I think that people have either used either mouse, eggs or cells which obviously are telomerase positive or else immortalized cell lines. I do not know of anybody who has actually tried to inject such a construct into a non-immortalized cell.

Bacchetti: I thought that William Brown did that. I thought he did not rescue any cell population and therefore there is no answer to the experiment.

Cooke: Very low efficiency, if anything.

Livingston: Well if you took a cell that had undergone fifteen population doublings and inject a yac blaring various telomere lengths is there a given DNA unit which will induce cell cycle arrest?

Wahl: It is not a yac but yes, we are doing it.

Livingston: The question is: Does a single suboptimal telomeric length bearing piece of DNA, also induce cell cycle arrest?

Wahl: Let me just clarify that, a nic does not lead to an arrest. A gap of a minimum of thirty nucleotides leads to an arrest. Double strand break leads to an arrest, the topology of end is not important and now the question is, if we add telomeric sequences of increasing length, can we prevent the linear from now inducing an arrest? That is the issue. The other way of looking at it is if we take a telomere of a certain size and we make it shorter, will it now induce the arrest? The substrates are being made, they are not easy to do. So, hopefully, within a short time we will know whether it is dominant.

Bacchetti: That question has been asked and answered in yeast where a single short telomere has an effect. But in mammalian cells we really do not know. I have a question about the telomere seeding and its cell dependence, I wonder if you want to elaborate on that.

Cooke: We do not know the basis of it. It is an observation we have to study a range of different cell types to find one that is particularly efficient at telomere seeding. We do not know whether it correlates with telomerase activity or any other aspect of the cells phenotype.

Bacchetti: Is that a matter of efficiency or you actually have cell lines where you have not been able to seed telomeres?

Cooke: We have cell lines where we have not been able to seed. So whether that is a question of very low efficiency or what, is impossible to say. But there are certainly many orders of magnitude difference in efficiency.

Mihich: Since we have some time, I dare to ask a philosophical question which is not very specific. We were talking yesterday about mutation and lability and today we are talking about the telomere length and the telomerase role in assuring an anti-senescent status to the cell. What is the relationship between the two? Is the tumor cell becoming progressive in its biological characteristics because it has time to become progressive given the so-called immortality phenotype? Or is it the other way round, that telomerase presence is already an expression of lability and of progression of the tumor?

Bacchetti: I consider telomerase as a facilitator of cell growth for a cell that has somehow become transformed or malignant rather than an initiator of that process. So I think I would go along with the idea that the presence of telomerase allows further evolution of the malignant cell phenotype.

Wahl: With regards to telomerase regulation, one thing that we have not heard about at this meeting is methylation and it has been reported in a number of instances that changes in methylation accompany immortalization. I am wondering if, has anybody tried an experiment where you use azacytidine to try to activate telomerase in a young cell?

Shay: In a simple answer, if you just treat a fibroblast that does not have telomerase, you do not get reactivation of telomerase. Even though there is an interesting history of, like you said, methylation changes—Peter Jones has published over the years—that there is less methylation with increasing age. There is about a five percent drop in global methylation. Again, there are no genes or molecular mechanisms to really translate those changes into a mechanism that would explain all the phenomena that we see as part of normal aging. While it is interesting, it is not clear if it is a cause or effect, or just a consequence of a lot of other things that are happening. Perhaps a more intriguing question is, why do mouse cells in general, or mice, have longer telomeres than humans? Some people have argued that if telomere erosion is such an important mechanism in regulating aging in human cells, why is it that mice, which are short lived organisms, have these very very long telomeres? I would be curious if Howard, or anybody else, would care to talk about those types of issues.

Cooke: Well, I think the thing to remember there is that different strains of mice have different length of telomeres and different sub-species of mice have different lengths of telomere. It suggests to me that actually the telomere length is not particularly critical. What is important is that it is regulated. Clearly, there are mice, in particular strains, which have gross differences between different isolates, but yet, as far as I am aware, no difference in spontaneous tumor rates or lifespan. It is clearly not important, as long as they have more than a minimum amount; presumably those mice are fine.

Shay: There is some evidence that, I do not know who has done the work, but at least I have heard that in many of these mouse strains that there are actually, internal large repeats of telomeres. Have you looked at any of that to see if there are any?

Cooke: That is not true in mice.

10

THE DNA DAMAGE CHECKPOINT

Leland Hartwell, Amanda Paulovich, and David Tocyzki

Fred Hutchinson Cancer Research Center
Seattle, Washington 98104
Department of Genetics
University of Washington
Seattle, Washington 98109

We use the yeast, *Saccharomyces cerevisiae*, to study how the cell cycle is controlled with a focus on the fidelity of genomic replication and transmission. This focus arises out of an interest in the contrasting instability of the genome in cancer cells in comparison to normal human somatic cells. Cancer hasgiven us a basic insight into an important evolutionary principle. Namely, that when an opportunity presents itself for an organism to invade a new niche, and the transition requires numerous genetic alterations, the selection for this demanding transition will favor organisms that are genetically unstable. Every population of cells or organisms will contain rare individuals that are genetically unstable by virtue of mutations in genes that normally function to maintain genomic integrity. By studying how the yeast cell controls the fidelity of the genome we hope to further our understanding of those cellular functions that work coordinately to ensure genomic integrity.

Three independent macromolecular processes are necessary for genomic replication and segregation: Assembly and segregation of the spindle poles, DNA replication, and assembly and dissasembly of the mitotic spindle. These processes are fundamentally independent of one another since in certain embryonic systems of in certain mutants we find that one of the three can be inhibited while the other two continue to occur. However, in normal somatic cells or in yeast cells these three processes are coordinated. That is, if one is directly inhibited by the agency of a specific inhibitor or a mutant protein, the other two also fail to occur. What is coordinating these three processes in normal cells?

Our first insight into the answer to this question came upon the discovery of the cyclin dependent kinases. The current paradigm is that a series of CDKs are activated and inactivated during the cell cycle to constitute the CDK cycle [Murray, and Hunt, 1993]. Each CDK in turn activates or inhibits a large number of proteins to assure the execution of the appropriate cell cycle event at the appropriate time. This paradigm may supply a sufficient solution to the ordering of cell cycle events in an unperturbed cell. However it is not a sufficient explanation when the cell is perturbed, either by extrinsic forces such as DNA damaging agents, or by intrinsic events, such as stochastic errors in DNA replication.

Genomic Instability and Immortality in Cancer
edited by Mihich and Hartwell, Plenum Press, New York, 1997

If the CDK paradigm were the whole story, then the only defects that could stop all aspects of the cell cycle would be defects that directly inhibited the CDKs of their transitions. We now know the biochemical defect in many cell cycle mutants (temperature-sensitive mutants that arrest all three processes of the cell cycle). As expected from the CDK paradigm, many of these mutations inactivate the CDKs themselves or affect the transitions in CDKs [e.g. mutations that prevent cyclin degradation]. However, many cell cycle mutants are defective in events that are peripheral to the CDK cycle, that is are in the biochemical components of events that are being activated or inhibited by CDKs. Examples include tubulin mutants (Neff, Thomas, Grisafi, and Botstein, 1982) and DNA polymerase mutants (Johnson, Snyder, Chang, Davis and Campbell, 1985). How can we explain the fact that mutations in the machinery for spindle formation or DNA replication prevent the other two processes of the cell cycle from occurring? The logical deduction is that defects in these "peripheral" events must signal back to CDKs and influence their ability to undergo subsequent transitions.

These facts led to the hypothesis that feedback controls, surveillance mechanisms, or checkpoints exist to detect defects in cell cycle events and when such defects are detected, to provide a signal that inhibits progression of the cell cycle (Hartwell, and Weinert, 1989).

One agent that was known to induce a cell cycle arrest at G2 in mammalian cells and yeast cells was exposure of cells to X-rays, presumably as a result of introducing breaks into DNA. Ted Weinert examined radiation sensitive mutants to see if any were defective in cell cycle arrest after exposure to X-rays and found that *rad9* mutants were defective (Weinert, and Hartwell, 1988). Other mutants that were more sensitive to X-rays than *rad9* mutants [e.g. *rad52*] were proficient in cell cycle arrest after X-irradiation. This result suggested that *RAD9* was a component of the checkpoint that monitored DNA damage and sent an inhibitory signal to the mitotic apparatus if breaks in DNA were detected.

The phenotype of the rad9 mutant accounted for the arrest of the cell cycle in response to extrinsically induced DNA damage. Was this gene also responsible for arrest of the cell cycle in response to intrinsic damage? A simple approach to answering this question was to construct *cdc rad9* double mutants. If the mutant failed to undergo cell cycle arrest at the restrictive temperature then the *RAD9* gene was also responsible for monitoring the types of intrinsic damage created by the *cdc* mutation in question. This analysis demonstrated that the *RAD9* gene was involved in arrest of the cell cycle for any defect that arrested the cell late in S phase [e.g. DNA ligase] but not for other defects, including those that arrested cells at the beginning of S phase such as hydroxyurea (Weinert and Hartwell, 1993).

One of the mutants that arrested in G2 in response to a shift to the restrictive temperature was *cdc13*. This mutant was further mutagenized and used in a screen to detect other genes that participated in this cell cycle checkpoint. A total of six genes were discovered: *RAD9, 17, 24, 53, MEC1* and *MEC3* (Weinert, Kiser and Hartwell, 1994). The pathway defined by these genes is called the DNA damage checkpoint.

Although some of these genes are essential for cell viability [MEC1 and RAD53] the others could be deleted with little or no effect on growth rate. This result means that the checkpoint is not necessary for a normal cell cycle. Thus, in an unperturbed cell cycle, mitosis is not coordinated with the completion of DNA replication through the DNA damage checkpoint. Perhaps the intrinsic timing of the CDK cycle is sufficient for such coordination.

However, even though a population of cells deleted for the *RAD9* gene grows at a rate indistinguishable from wild type, there is a consequence of the *rad9* deletion. The frequency of chromosome loss in the population is elevated about 20 fold (Hartwell and Weinert,

1989). We interpret this to mean that intrinsic, stochastic damage occurs in rare cells in the population and when such damage occurs, the DNA damage checkpoint is necessary for the damaged cell to successfully delay cell cycle progress and repair the damage.

What is the DNA structure that the DNA damage pathway recognizes as damage? The *cdc13* mutant has provided a clue. Mutants defective in DNA replication stimulate mitotic recombination at the restrictive temperature, presumably because the recombinational repair pathway is activated. Most mutants induce DNA recombination all along the chromosome, but *cdc13* mutants induce recombination specifically near the ends (Garvik, Carson and Hartwell, 1995). This observation provoked us to look for unusual DNA structures near the ends and we found that one of the two DNA strands was being extensively degraded in *cdc13* mutants at the restrictive temperature. The polarity of strand degradation was the same as that observed during double strand break repair [Sugawara and Haber, 1992]. These results suggested that *cdc13* mutants might be defective in masking telomeres, so that in *cdc13* mutants the cell acts on telomeres as if they were double strand breaks. Moreover, since *cdc13* mutants generate a signal that is more intense than that of other known *cdc* mutants, this result suggests that single stranded DNA might be the structure that generates the signal. Work with Xenopus extracts had previously demonstrated that single stranded DNA generated the strongest signal for the DNA damage checkpoint in that system [Kornbluth, Smythe, and Newport, 1992]. Recent work has shown that some of the checkpoint genes participate in the generation of single stranded DNA in the *cdc13* mutant at the restrictive temperature suggesting that some of these mutants are involved in processing broken DNA to generate the single stranded DNA signal [Lydall, and Weinert, 1995].

Yeast cells undergo a relatively brief delay at the G1/S boundary in response to UV damage and this delay is dependent upon many of the DNA damage checkpoint genes [Siede, Friedberg, and Friedberg, 1993]. This result demonstrated that the DNA damage checkpoint was not restricted to acting on the G2/M transition.

Recently, we have discovered that the DNA damage checkpoint genes also control the rate of DNA replication over DNA that has received alkylation damage (Paulovich, and Hartwell, 1995). Wild type cells exposed to methylmethanesulfonate (MMS) exhibit a very slow S phase and, at appropriate concentrations of MMS, retain their viability. In contrast, *mec1* and *rad53* mutants undergo a rapid S phase in the presence of MMS damage and die. These results suggest that the DNA damage checkpoint is responsible for recognizing alkylation damage and reprogramming DNA replication so that it proceeds in a different way over damaged DNA than over undamaged DNA. Interestingly, the other DNA damage checkpoint genes, *RAD9*, *RAD17*, *RAD24*, and *MEC3* also play a role in DNA replication over alkylated DNA although a more subtle role than the other two genes. In these latter mutants, cells replicate about half their genome rapidly and about half slowly. They also die. This result is reminiscent of the fact that these same four mutants do not respond to blocks early in DNA replication (like *mec1* and *rad53* mutants) but do respond to blocks late in DNA replication. The nature of these distinctions is completely mysterious.

One attractive hypothesis for how DNA replication is altered during exposure to alkylation damage has to do with sister chromatid exchange (SCE). SCE is induced in mammalian cells in response to bulky lesions in DNA and occurs only during replication over damaged DNA. We have shown that *S. cerevisiae* also has replication dependent SCE (Kadyk and Hartwell, 1993). One hypothesis would be that SCE reflects the ability of the replication fork to switch templates from a damaged template to the newly synthesized sister strand of the same polarity, thereby avoiding replication over damaged bases. Such strand switching might considerably slow replication. The fact that DNA damage checkpoint mutants lack the ability to replicate damaged DNA slowly could be interpreted to mean that they are responsible for altering the replication complex in such a way as to permit strand switching.

Many signal transduction systems adapt. That is they change their sensitivity in the face of continuous signal so that they no longer elicit the same behavioral response as they do in the face of new signal. The DNA damage checkpoint also adapts. This fact was first demonstrated by Lisa Sandel and Virginia Zakian [Sandell, and Zakian, 1993]. We have used the system devised by these workers to screen for mutations that block adaptation. We find such mutations. The phenotype of these mutants is that the cells remain permanently arrested in response to an unrepaired double strand break whereas wild type cells go on to divide and transmit the broken chromosome. The adaptation process can be thought of as the "window" to damage transmission. Cells that cannot adapt do not transmit broken chromosomes to progeny whereas wild type cells do. The adaptation defective yeast mutants are analogous to mammalian cells that undergo apoptosis in response to DNA damage.

One of the important and amazing discoveries that came out of the CDK field has been the high degree of functional conservation between CDKs among all organisms. Is the DNA damage checkpoint conserved between human cells and yeast? Certainly the functional responses are conserved. Mammalian cells show response to DNA damage at the G1/S, G2/M and during S phase similar to those of yeast. Moreover, one gene known to be involved in all three of these responses, the gene defective in the cancer prone syndrome, ataxia telangiectasia (ATM) is structurally homologus to the yeast gene, MEC1 that also plays a role in these three processes (Keith, and Schreiber, 1995). These proteins are thought to encode DNA dependent protein kinases. Whether the other genes are also conserved and whether the adaptation pathway bears an evolutionary relationship to the apoptotic pathway remains to be seen.

REFERENCES

Garvik, B., Carson, M. and Hartwell, L. 1995. Single-stranded DNA Arising at Telomeres in *cdc13* Mutants May Constitute a Specific Signal for the *RAD9* Checkpoint. Molec. & Cell Biol. *15*:6128–38.

Hartwell, L. and Weinert, T. 1989. Checkpoints: Controls that ensure the order of cell cycle events. Science *246*:629–634.

Johnson, L. M., Snyder, M., Chang, L. M., Davis, R. W., Campbell, J. L. 1985. Isolation of the Gene Encoding Yeast DNA Polymerase I. Cell *43*:369–377.

Kadyk, L. C. and Hartwell, L. H. 1993. Replication-Dependent Sister Chromatid Recombination in *rad1* Mutants of *Saccharomyces cerevisiae*. Genetics *133*:469–487.

Keith, C. T., and Schreiber, S. L. 1995. PIK-Related Kinases: DNA Repair, Recombination, and Cell Cycle Checkpoints. Science *270*:50–51.

Kornbluth, S., Smythe, C., and Newport, J. W. 1992. *In vitro* Cell Cycle Arrest Induced by Using Artificial DNA Templates. Molec. Cell Biol *12*:3216–3223.

Lydall, D., and Weinert, T. 1995. Yeast Checkpoint Genes in DNA Damage Processing: Implications for Repair and Arrest. Science. *270*:1488–1491.

Murray, A., and Hunt, T. 1993. "The Cell Cycle, An Introduction." W. H. Freeman and Company, New York.

Neff, N. F., Thomas, J. H., Grisafi, P. and Botstein, D. 1982. Isolation of the Beta-Tubulin Gene from Yeast and Demonstration of its Essential Function *in vivo*. Cell *30*:211–219.

Paulovich, A. G. and Hartwell, L. H. 1995. A Checkpoint Regulating the Rate of Progression Through S Phase in Response to DNA Damage. Cell *82*:841–7.

Sandell, L., and Zakian, V. A. 1993. Loss of a Yeast Telomere: Arrest, Recovery and Chromosome Loss. Cell *75*:729–739.

Siede, W., Friedberg, A. S., and Friedberg, E. C. 1993. RAD9 Dependent G1 Arrest Defines a Second Checkpoint for Damaged DNA in the Cell Cycle of *Saccharomyces cerevisiae*. PNAS *90*:7985–7989.

Sugawara, N. and Haber, J. E. 1992. Characterization of Double-Strand Break-Induced Recombinatoin: Homology Requirements and Single-Stranded DNA Formation. Molec. Cell Biol. *12*:563–575.

Weinert, T. A. and Hartwell, L. H. 1988. The *RAD9* Gene Controls and Cell Cycle Response to DNA Damage in *Saccharomyces cerevisiae*. Science *241*:317–322.

Weinert, T. A. and Hartwell, L. H. 1993. Cell Cycle Arrest of *cdc* Mutants and Specificity of the RAD9 Checkpoint. Genetics *134*:63–80.

Weinert, T. A., Kiser, G. L. and Hartwell, L. H. 1994. Mitotic Checkpoint Genes in Budding Yeast and the Dependent of Mitosis on DNA Replication and Repair. Genes and Development *8*:652–665.

DISCUSSION

Bernards: I am sure there are going to be a lot of questions. The question that I think will be inevitable is, how do you know what bad1 looks like?

Hartwell: This is actually a very difficult gene to clone. We have cloned two genes which in single copy will complement the bad1 defect. One is an essential G2 phosphatase, one is an essential G2 kinase. But we are not sure that either of those is the bad1 gene and so that, at present, is still undefined.

Wahl: When you do the MMS experiments, is that continuous exposure or is it just one hit of MMS?

Hartwell: That is continuous exposure. We find we get the clearest results by just leaving it in because it can be processed back out.

Wahl: So the lesions are processed very rapidly. Is it not possible to use a pulse because the number of lesions that would be introduced would be insufficient to slow down the S-phase to an extent that could be measured accurately?

Hartwell: It is not as clean if you do it in a pulse. We can do pulse experiments like the UV experiment because you can take out the excision repair and then you see comparable phenomenon due to unexcised UV damage. But with the MMS, because there are so many lesions and because we do not know which are the important ones yet, we just leave it in.

Hoeijmakers: I am very intrigued by this adaptation process in yeast cell. There must be a time clock which is ticking and at a certain moment decides well, let us continue anyway. Do you know whether such a system exists in mammalian cells? Of course, in mammalian cells, there is also another potential outcome: to take the apoptotic pathway? Is there anything known about the parallel genes in mammalian cells?

Hartwell: No, I do not think there are any comparable mutants in mammalian cells that are defective for adaptation, but it is quite clear from the thing that I described that most of these checkpoints will show some kind of adaptation, so like the spindle poison experiment. In yeast Andrew Murray has just reported that they have a gene that affects the spindle arrest checkpoint in yeast, that also blocks the adaptation of that. So the answer is no, we do not know about mammalian cells yet, but I strongly suspect that they are going to be related to the apoptotic decision because I think the system has evolved differently in different tissues in mammalian cells and in yeast to different ends. And the length of the delay depends upon the kind of damage that the cell is seeing. For example, the delay is very short in a G1 to S experiment. Even if you cannot remove the UV damage it is short, the cells go ahead and go into S. Whereas for a single double strand break, the cell tries very hard to repair it, it sits there for about ten hours. So I think it is a sophisticated response, decision making process.

Hoeijmakers: Another phenomenon which was very interesting is the difference in the effects of the RAD9 and RAD17, RAD24 and MEC1 response versus the MEC3 response in S-phase replication of a damaged template, where you see in one case all replication is inhibited, in the other case, it is about fifty percent. You mentioned that there are several models to explain this observation. One option that is appealing to me is the following: A lesion in a leading strand may have other effects or require other factors to bypass than a lesion in a lagging strand. So, do you have any evidence that these four genes which cause a partial deficiency in replication, are related to problems on one or the other strand? Or is it related to active versus inactive DNA.

Hartwell: Yes, we do not know. I think that it is a nice model, it could be transcribed versus non-transcribed. We have also thought in terms of a model of a replication dependent repair which is analogous to a transcription coupled repair system. You could use that to explain this. You could also imagine that the subtle genes, as we call them, are responding to only a certain class of damage, whereas the MEC1 is responding to all kinds of damage. I think there are various categories, you can imagine, some of which are testable and we just do not have any data on it yet.

Hoeijmakers: Finally, you mentioned the homology of the ATM gene with yeast, TEL1 and other genes within the same family. You did not talk about RAD9, RAD17 and RAD24 which are very interesting genes from that perspective, as well. Do you know anything about functional or structural homologues in mammalian cells?

Hartwell: I do not. To my knowledge, there is no RAD9 homologue known as yet.

Hoeijmakers: Do you know whether they are strongly conserved in various yeasts?

Hartwell: There are homologues but with different numbers; this whole system has been quite extensively worked in. There certainly are some genes for the three to five prime exonuclease which is RAD17 in *cerevisiae.*

Hoeijmakers: That is already quite an evolutionary distance that they have overcome. There must be mammalian homologues.

Hartwell: They are undoubtedly there, yes.

Wahl: Jan, I just wanted to comment that we have looked carefully at the response to double strand breaks in human fibroblasts: I want to make the point that maybe human fibroblasts do not react like every other mammalian cell. So let us just restrict it to that. In response to a small number of breaks in G1, the cells will arrest for about thirty-six hours and then there are two classes of cells that come out of that. Sixty percent will remain arrested and senescent the other forty percent will go into S-phase. The forty percent that go in, go in with breaks. It looks like adaptation. The reason we know that they go in with breaks is that they arrest in G2 and then if they happen to make it through that G2 they will get into the next G1 and about another sixty or seventy percent will be eliminated, and so on. So it looks like there is an adaptive response, p53 seems to be part of that, because if you eliminate p53 a hundred percent of the cells now go in without delay. But they will then delay in G2 and some of those will be eliminated. So there may actually be two adaptive phases and G2 would be the phase that would be primarily responsible for repair of double strand breaks.

Nasmyth: Do you know whether the adaptation involves the cells getting bigger? That is, if you artificially were to make cells bigger, which would require protein synthesis, then could you reduce the adaption periods have you done any experiments along those lines?

Hartwell: We have thought about that; there are a lot of experiments to do here particularly related to specificity. For example, does this effect the spindle checkpoint, and it would be the same experiment because one could hold cells to allow them to get bigger. We have not been able to do those experiments yet, because we are not yet sure we have the cloned gene and we cannot do the constructs that we want to do. It is very difficult to move this gene around and keep the constructs.

Nasmyth: But you could look up the adaptation period, say you take three mutations which triples the size of the cell, nearly, and just ask is the adaptation period shorter?

Hartwell: That we could do and have not done. To reply to a sort of related question that you asked me last week; is this really a specific phenomenon that is set up to deal with this specific issue or it is just that when a cell gets big enough it eventually sort of leaks through for some other non-specific reason. And certainly specificity with the checkpoint is important. Whether the bad1 mutant, for example, just after sitting there for so long dies, for some non-specific reason we cannot argue against at the moment. What you want to do is to allow the cells to sit there for ten hours when the wild type would have recovered and the mutant does not. And then turn on the double strand break repair and remove the signal and ask if the cell can then get out of it. Because then you would know that is still O.K. So that is the right experiment and we have not done it yet.

Nasmyth: And ultimately, if you saw the mRNA or in some way the gene product induced in response to the damage, then that would obviously give a clue. I had another question regarding the connection of the ATM and MEC1. I mean, I was quite surprised by Sheila's claim that ATM is actually a cytoplasmic protein and yet for all this sort of checkpoint control, the damage originates in the nucleus and the cell cycle machinery that has got to impact on it is in the nucleus. Do you know if in yeast these protein kinases are unclear or not?

Hartwell: I do not know either. I do not think there has been much done by way of biochemistry. At the ATM meeting, people were very confused about the antibody localization results with ATM. I am not sure that story is clear at all.

Stark: I wanted to comment a little bit more, maybe add a little bit of information about the status of mammalian cell mutants defective in DNA dependent protein kinase, ATM or poly ADP ribose polymerase. Some of the work is from our lab and some of the work from other labs. We have looked at cells deficient in poly-ADP ribose polymerase and I guess, everyone knows that that knockout mouse has been made and as far as I am aware, shows no obvious defect, perfectly normal happy mouse.

Hartwell: Also in response to damage?

Stark: Yes. And we have looked at ADPRT-minus cells in a couple of ways, normal cell cycle arrest in response to damage and ability, or not, to undergo gene amplification and induction of p53. And although there is some quantitative defect in the amount of p53 that is induced, ADPRT-minus cells still induce p53 in response to damage and, as far as all the other assays go, in our hands, they are indistinguishable from the ADPRT-plus cells. So, if they have any defect at all, it is only a very partial defect. I think similarly for the other two lesions, I know David Lane has done some work with the ATM cells, looking at the p53 response. Michael Kastan had done some previously as well. There again I think there is a partial defect to delayed response, perhaps quantitatively not as great, but p53 is induced and the cells have a more or less normal phenotype. I think, somebody here who may know better can correct me, but, I have heard, kind of indirectly from Steve Jackson, who has looked at this issue in cells that are deficient in DNA dependent protein kinase and again it seems to be pretty much the same story. If anybody knows better please add information, but, to the best of my knowledge, none of these mutations singly have a very important phenotype either with respect to the induction of p53 or with respect to the normal response to DNA damage. It leads me to think that either none of these are really involved at all or perhaps, more likely, all of them are involved partially and in order to see a full defect we are going to have to create double and perhaps triple mutants.

Hartwell: The ATM phenotype from human cells is quite dramatic. The claim there from Mike Kaston is that there is a defect in p53 induction but with respect to this control of the cell cycle, which is ultimately the important thing, they are quite defective in all three of these processes, G1 S and G2.

Gatti: I am a little bit confused. If you irradiate, for instance, mammalian cells, or any kind of cells, after one hour perhaps you can find many chromosome breaks in mitosis. So, whereas from what has been said here it appears that yeast cell arrest for a long time and also I think, mammalian fibroblast do arrest for a long time. But how can you reconcile these findings with the fact that with x-ray radiation you find chromosome breaks within two hours, even after one hour?

Hartwell: Are these cell lines?

Gatti: Cell lines, lymphocytes, Drosophila cells, plants, any kinds of cells. If you radiate them with x-rays you find breaks after a short time.

Hartwell: Yes, I think the question is, quantitatively and whether the cells are normal or not. But for normal cells, at least some of the data that I can remember is, that if you look by premature chromosome condensation, PCC, that there are many many breaks in the interphase cell, most of which get repaired, and a few of which will appear in mitosis. So how effective this system is, is what we really need to ask and also how normal are the cells we are looking at. There are probably people here who are much better to comment on that than me.

Gatti: People who work on lymphocytes from normal individuals, even if they radiate them with very small doses, they find aberration after one hour or two. There is a certain mitotic delay but it is very short. So we should probably assume that those breaks are repaired in some way, so that the broken end of the chromosome actually are not the real double strand breaks but they are different from the kind of breaks which are induced with this long delay.

Hartwell: Do we have any comments from the mammalian people here?

Wahl: Radiobiologists have defined many types of lesions induced by radiation. There is one type called the "multiple locally damaged site" by John Ward and this is supposed to be some type of lesion that is very difficult, if not impossible, to repair. So, for any dose of radiation you will get lesions that can be repaired, some can be repaired very rapidly, some can be repaired a little bit more slowly and others cannot be repaired at all. Those are infrequent but from our microinjection experiments we would say that if the cell has one of that type, it will be removed from the dividing population. For the other types, liquid holding experiments indicate that you can repair many types of lesions, but you do not repair all types at the same rate. So I do not see any inconsistency with the fact that irradiation generates multiple types of DNA damage, and so might breakage of dicentric chromosomes, or that induced by drugs, etc.

Gatti: What I am saying, if you irradiate cells and you fix them right away, you find many breaks, many chromatid breaks, isochromatid breaks, exchanges all sorts of chromosome aberrations.

Wahl: You are saying that there just has not been enough time to go through the cycle to generate that kind of lesion? Is that what you are saying?

Gatti: What I am saying is that there is not such a long delay. So that means that this cell has not been delayed for a long time unless the kind of breaks we see are repaired in such a way that they do not trigger this long delay.

Livingston: Just two questions: One, in a simple bad1 mutant, are there any changes in cell cycle intervals?

Hartwell: Well, the kind of experiments that we have done are these where we look for the effects on chromosome transmission and in a bad1 mutant that is RAD9 and RAD52+ I showed you the data for that and what happens is that it looks pretty much like wild type except that wild type will fail to repair some chromosomes and some percentage, I forget, somewhere between ten and twenty percent of the cells will actually transmit a broken chromosome in the wild type. In the bad1 mutant that category is gone, it is filtered out and they appear in the dead category. But the same fraction of cells repair and go on in the bad1 and the wild type.

Livingston: If one does not irradiate and tests cell cycle intervals, what happens? Do the cells sporulate?

Hartwell: Yes we find no differences in the cell behavior without radiation.

11

A EUKARYOTIC CELL CYCLE[*]

Kim Nasmyth

Research Institute of Molecular Pathology (IMP)
Dr. Bohr-Gasse 7
A-1030 Vienna, Austria

1. PROLOGUE

Cell proliferation depends on the fission (or budding) of cells that have duplicated all of their constituents and divided them more or less equally between daughter cells. Most constituents are synthesised continuously throughout the cell cycle and their segregation does not require great care either because they are sufficiently numerous or because they can be synthesised de novo using directions supplied by the genome. Chromosomes and possibly also microtubule organizing centres are exceptions to this rule. Each gene within the genome must be replicated using itself as a template once and only once per cycle and each sister chromatid must be segregated away from each other to opposite poles of the cell.[1] Because most of the instructions for building a cell reside within chromosomes, their duplication and segregation must occur with a fidelity which far exceeds that of other biosynthetic processes. Nuclear division must be avoided when sister chromatids have not properly aligned on the mitotic spindle, when one chromatid has been damaged and needs repairing using its undamaged sister, or when DNA replication has not yet been completed. Cells therefore possess surveillance mechanisms[2] that detect these sorts of accidents and induce inhibitors of chromosome alignment on the mitotic spindle (metaphase) and/or disjunction of sister chromatids (anaphase).

Two other salient features characterize the chromosome cycle of eukaryotic cells. Cells must duplicate and segregate their chromosomes at the same rate at which they duplicate all other cell constituents (i.e. mass doubling) if they are to maintain a constant ratio of cytoplasm to nucleus. They achieve this goal by restricting key transitions of the chromosome cycle, either G1/S or G2/M, to cells that have attained a critical mass.[3,4] Finally, cells must not initiate re-duplication of their chromosomes before they have segregated the previous set of sister chromatids at anaphase. This last control is important if cells are to maintain the constant ploidy needed for sexual life cycles involving meiosis; it may also be essential for avoiding ambiguities when deciding which pairs of chromatids are "sisters" that need to be disjoined at anaphase.[1]

---

[*] A related article on this subject was published in *Trends in Genetics* in October 1996.

Genomic Instability and Immortality in Cancer
edited by Mihich and Hartwell, Plenum Press, New York, 1997

It may now seem obvious that the mechanisms regulating cell division would be found to be a highly conserved feature of eukaryotic cells. This was less clear twenty years ago when the pioneering genetic studies of the cell cycle were initiated.[5] This article presents one view as to what lies at the heart of the budding yeast cell cycle. It is written on the premise that most of the key players such as cyclin dependent kinases, the anaphase promoting complex, ORC, and Mcm proteins were performing similar functions in the common ancestor of yeast and man. Ideas about the budding yeast cell cycle might therefore have universal significance.

2. CDKS, THEIR SUBSTRATES, AND THE APC

Cell cycle studies over the last 20 years, whether genetic ones on yeast or biochemical ones on frogs, have mainly been concerned with identifying regulatory molecules whose changes in activity are responsible for driving key chromosome cycle transitions. The "Rome" to which all avenues have until recently led, as it were, is a family of protein kinases whose activity is dependent on regulatory subunits called cyclins, known as Cdks.[6] S. cerevisiae posesses at least four Cdks: Cdc28,[7] Pho85,[8] kin28,[9] and Srb10,[10] of which only Cdc28 has a clear role in regulating the chromosome cycle; the others are all involved in regulating transcription. Cdc28 (or Cdk1 as I shall refer to it) performs the tasks performed by cdc2 in S. pombe and those which are shared between Cdk4, Cdk2, and Cdk1(cdc2) in mammalian cells. Its various functions are performed by varieties of the kinase that differ mainly if not solely in their cyclin subunit. Cln3/Cdk1 activates transcription in late G1,[11] Cln1 and Cln2 turn off proteolysis of B-type cyclins, turn on proteolysis of a cyclin B/Cdk1 specific inhibitor, turn off the ability of haploid cells to respond to mating pheromones, and trigger the polarization of the cytoskeleton needed for bud formation,[11-13] Clb5 and 6 trigger DNA replication,[12] Clbs3 and 4 trigger the formation of mitotic spindles,[14] and Clbs 1 and 2 trigger nuclear division[15] and isometric bud growth during G2.[13] It is presumed but it has rarely been demonstrated that different cyclins determine either Cdk1's location or association with other proteins and/or its substrate specificity.

The notion that different Cdk sub-types catalyse different cell cycle transitions is one of the prevailing paradigms of the cell cycle field.[16] Though largely correct, it is not true that cell cycle position is defined simply by the state of activity of different Cdk1 forms. Activation in G1 cells of the supposedly "mitotic" Clb2/Cdk1 kinase (instead of the S phase promoting Clb5/Cdk1 kinase) triggers yeast cells to enter S phase;[17] that is, Clb2/Cdk1 induces S phase in G1 cells but nuclear division in G2 cells! Thus, Cdks do not also instruct cells how to respond to kinase activation and the cell cycle programme is not simply a prescription for activating and inhibiting different Cdk sub-types in a given order. As previously hinted by cell fusion experiments,[18] how cells respond to the activation of particular Cdks depends on the state or presence/absence of their substrates. How these are regulated is therefore an equally important aspect of chromosome cycle regulation.

The induction of S phase by the unscheduled activation of a supposedly mitotic Cdk implies that there is considerable overlap in the activities of different cyclins in yeast. S phase is normally triggered by activation of Clb5 or Clb6/Cdk1 (because these are the first Clb/Cdk1 subtypes to appear), but in their absence, S phase is triggered by the later appearance of "mitotic" Clb/Cdk1 subtypes (Clbs1–4).[12] Likewise, Clb5 can assume the functions of Cln1 and Cln2 when expressed ectopically. Nevertheless, specificity in the activity of different cyclins does exist; Clb5/Cdk1, for example, is active during G2 but it cannot trigger nuclear division in the absence of Clbs1 and 2.[19]

It has also become clear only recently that certain key transitions might be triggered by changes in the acitivity of factors that are not Cdks. It was thought that the metaphase to anaphase transition, an irevocable step in the chromosome cycle, was triggered by the destruction of cyclin B/Cdks.[20] This seems not to be the case. Expression of high levels of non degradable B-type cyclins fails to block anaphase in yeast,[21] frogs,[22] and flies.[23] Nevertheless, many of the genes specifically involved in cyclin proteolysis are necessary also for the onset of anaphase.[24] These genes encode components of a 20S particle called the Anaphase Promoting Complex, conserved between yeast and frogs, whose function is to catalyse the ligation of multiple ubiquitin molecules to cyclins and thereby target them for proteolysis by the 26S proteosome.[25–27] The implication is that the separation of sister chromatids at anaphase depends on the ubiquitination by the APC (and consequent destruction) of proteins other than cyclins.[22,24] Whether changes in the activity of the APC actually triggers anaphase and what are its key substrates is as yet unknown.

3. FIRING ORIGINS ONCE PER CYCLE

Entrusting the capacity to direct the synthesis of enzymes to a genome that is replicated with high fidelity must have been one of the key developments in the evolution of life on this planet. Part of this package was a mechanism that ensured the equal replication of all genes. Many bacteria achieve this goal by the simple expedient of having their entire genome encoded by a single chromosome that is replicated from a single origin. Eukaryotic cells distribute their genome between multiple chromosomes, each of which is replicated from multiple origins, some of which fire early during S phase and some of which fire late, but none of which fire more than once between successive rounds of sister chromatid disjunction. Sequences sufficient to confer the bidirectional replication of budding yeast chromosomes consist of an essential core sequence bound by the six subunit Origin Recognition Complex[28] (ORC) flanked by less well specified sequences containing binding sites for transcription factors such as Abf1.[29] Firing is triggered by the activation in late G1 of cyclin B/Cdk1 kinases, normally Clb5 and Clb6.[12]

3.1. S Phase Initiation Involves Two Steps

All six Clb/Cdk1 kinases capable of triggering S phase in a G1 cell remain active until the onset of anaphase and yet origins fire only once during this period. The lack of firing is not due to a lack of ORC binding, for this is bound to origins throughout the cell cycle.[30] Changes in the activity or location of two other sets of initiation factors may be responsible. The sequences adjascent to those bound by ORC are exposed (to exogenously added nuclease) during G2 and M phases but are protected during G1.[31] The formation of this G1-specific footprint as cells exit from M phase depends on the de novo synthesis of an unstable protein encoded by the *CDC6* gene whose sequence is related to the Orc1 subunit.[32] Cdc6 is necessary for maintaining the footprint during G1 and for the subsequent initiation of DNA replication. This suggests that the formation during G1 of a "pre-replicative complex" containing ORC and Cdc6 is a pre-condition for the firing of origins upon the subsequent activation of Clb/Cdk1 kinases.

Another set of proteins important for origin firing are those encoded by Cdc46, Cdc47, Mcm2, Mcm3, and Mcm4.[33,34] These encode related proteins all of which are essential for initiation and have homologous counterparts in animal cells. Many of the Cdc46/Mcm proteins change their location during the cell cycle;[35] the yeast proteins accumulate within the nucleus during G1 but in the cytoplasm during G2, whereas animal counterparts are found within nuclei at all stages of the cell cycle but are only bound to chromatin during G1.[36,37] The regulation of Mcm nuclear accumulation (which might be a reflection of other changes in the properties of these proteins) resembles that of the transcription factor Swi5, whose accumulation in the cytoplasm during G2 and M phases is due to Clb/Cdk1 activity.[38] Destruction of Clb kinases during anaphase might therefore trigger accumulation within nuclei not only of Swi5 but also of Mcm proteins; it might also alter their ability to bind to chromatin.

3.2. A Dual Function for S Phase Promoting Cdks

Clb/Cdk1 kinase inactivation during anaphase might also trigger the appearance of pre-RCs at origins, which occurs simultaneously with the accumulation of Mcm proteins in nuclei. Inactivation of Clb/Cdk1 kinases in nocodazole blocked cells (due to ectopic expression of the Clb/Cdk1-specific inhibitor p40Sic1) induces the formation of pre-RCs and permits re-replication in the absence of any chromosome segregation when Clb/Cdk1 kinases are re-activated.[39] Thus, the very same Clb/Cdk1 kinases that trigger replication in G1 cells are responsible for preventing formation of the pre-RCs needed for another round. Furthermore, this seems true not only for Clb5, whose normal role is to promote S phase, but also for Clb2, whose normal role is to promote nuclear division.[39] The dual replication functions of most if not all Clb/Cdk1 kinases is another instance where their activities seem to overlap. The initiation of DNA replication can therefore be seen as a two step process: first, formation of pre-RCs (accompanied perhaps by the activation of Mcm proteins) and second, origin firing triggered by Clb/Cdk1 activation.[39] Because these two steps cannot occur simultaneously, origin firing depends on a cycle of Clb/Cdk1 activation: a period of low kinase activity that permits pre-RCs to form followed by a period of high kinase that permits origins that have previously formed pre-RCs to fire.

Such a mechanism provides possibly for the first time a coherent hypothesis for how cells prevent origins from firing more than once during the cell cycle. Activation of Clb/Cdk1 kinases in late G1 simultaneously triggers firing of origins that had formed pre-RCs during the preceeding period of low kinase activity while at the same time terminates formation of further pre-RCs.

To explain fully the equal replication of all genomic sequences, one must also propose that the act of initiation destroys pre-RCs that have just given birth to replication forks and that passage of a replication fork destroys pre-RCs that had not yet been fired. The hypothesis also suggests that whether an origin fires early or late is due not to differences in the timing of pre-RC formation but due to differences in the rate at which Clb/Cdk1 kinases trigger initiaton from pre-RCs.

Because Cdc6 is an unstable protein[40] needed for the formation of pre-RCs, the timing of its synthesis relative to Clb/Cdk1 activation should be crucial. Individual Clb/Cdk1 kinases have different profiles of activity during the cell cycle, largely due to differences in gene transcription. Clbs 5 and 6 appear in late G1, Clbs3 an 4 during S phase, and Clbs1 and 2 during G2. All kinases remain active until the onset of anaphase when increased proteolysis and the accumulation of p40Sic1 generate a state in which all six kinases are inactive. *CDC6* transcription occurs in two bursts, one as cells exit from mitosis and a second during late G1. Significantly, both bursts occur during the period of low Clb/Cdk1 kinase; that is, between exit from anaphase and destruction of p40Sic1 in late G1.

As predicted by this view of the replication cycle, pre-RCs cannot be formed and cells fail to replicate DNA when Cdc6 synthesis is delayed until after Clb/Cdk1 kinases have been activated. There exists, as it were, a "point of no return" after which Cdc6 synthesis fails to support an S phase in cells that had exited from anaphase in the absence of Cdc6 synthesis.[41] This point occurs around the time that cells should have initiated DNA replication; i.e around the time of activation Clb5/Cdk1 kinase.

The notion that a drop in the activity of Clb/Cdk1 kinases is normally necessary for the formation of pre-RCs at future replication origins provides an explanation for how re-replication might be linked to anaphase. Kinase inactivation (see below) is due largely to proteolysis of B-type cyclins, which occurs because of their ubiquitination mediated by the APC.[24,25] Sister chromatid separation also depends on the APC, though what proteins need to be degraded for this transition (anaphase blockers) is not known. The APC may be regulated in such a way that it cannot ubiquitinate B-type cyclins before it ubiquitinates the putative anaphase blockers. In the simplest situation, APC activation at the metaphase to anaphase transition might trigger simultaneously destruction of proteins blocking sister chromatid separation and that of B-type cyclins blocking the formation of pre-RCs.

4. REGULATION OF CLB/CDK1 KINASES

An oscillation in the activity of Clb/Cdk1 kinases is a key feature of the chromosome replication cycle outlined above. How do cells switch from low to high kinase states, maintain high kinase levels while chromosomes are aligned on the mitotic spindle and finally switch back to the low kinase state needed for the next round of DNA replication as cells undergo anaphase?

The switch between low and high Clb/Cdk1 kinase state is characterized by changes in the transcription of cyclin B genes, by reductions in the rate of cyclin proteolysis, and by disappearance due to proteolysis of the p40Sic1 Clb/Cdk1-specific inhibitor. High rates of Clb proteolysis, in particular that of Clbs1–4, and low rates of *CLB* gene transcription play an important part in maintaining low Clb levels in early G1 cells, but more important still is their high level of p40Sic1, which is sufficient to inhibit all six Clb/Cdk1 kinases even when cyclins are allowed to accumulate.[12] Cells can be nudged out of their low kinase state by changes in any of these processes, suggesting that all three are important and interlocking.[17,19]

The first event leading to Clb/Cdk1 activation is the sudden activation of *CLB5* and *CLB6* gene transcription when G1 cells reach a critical size. This event is part of a large programme of gene activation mediated by a pair of related transcription factors, called SBF and MBF.[42] Two other key genes turned on by SBF and MBF are *CLN1* and *CLN2*, which encode a pair of related cyclins[43] that have a key role in activating Clb/Cdk1 kinases. The accumulation of Cln1/ and Cln2/Cdk1 kinase activity in late G1 turns off APC activity,[11,17] which facilitates the accumulation of Clb proteins, and switches on proteolysis of p40Sic1, which releases Clb/Cdk1 kinase complexes from inhibition. A reciprocal switch in the rates of Clb and p40Sic1 proteolyis triggered by *CLN1/2* transcription is therefore the key step that triggers entry into a high Clb/Cdk1 kinase state.

Proteolysis of p40Sic1 depends on an ubiquitin conjugating enzyme encoded by *CDC34;* it also requires *CDC4* and *CDC53*.[12] Ubiquitination, which targets proteins for destruction by the proteosome, might be triggered by phosphorylation of key residues within p40Sic1 caused by the rise of Cln/Cdk1 kinases in late G1. Mutations that eliminate potential phosphorylation sites within p40Sic1 prevents its proteolyis in vivo, whereas phosphorylation of p40Sic1 by Cln/Cdk1 kinases promotes its ubiquitation by Cdc34p in vitro (T. Böhm and R. Deschaies personal communication). Thus, the switch that triggers p40Sic1 proteolysis in vivo might be driven by changes in the substrate (i.e. phosphorylation by Cln kinases) rather than by changes in the proteolytic machinery itself (i.e. Cdc4, 34, 53). Indeed, Cdc34 is also involved in the proteolysis of several proteins whose degradation is not apparently cell cycle regulated and which have little or no role in cell cycle control. How Cln1/ and Cln2/Cdk1 kinases simultaneously turn off the APC mediated ubiquitination of Clb proteins is not understood but it might be due to changes in the APC itself.

How cells switch from the high back to the low Clb/Cdk1 kinase state is less clear. There is a large collection of gene products (which I will call the Cdc15 group) required for this switch[21] but not needed for maintaining the low kinase state once cells have entered G1.[24] These include kinases like Cdc15, Cdc5 (Polo), and Dbf2,[44] a ras-like GTPase Tem1,[45] a nucleotide exchange factor Lte1, and a phosphatase Cdc14.[46] Mutants with defects in any one of these proteins initiate anaphase, but arrest with post-anaphase mitotic spindles and high Clb/Cdk1 activity (due to low p40Sic1 and high Clb cyclin levels), and do not re-replicate their genomes. They also fail to undergo cytokinesis. Elevating (artificially) the synthesis of p40Sic1 allows cells lacking Cdc15 or Dbf2 to proliferate, albeit poorly.[47] One suspects therefore that at least some of the Cdc15 group of genes are specifically concerned with switching cells from the high to the low kinase state. How they effect this transition and how they do so at the appropriate time (at or soon after separation of sister chromatids) is not understood.

In summary, then, the replication cycle in budding yeast is driven by sudden transitions between low and high Clb/Cdk1 kinase states. Both the rapidity or suddenness of these transitions and the persistence of the high kinase state once established are due to multiple mechanisms whereby Clb/Cdk1 kinases promote their own activity. This general principle, the crucial targets of Clb/Cdk1 kinases (e.g. ORC, Cdc6, and Mcm proteins), and the mechanisms by which Clb/Cdk1 kinases maintain their own activity until anaphase has been triggered may turn out to be highly conserved between eukaryotic lineages.

5. SIZE CONTROL

Budding yeast cells must grow to a critical size before they can initiate a new chromosome cycle.[48] Conversely, blocking the chromosome cycle at almost any stage has little or no immediate effect on cell mass accumulation. Thus, coordination between cell growth (mass accumulation) and the chromosome cycle is achieved by a dependence of chromosome replication on cell growth and not vice versa. In cells growing in rich media, daughter cells inherit at least some pre-RCs formed at the end of the previous mitosis but they remain in a low Clb/Cdk1 state and do not form buds until they have grown to a critical size.[12] The key event that triggers entry into the chromosome cycle and simultaneously initiates budding is activation of the Cln1/ and Cln2/Cdk1 kinases. The *CLN1* and *CLN2* genes are silent in newly born daughter cells but along with many other genes involved in DNA replication, are suddenly activated when cells reach a critical size. Two related transcription factors, SBF and MBF, are responsible for this late G1-specific programme of gene activation[49] and their activation at a given cell size depends on the Cln3/Cdk1 kinase.[11] Deletion of the *CLN3* gene delays the entire transcriptional programme for several hours, whereas its duplication advances it,[50,51] suggesting that small changes in Cln3 levels might be one of the factors that activate SBF/MBF. Cln3 is an unstable protein and is synthesised continuously in cells growing towards the critical size. Cln3's rate of synthesis and amount per cell reflects therefore the cell's total rate of protein synthesis, which is proportional to the cell's mass. How Cln3 activates gene expression only when cells achieve a critical mass remains, however, a mystery. SBF is already bound to the *CLN2* promoter in small unbudded cells, suggesting that Cln3/Cdk1 somehow facilitates activation of RNA polymerase by SBF molecules that have previously bound to the promoter.[52]

6. A RECIPROCATING ENGINE AT THE HEART OF THE EUKARYOTIC CELL CYCLE

A key requirement for the mechanism that drives chromosome duplication is that it should permit replication origins to fire once and once only once between succeeding rounds of sister chromatid segregation. The momentum of the device used by budding yeast seems to be maintained by fluctuations in the activity of Clb/Cdk1 kinases. It is a reciprocating device in which motion (i.e. rounds of replication) is driven by the alternate inactivation and re-activation of Clb/Cdk1 kinases. In this sense, it is deeply analogous to the reciprocating steam engine, whose steam corresponds to kinase activity and whose piston corresponds to origins. Steam, i.e. kinase, can only do work by driving the piston upwards (i.e. drive replication) if it (i.e. origins) has previously returned to the down state (i.e. contains pre-RCs). Passage from the up state (an origin that has initiated replication) depends on evacuation of the steam (i.e. inactivation of kinases) from the chamber containing the piston ((i.e. the cell). The essence of the reciprocating steam engine is that entry of steam into the chamber only performs work if the piston has been returned to the down state by previous evacation of steam from the chamber. Likewise Clb/Cdk1 kinases only drive replication when their prior inactivation has permitted the formation of pre-RCs. Just as the piston can only move once during a cycle of expansion and contraction, so too can origins only fire only once during a cycle of kinase activation and inactivation. The valves that control the entry and exit of steam from the chamber are an essential aspect of any steam engine and so are the molecules like Cln/Cdk1 kinases that switch cells between low and high Clb kinase states. As indeed are the mechanisms that ensure that they open and shut at the correct stage of the cycle. Elaborate mechanisms are needed to ensure that evacuation of steam from the chamber (i.e activation of the APC) is linked to movement of the piston to the upstate (i.e. completion of replication). If then, there exists a cell cycle engine,[53] these aspect of the cell cycle must surely be a key part of it. The metaphor of a reciprocating engine might even apply to the DNA replication cycle of eubacteria, in which the initiation signal (whose identity is still unclear) both triggers replication and sequestration of origins in membranes, where they are temporarily sheltered from the initiation signal and thereby avoid unscheduled re-replication.[54]

REFERENCES

1. Nasmyth, K. (1995) Phil. Trans. R. Soc. Lond. B 349, 271–281
2. Hartwell, L.H. and Weinert, T.A. (1989) *Science* 246, 629–634
3. Killander, D. and Zetterberg, A. (1965) *Exp Cell Res* 40, 12–20
4. Nurse, P. (1975) *Nature* 256, 547–551
5. Hartwell, L.H. (1993) in *The early days of yeast genetics* (Hall, M.N. and Linder, P., eds), pp. 307–314, Cold Spring Harbor Laboratory Press
6. Nurse, P. (1990) *Nature* 344, 503–8
7. Nasmyth, K. (1993) *Curr Opin Cell Biol* 5, 166–79
8. Cross-F. (1995) *Trends-Genet.* 11, 209–11

9. Valay-JG *et al.* (1995) *J-Mol-Biol.* 249, 535–544

10. Liao-SM *et al.* (1995) *Nature* 374, 193–6

11. Dirick, L., Bohm, T. and Nasmyth, K. (1995) *EMBO-J.* 14, 4803–4813

12. Schwob, E., Bohm, T., Mendenhall, M.D. and Nasmyth, K. (1994) *Cell* 79, 233–44

13. Lew, D.J. and Reed, S.I. (1993) *J Cell Biol* 120, 1305–20

14. Fitch, I. *et al.* (1992) *Mol Biol Cell* 3, 805–18

15. Surana, U. *et al.* (1991) *Cell* 65, 145–61

16. Hayles, J., Fisher, D., Woollard, A. and Nurse, P. (1994) *Cell* 78, 813–22

17. Amon, A., Irniger, S. and Nasmyth, K. (1994) *Cell* 77, 1037–50

18. Johnson, R.T. and Rao, P.N. (1971) *Biol. Rev.* 46, 97–155

19. Schwob, E. and Nasmyth, K. (1993) *Genes Dev* 7, 1160–75

20. Murray, A.W. and Kirschner, M.W. (1991) *Sci Am* 264, 56–63

21. Surana, U. *et al.* (1993) *Embo J* 12, 1969–78

22. Holloway, S.L., Glotzer, M., King, R.W. and Murray, A.W. (1993) *Cell* 73, 1393–402

23. Rimmington, G., Dalby, B. and Glover, D.M. (1994) *J. Cell Sci.* 107, 2729–2738

24. Irniger, S., Piatti, S., Michaelis, C. and Nasmyth, K. (1995) *Cell* 81, 269–277

25. King, R.W. *et al.* (1995) *Cell* 81, 279–288

26. Sudakin, V. *et al.* (1995) *Molec. Biol. Cell* 6, 185–198

27. Zachariae, W. and Nasmyth, K. (1996) *Mol. Biol. Cell* 7, 791–801

28. Bell, S.P. and Stillman, B. (1992) *Nature* 357, 128–34

29. Marahrens, Y. and Stillman, B. (1992) *Science* 255, 817–23

30. Diffley, J.F. and Cocker, J.H. (1992) *Nature* 357, 169–72

31. Diffley, J.F., Cocker, J.H., Dowell, S.J. and Rowley, A. (1994) *Cell* 78, 303–16

32. Cocker, J.H. *et al.* (1995) *Nature* 180–182

33. Hennessy, K.M., Lee, A., Chen, E. and Botstein, D. (1991) *Genes Dev* 5, 958–69

34. Tye, B.-K. (1994) *Trends in Cell Biol.* 4, 160–165

35. Hennessy, K.M., Clark, C.D. and Botstein, D. (1990) *Genes Dev* 4, 2252–63

36. Kubota, Y. *et al.* (1995) *Cell* Cell, 601–609

37. Madine, M.A. *et al.* (1995) *Current Biol.* 5, 1270–1279

38. Moll, T. *et al.* (1991) *Cell* 66, 743–58

39. Dahmann, C., Diffley, J. and Nasmyth, K. (1995) *Current Biol.* 5, 1257–69

40. Piatti, S., Lengauer, C. and Nasmyth, K. (1995) *EMBO J* 14, 3788–3799

41. Piatti, S. *et al.* (1996) *Genes and Devel.* 10, 1516–1531

42. Koch, C. and Nasmyth, K. (1994) *Curr Opin Cell Biol* 6, 451–9

43. Hadwiger, J.A. *et al.* (1989) *Proc. Natl. Acad. Sci. U.S.A.* 86, 6255–6259

44. Toyn, J.H. and Johnston, L.H. (1994) *Embo J* 13, 1103–13

45. Shirayama, M., Matsui, Y. and Toh, E.A. (1994) *Mol Cell Biol* 14, 7476–82

46. Wan, J., Xu, H. and Grunstein, M. (1992) *J Biol Chem* 267, 11274–80

47. Donovan, J.D., Toyn, J.H., Johnson, A.L. and Johnston, L.H. (1994) *Genes Dev* 8, 1640–53

48. Hartwell, L.M. and Unger, M.W. (1977) *J. Cell. Biol.* 75, 422–435

49. Koch, C. *et al.* (1993) *Science* 261, 1551–7

50. Cross, F.R. (1988) *Mol. Cell. Biol.* 18, 4675–4684

51. Nash, R. *et al.* (1988) *EMBO J.* 7, 4335–4346

52. Koch, C., Schleiffer, A., Ammerer, G. and Nasmyth, K. (1996) *Genes & Dev.* 10, 129–141

53. Murray, A.W. and Hunt, T. (1993) *The cell cycle — an introduction*, W.H. Freeman and Company, New York.

54. Crooke, E. (1995) *Cell* 82, 877–880

DISCUSSION

Wahl: The experiment with regulated cdc6 expression was very interesting and I am curious about whether there is a second "licensing system", for want of a better word, that enables late origins to fire. Do you think that cdc6 is not part of the prereplication complex for late origins? Is there another one, or do you think that late origins actually initiate early, but are prevented from elongating until later?

Nasmyth: You are getting at a very key problem. What is determining replication timing? According to the model that I have given you, one would be tempted to believe that the thing that causes an origin to be late is not when you form prereplication complex but rather when you allow the kinase to get at those preformed prereplication complexes. Whether that is true, only experiments will be able to distinguish. But that is certainly a prediction of this model because once you have activated that kinase, that should be a global property of the nucleus. According to this model you should not be able to form prereplication complex. Unless, of course, the kinase is regulated, as it were, at a local level. There are domains within the nucleus which are not accessible to the kinase and, while they are not accessible, you can still continue to form prereplication complex. I mean it could be pretty complicated. Naively I would think that you form prereplication complexes all at the same time, maybe, and then it is the time it takes for the kinase to get to them and trigger them to initiate that would explain the difference between early and late.

Livingston: You increase p40, shut down Clb cdk1, the prereplication intermediate forms (you show it by footprint), back comes kinase activity and does the origin fire?

Nasmyth: Yes it does. We have done that experiment. I went through it, but probably too fast. You take a G2 cell, it has high kinase, you kill the kinase by inducing artificially high levels of p40 which overcomes the proteolytic system. That is sufficient to form prereplication complexes by the footprinting assay. You will not refire while you leave p40 levels high. But then if you turn off synthesis of p40 it gets degraded, you reactivate the kinase and you get a full round of prereplication in the absence of any mitosis. There are data from Paul Nurse's lab where they picked up rum1 which is a kind of homologue of p40, which indeed has that property of allowing multiple rounds of pre-replication.

Gatti: When you draw the Pds1p sketch, you suggested that the protein holds together the sister chromatids. What is the meiotic phenotype of those mutants?

Nasmyth: I cannot tell you what the meiotic phenotype is. In yeast it is known that in metaphase, as in mammalian cells, the sister chromatids were attached throughout. They are not just attached at kinetochores. Now there is a huge literature suggesting that the attachments that hold kinetochores together are qualitatively different from those that attach the arms. Most of the evidence can be interpreted by saying that it is harder to get the kinetochores apart in the arms. But I think in most normal cells the arms are held together as well. That is clearly the case in yeast. It has been very vividly demonstrated by FISH analysis in yeast. So in the cdc20, if you knockout the Apc where there is no anaphase then sister chromatids are associated along the entire chromosome, not just at the kinetochore.

Hoeijmakers: It is a very fascinating story how all these proteins are being regulated in terms of their degradation and in terms of the involvement of the proteolysis machinery which is there all the time, but has to be put into action at the right moment. Do you know how this proteolysis machinery is activated? You mentioned factors which may be involved, which are not part of the proteolysis apparatus.

Nasmyth: You have hit on what I think is the real Achilles' heel of the cell cycle field. I think we are beginning to get a fairly good idea of what happens at the G_1/S transition. We have for several years had a fairly good feeling about what happens at the G2 M transition. We are now beginning to get an idea of what actually happens at the M phase to anaphase transition. The one area that we really have not got a clue, is the relationship between activating the cylin B CDK kinase and triggering what we suspect is activation of the anaphase promoting complex. That is, in my view, the biggest black box in understanding how you work your way through these various states of the cell cycle, in the eukaryotic cell cycle at the moment. That is how the activation of the mitotic kinase leads to triggering anaphase. We know that in the frog system you can take these extracts and you can drive that process, you can, as it were, turn on the proteolysis machinery by adding to them the cyclin B kinase. But there is a long lag phase, so there must be a lot of things happening in between. Some of those genes that I put up there, some of them are kinases, some of them are phosphatases, etc. It is really a key area which we are fundamentally ignorant about, is how activating that kinase, not only causes chromosome alignment, but then with a lag period, as long as all the chromosomes are aligned and no checkpoint mechanisms shut off the process triggering sister chromatid separation. What is going on in the lag between alligning and triggering sister chromatic separation and what control circuits are going on and how the checkpoint control mechanisms impact on that is really a key area for future investigation.

Hoeijmakers: One other question that I had, which was not very clear to me, but maybe you could explain it is that origin of replication is protected by cdc6 which starts at the beginning of G1 and then cdc6 is degraded at the beginning of S-phase. You mentioned also that the cyclins are lowered at that point so what is then the need to protect the origin if you have already a low cyclin content. Is this a second safety mechanism?

Nasmyth: No, I think the protection is merely a reflection of you building a complex of origins at preinitiation complex, and this is the enzymatic machinery which you must build. Having built that complex, you then activate the kinase, the kinase now catalyzes that complex to allow polymerase and primase and all of that to catalyze the initiation of DNA replication. So, we think you need the low kinase to permit the formation of this multicell unit complex. You then need the high kinase to trigger that complex to catalyze the initiation of DNA replication. And I think that this concept of the kinase having dual functions has now been accepted. For instance, promoting kinase can inhibit the ability to prevent multiple rounds of initiation. Likewise in the frog system where there has been a lot of work on this.

Hartwell: I have two questions and I am going to ask them one at a time, so I do not get confused. One relates to this issue of whether there is this sort of two S-phases, I refer to various anecdotal things. Steve Reid at one time reported that certain combinations of Clbs activated S, but the cells only went about half way through. Have you seen anything like that? Do you have any thoughts about that?

Nasmyth: We have not seen anything llike this. There is one observation that I think would be consistent with that, and that is that there are six cyclin Bs, any one of which, I think, can probably drive replication. But obviously, the ones that come on earliest would be the ones that are normally doing it. And those two are Clb5 and Clb6. Now, of those two, it looks like Clb5 is the important one, at least in mitotic cells. If you knock out Clb5, then cells are probably slow to initiate and then they also spend a long time in S-phase, as if many origins failed to fire. Now it turns out that if you knock out both Clb5 and Clb6, then they are slow to initiate but once they initiate they do not get stuck in S-phase and the implication is that Clb6 is doing a bad job at many origins. And this maybe very, very slow to activate origins and is actually getting in the way. So maybe Clb5, at least in mitotic cells, is the guy that is normally driving it, and Clb6 is not very good.

Hartwell: The second question is: as you know, this sic degrading activity, these three mutants have very unusual phenotypes in that they sit there and they bud periodically with roughly cell cycle interval. Do you have insight into what is going on there?

Nasmyth: No. I think that that phenotype is due to the low cyclin B kinase. Because if you knock out all of the sic cyclin B genes, you get the same phenotype. They arrest in G1 and they undergo what looks like multiple budding cycles. The cell cycle periods just have never been looked at. So it seems that if you keep the cyclin B kinase low, but the Cln1 and Cln2 kinases high, then that is what you get. Now, why you do it once, at one point in time, and then you wait a bit, and then you do it other time, it smacks of "once and only once" again, and I wonder whether some similar principle is going on. It is fundamentally a reciprocating system and the chamber of the steam engine, as it were, is the cell, and the steam that you put into it that does work, is the kinase. So the key thing is that you start off with the chamber empty of kinase and you put kinase in and for it to drive replication, you must previously have had the piston down, that is that you must have formed replication complexes. You put steam into the chamber, that is the kinase activation, drives the piston up. You can go on pouring steam into the chamber, it will do no further work. You will not get another round. In order to get a second round, you have got to get the steam out of the chamber, the piston can come down and then when you put steam back in, you get another round. This analogy is quite good so far, but it gets better because the key to functioning the steam engine is the valves that allow the steam in and out. The valves of the replication engine are, as it were, the enzymes that trigger the transitions from low kinase to high kinase and from high kinase back again. The Cln cyclins, as it were, are the valves for letting the steam in. That is the way of thinking. The steam is the Clb kinase and in a mammalian cell it would be cyclin E and cyclin A. And cyclin D and the Clns is, as it were, the valves for that. Now what is the valve for turning it off, we do not know. You have raised the issue for those genes that are required to activate the APC and those would, as it were, be the valves for getting the steam out of the chamber.

Pagano: Have you checked what is the state of phosphorylation of p40?

Nasmyth: Yes, this is a long sad story. We do see that p40 on a one-dimentional SDS gel. In G1 it looks as if it is hypophosphorylated and at the G1/S-phase transition it becomes hyperphosphorylated. p40 is a phosphoprotein. We also know that it is the activation of the Cln1–Cln2 kinase that triggers p40 proteolysis. But if you knock out those two kinases, then we delay p40 proteolysis. We have then tried to guess what the phosphorylation sites are and we have looked for sites that look like cdc-cdk phosphorylation sites. You make mutations in those sites and you get a wonderful phenotype, p40 is now stable as a rock. Everything looks great until you try to demonstrate that the residues that we have changed are actually phosphorylated in vivo. Thus far one of the key sites has a huge effect on the proteolysis of p40. We have not yet been able to demonstrate that that site is actually phosphorylated in vivo. The answer is that we do not know for sure what is going on.

12

THE INTEGRATION OF SIGNALLING PATHWAYS IN MAMMALIAN CELLS

Gerard I. Evan, Andrea Kauffmann-Zeh, Eugen Ulrich, Trevor Littlewood, David Hancock, and Elizabeth Harrington[*]

Imperial Cancer Research Fund Laboratories
44, Lincoln's Inn Fields
London WC2A 3PX, England

INTRODUCTION

Cancer affects one in three persons in the Developed World of whom half will die from their disease. Moreover, the half that survive suffer an unpleasant mixture of surgery and radio- and chemo-therapy, all of which can carry severe side effects and are of limited efficacy. Cancer is therefore both a very common disease and a frequent cause of death. Tumours are, however, clonal in origin: they arise from single mutant cells that progressively evolve an ensemble of mutations in critical genes that govern cell proliferation, survival, cell adhesion and cell mobility. Thus, if cancer affects one in three people, this means that the individual cancer cell only arises and propagates in one in three persons. As the human body comprises some ten thousand billion (10^{14}) cells, that undergo a total of some 10^{16} cell divisions during the course of a human life, the cancer cell becomes extremely rare.

The classical explanation for the rarity of the cancer cell is that it takes a significant number of sequential mutations to turn a normal cell into a cancer cell: the combined probability of such an ensemble of mutations is very low with the result that the cancer cell is very rare. This multistage carcinogenesis model is endorsed by numerous studies indicating that cancers arise by progressive acquisition of mutations. However, there are two significant problems with this explanation for the rarity of cancer cells. The first is that it ignores any clonal expansion that occurs between oncogenic mutations. As most oncogenic lesions compromise control of proliferation in affected cells, clonal expansion is very likely and this would significantly increase the number of mutant cells available as targets for further oncogenic lesions. In effect, growth deregulating lesion promote the acquisition of further oncogenic lesions. The second problem is one of information. In

---

* Present address: Laboratory of Molecular Oncology, Massachusetts General Hospital, Building 149, 13th Street, Charlestown, Massachusetts 02129.

many ways, a tumour cell can be considered as a normal cell whose proliferative mechanisms have been activated by mutation rather than by extracellular signals. It follows that the more mutations are required to generate a fully malignant cell, the more independent extracellular signals must be required for the normal counterpart of that tumour cell to propagate. This presents a biological conundrum: how can it be relatively easy for a normal cell to proliferate, but be so difficult for any cell to accomplish the same through sequential mutation?

Recent discoveries suggest effective solution to this puzzle. Evidence will be presented here that three critical regulators of cell proliferation and survival, the oncogenes *myc*, *ras* and *bcl*-2, possess dual functionality as both promoters and suppressors of cell propagation. Thus, the normal functions of these proto-oncogenes are obligatorily interlocked and it is only when they are all activated together that the negative growth aspects of one are suppressed by the positive growth properties of the others. In effect, the system works like a combination lock in which out-of-sequence activation of the components triggers the death or growth arrest of the affected cell.

THE c-*myc* ONCOGENE

It is well established that proto-oncogenes encode essential components of signal transduction pathways that regulate important aspects of cell behaviour and whose inappropriate activation promotes cellular transformation. The *myc* oncogenes were originally identified by homology with the transforming sequence of the avian myelocytomatosis virus that causes acute myelocytic leukaemias and carcinomas in affected chickens. The viral *myc* gene is derived from the cellular or c-*myc* gene, but at least three other *myc* genes are present in the human genome of which three, c-*myc*, N-*myc* and L-*myc* have been implicated in human neoplastic disease. Most notably, c-*myc* is activated and deregulated by chromosomal translocation in non-AIDS Burkitt's lymphoma and N-*myc* is amplified and over-expressed in the more serious incidences of childhood neuroblastoma.

In man, c-*myc*, is expressed in virtually all proliferating cells. Quiescent or postmitotic cells express none. A large body of data indicates that expression of c-*myc* is both necessary and, at least in the case of many haematopoietic, lymphoid and mesenchymal cells, sufficient for cell proliferation. For example, introduction of a constitutively expressed c-*myc* gene renders mesenchymal cells independent of external mitogens for proliferation and refractory to growth inhibitory effects of cell-cell contact, TGFβ and interferons [1, 2].

c-*myc* encodes a short-lived sequence-specific DNA-binding phosphoprotein possessing an N-terminal domain transcriptional-activation domain and a C-terminal DNA-binding/dimerisation bHLH-LZ domain [3] akin to that present in several known transcription factors [4–8]. Myc proteins act via dimerisation with the stable bHLH-LZ protein Max [5, 7–10]: Max is also the partner for the Mad bHLH-LZ proteins which compete with Myc proteins for Max and antagonise Myc function [11]. The Myc:Max:Mad proteins therefore comprise an extended and inter-dependent growth regulatory network.

Mutagenesis indicates that ability of c-Myc to drive cell proliferation requires integrity of all of both its transcriptional activation and bHLH-LZ dimerisation and DNA binding domains (reviewed in [3]), strongly suggesting that c-Myc promotes cell proliferation through its action as a transcription factor. However, with very few exceptions [2, 12, 13], target genes regulated by c-Myc have remained elusive.

c-*myc* in Leukaemia and Tumour Formation

Deregulated or altered expression of *myc* proto oncogenes is observed in virtually every tested human cancer, including leukaemia and lymphomas. The mechanisms responsible for *myc* gene deregulation include gene amplification (e.g. in myeloid and plasma cell leukaemias as well as in a variety of carcinomas), translocation (e.g. the classical translocation of c-*myc* to the Ig loci in Burkitt's lymphoma and acute lymphocytic leukaemias), and mutations in the regulatory region and structural regions of the gene. The net effect of almost all of these alterations is to compromise the normally tight regulation of *myc* expression such that the gene and its activity can no longer be shut down. Given the ability of *myc* genes to drive cell proliferation, it appears clear why such lesions are oncogenic.

However, although dominant oncogenes like *myc* are very effective inducers of *tumours* in animals, they are very inefficient inducers of *tumour cells*. For example, transgenic deregulation of *myc* in many mouse tissues almost invariably generates a tumour. However, the long latency in tumour formation and the clonality of the eventual neoplasm both indicate that even though *myc* was transgenically expressed in millions of cells, tumours arise as rare somatic events that depend on secondary mutations. The nature of such secondary (co operating) mutations can be partly revealed by a number of techniques, most notably by retrovirus insertional mutagenesis tumour acceleration assays [14, 15]. In other cases, oncogenic co-operation has been directly demonstrated between c-*myc* and specific other oncogenes (e.g. Ha-*ras*, v-*abl* and *bcl*-2) by co-transgenic expression. Until recently, it has not been clear quite why oncogenes such as c-*myc*, which are very efficient inducers of cell transformation *in vitro*, are so inefficient at inducing tumour cells *in vivo*. One clue is provided by the dramatic oncogenic synergy between c-*myc* and the anti-apoptotic oncogene *bcl*-2: it appears that growth deregulating lesions such as c-*myc* activation are not only good inducers of cell proliferation but also efficient triggers of programmed cell death or apoptosis [1].

c-Myc-Induced Apoptosis

Induction of apoptosis by c-Myc is most evident under conditions of growth factor deprivation [1, 16] or following physical or genotoxic insult [1]. Intriguingly, the more c-Myc is expressed in cells, the greater their tendency to undergo apoptosis: at extremely high levels of c-Myc expression—for example the level seen in many haematopoietic and epithelial tumour cells (20–50,000 molecules of c-Myc protein per cell)—it is very difficult to maintain cell viability *in vitro* and cultures of such cells almost invariably involute.

The cytotoxic attribute of c-Myc is genetically indissociable from its mitogenic property. Identical regions of the c-Myc protein are required for both growth promotion and induction of apoptosis. Both the N-terminal *trans*-activation domain and the C-terminal DNA-binding and dimerisation bHLH-LZ domain [1] mediating interaction with the heterologous partner protein Max [7] are necessary for both transforming and apoptotic functions of c-Myc. Moreover, physical interaction with Max is absolutely required for induction of both proliferation and apoptosis [8]. These observations all strongly suggest that c-Myc induces apoptosis, like proliferation, via a transcriptional mechanism—presumably through the modulation of appropriate target genes. Thus, c-Myc can drive two completely contradictory processes within a cell: proliferation and death.

Induction of Apoptosis Is a Ubiquitous Property of Growth-Deregulating Lesions in Cancer

The "paradoxical" state of affairs in which c-Myc, a potent inducer of cell proliferation, is also a potent trigger of cell death, has since been generalised to include virtually all other promoters of cell proliferation. Examples include the adenovirus oncoprotein E1A [17], the G1 progression transcription factor E2F1 [18–20] and the Cdc25a phosphatase [21] responsible for triggering activation of the cell cycle cyclin-dependent kinase machinery, all of which have also been shown to be effective inducers of apoptosis under certain conditions. It seems that any lesion promoting deregulated mammalian cell proliferation also obligatorily opens a second programme for cell suicide. This Dual Programme concept (Figure 1) has fundamental implications for cancer. All tumour cells possess growth-deregulating lesions and must *ipso facto* be sensitised to the induction of apoptosis by those lesions. Thus, cells that express deregulated oncogenes such as c-*myc* [1], E1A or *ras* [22] exhibit greatly increased sensitivity to induction of apoptosis in response to a wide range of insults such as growth factor withdrawal, genotoxic insult, cytostatic agents, nutrient privation or exposure to interferon γ.

Clearly, factors that influence the apoptotic arm of the Dual Programme are therefore likely to be of great importance in determining the survival and expansion of tumour cells. Indeed, accumulating evidence suggests that many of our existing anti-cancer therapies, particularly genotoxic agents, work by inducing apoptosis in target tumour cells that are sensitised to its by the oncogenic lesions they possess. It has therefore become vital to determine the molecular mechanism by which such sensitisation to apoptosis is conferred. Unfortunately, however, almost nothing is known of the molecular mechanical linkage between cell proliferation and cell death.

The further possibility emerges that a major way by which tumours may evade treatment is through the evolution of mechanisms that suppress apoptosis. Indeed, a number of such mechanisms have been uncovered that mitigate the pro-apoptotic action of both

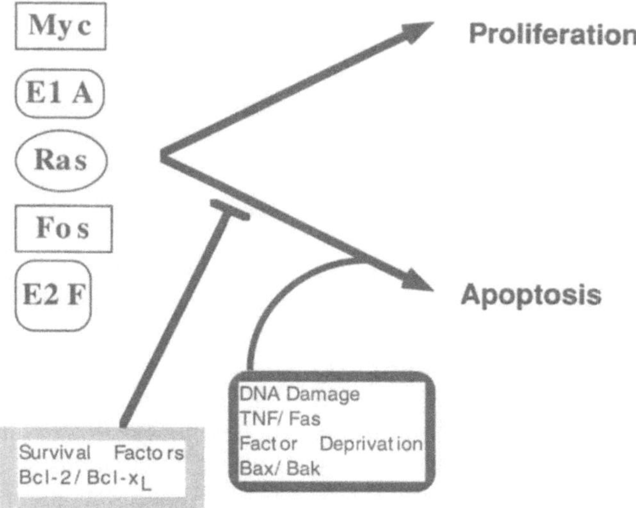

Figure 1. Dual Programme model for the coupling between cell proliferation and apoptosis.

oncogenes and cytotoxic agents. Examples of anti-apoptotic mechanisms in tumours include the mal-expression of certain gene products that inhibit apoptosis and the unscheduled activation of specific signal transduction pathways that promote cell viability. The idea of tumour cells with compromised apoptotic machinery presents a particularly worrying problem in cancer therapy. Such "deathless" cells would tend to survive therapy *despite* the damage inflicted upon them. Given that most existing anti-neoplastic agents are mutagenic, cells that survive their effects would probably sustain mutations which would drive the genetic diversity of the surviving tumour.

MECHANISMS OF SUPPRESSION OF APOPTOSIS IN CANCER

Anti-Apoptotic Oncogenes

One well characterised example of an anti-apoptotic lesion in cancer is the deregulated expression of the proto-oncogene *bcl*-2. *bcl*-2 maps to the site on chromosome translocation of the t(14,18) translocation found in many follicular B cell lymphomas in man and encodes polypeptide localised to most intracellular membranes [23, 24] Although widely expressed during development, *bcl*-2 expression in the adult is confined to immature and stem-cell populations and long-lived cells such as peripheral sensory neurones and resting B lymphocytes [25]. T cells expressing a *bcl*-2 transgene are markedly resistant to the cytocidal effects of glucocorticoids, radiation and anti-CD3-triggered apoptosis. Targeted expression of Bcl-2 to lymphoid cells in transgenic mice leads to increase in numbers of mature resting B cells and potentiates their longevity. In addition, *bcl*-2 exhibits dramatic synergy with c-*myc* when both are targeted to lymphoid cells in transgenic mice (reviewed in [26]). As outlined above, this synergy arises because *bcl*-2 suppresses the apoptotic programme implemented by c-*myc*, leaving c-*myc* free to drive uncontrolled cell proliferation [27–29]. Expression of *bcl*-2, or other anti-apoptotic members of the *bcl*-2 family, has now been shown to be an effective suppressor of apoptosis induced by many diverse triggers, including oncogene deregulation, growth factor deprivation, activation of the Fas/CD95 and TNF killing signal transduction pathways, and physical and genotoxic insult [30].

Anti-Apoptotic Signalling Pathways

Signal transduction pathways that suppress apoptosis are also candidate sources for anti-apoptotic lesions. A number of such pathways have been defined, although the precise molecular basis for their anti-apoptotic actions remain obscure. In our own laboratory we have studied in detail the Insulin-like Growth Factor-I (IGF-I) signalling pathway because IGF-I is a pervasive survival cytokine active on most haematopoietic, mesenchymal, epithelial and neuroectoderm-derived cells. Moreover, IGF-I signalling is implicated in the growth of many, if not most, tumours [31].

IGF-I binds the Insulin-like Growth Factor Receptor-I, IGF-IR, a trans-membrane receptor tyrosine kinase that also binds the ligands IGF-II and insulin (albeit at supraphysiological concentrations). A role for IGF-I in inhibiting apoptosis was first demonstrated by its ability to suppress apoptosis following cytokine withdrawal in IL-3-dependent haematopoietic cells [32] and after serum withdrawal in Rat-1 cells transformed by the c-*myc* oncogene (Rat-1/Myc cells) [33]. In the latter system, IGF-I provided the most potent inhibition of c-Myc-induced apoptosis out of a range of cytok-

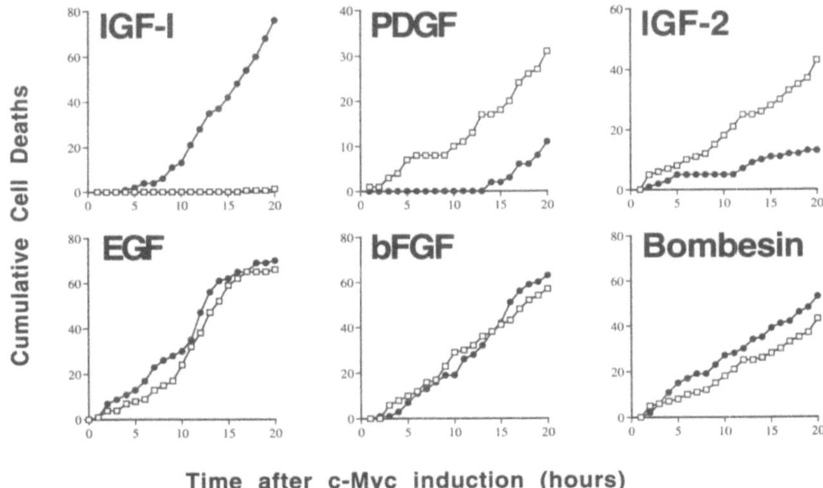

Figure 2. Suppression of c-Myc-induced apoptosis by a discrete group of cytokines. Cultures of Rat-1 fibroblasts expressing a 4-hydroxytamoxifen (OHT)-dependent conditional form of the c-Myc oncoprotein (Littlewood, 1995 #3574) were cultured in medium containing 0.1% serum. At time 0, c-Myc was activated by addition of OHT and 100 individual cells followed by the technique of time-lapse video microscopy (Evan, 1992 #2411; Fanidi, 1992 #2358; Harrington, 1994 #3076). Each apoptotic cell death was scored and cumulative deaths are shown plotted against time. Each graph represents a paired experiment in which c-Myc was activated in the absence (□) or presence (●) of the cytokine shown.

ines tested (Figure 2). IGF-1 also protected Rat-1/Myc cells in the presence of cyclohexi-mide, indicating that suppression of apoptosis does not require *de novo* synthesis of proteins. IGF-I is also a recognised neurotrophic agent that supports survival of CNS-derived cells [34].

Mechanism of Suppression of Apoptosis by IGF-I

To determine the molecular basis for IGF-I suppression of apoptosis, we have used a series of IGF-I receptor mutants to define domains required for transmitting the anti-apoptotic signal from the receptor. An IGF-I receptor lacking a functional ATP-binding site provided no protection from apoptosis in dictating the requirement for tyrosine kinase activity. However, mutations at tyrosine residue 950 or in the tyrosine cluster (1131, 1135, and 1136) within the kinase domain, remained capable of suppressing apoptosis, even though such mutations inactivate the transforming and mitogenic functions of the receptor. In the C-terminus of the IGF-IR two mutations, one at tyrosine 1251 and one that replaces residues histidine 1293 and lysine 1294, abolished the anti-apoptotic function, whereas mutation of the 4 serines at 1280–1283 did not. Furthermore, receptors truncated at the C-terminus exhibited enhanced anti-apoptotic function that was constitutive and no longer required ligation of IGF-I ligand. Finally, mutation of Y950, a residue important for inter-action of IGF-IR with the major intracellular docking protein IRS-1, to phenylalanine generated a receptor which could still confer protection from c-Myc induced apoptosis, albeit at reduced level. These studies demonstrate that the domains of the IGF-I receptor required for its anti-apoptotic function are distinct from those required for mitogenesis or transformation.

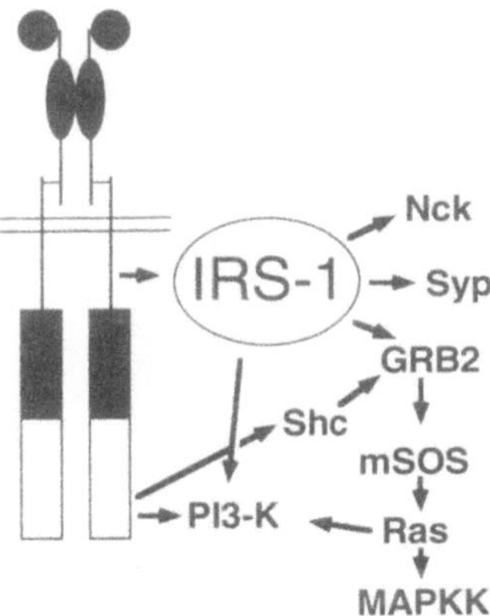

Figure 3. Schematic diagram of some of the known signalling pathways downstream of the mammalian IGF-I receptor.

Dissection of the IGF-I Anti-Apoptotic Signalling Pathway

In cells of neuronal origin, recent studies implicate the intracellular signalling enzyme phosphatidylinositol 3-kinase (PI3K) in transduction of survival signals in neuronal cells [35]. PI3K is potentially activated by IGF-IR via three independent routes—via IRS-1, via the C terminal of IGF-IR itself, and via Ras (Figure 3). A critical role for PI3K in IGF-I survival signalling is indicated by the fact that the well-characterised inhibitors of PI3K, wortmannin and LY294002, are both effective at blocking IGF-I-mediated protection from apoptosis induced by c-*myc* or ultraviolet irradiation in fibroblasts. We therefore investigated whether PI3K is directly involved in IGF-I survival signalling by assessing the effect of a constitutively activated PI3K mutant on c-Myc-induced apoptosis in fibroblasts. At the same time, we also examined the effects on apoptosis of Ras, a key upstream effector of PI3K [36], as well as two of the known downstream effectors of PI3K—the protein kinases PKB/Akt [37, 38] and p70^{S6K} [39].

The canonical PI3K activity comprises a regulatory p85 subunit and a catalytic p110 subunit that phosphorylates PtdIns(4,5)P$_2$ to create the second messenger PtdIns(3,4,5)P$_3$. Mutation of the lysine reside at position 227 of p110 to a glutamic acid (K227>E) generates a constitutively activated PI3K. Accordingly, this mutant was expressed in Rat-1 fibroblasts and assayed for its ability to affect c-Myc and UV-induced apoptosis. The (K227>E) PI3K mutant afforded excellent protection from apoptosis, and this effect was blocked by either wortmannin and LY294002, indicating the involvement of active PI3K in the protection [22]. Known downstream effectors of PI3K are the serine/threonine protein kinases p70^{S6K} and PKB/AKT. We were able to exclude p70^{S6K} effector from survival signalling because the macrolide rapamycin, which induces dephosphorylation and concomitant inactivation of p70^{S6K} [40], had no detectable inhibitory effect on PI 3-kinase-mediated suppression of apoptosis. Another candidate downstream effector of PI 3-kinase is the serine/threonine protein kinase PKB/Akt which is activated by both the constitutively active (K227>E) mutant of p110 PI3K and by the PI3K-specific activating mutant Ras V12 C40 [22]. Expression of a constitutively activated Akt/PKB mutant (*gag*-PKB) afforded excellent protection from apoptosis [22]. Thus, it is likely that PI 3-kinase-mediated activation of the PKB/Akt kinase is a significant component of the anti-apoptotic signalling pathway triggered by IGF-I binding to its receptor (Figure 4).

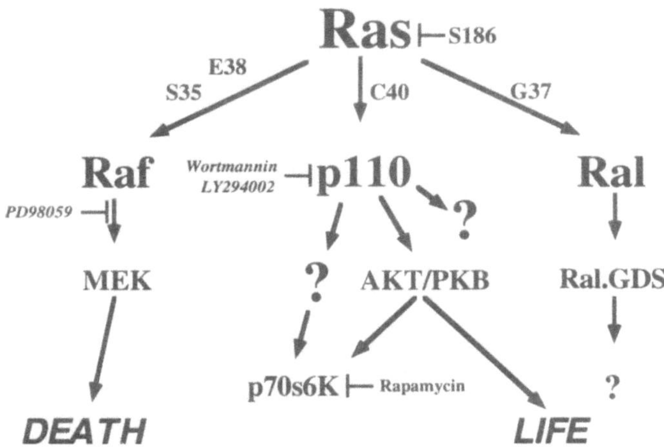

Figure 4. Schematic representation of three signalling pathways downstream of Ras. Effector mutants described in the text are shown associated with their relevant effector pathway. Inhibitors with known targets are shown in italics.

Ras Transduces Both Survival and Killing Signals

The point mutation in the constitutively active p110 K227>E mutant lies in the domain of p110 that interacts with Ras (p21^{ras}) and is thought to induce a conformational change that mimics that caused by binding to activated Ras during normal PI 3K signal transduction. This implies that Ras lies upstream of PI3K in the IGF-I survival signalling pathway: a surprising conclusion because our earlier work had demonstrated that a constitutively activated oncogenic V12 Ras mutant *enhances* rather than suppresses apoptosis induced by c-Myc or UV. This poses the paradox that Ras promotes apoptosis yet appears to be a necessary component of anti-apoptotic signalling.

A clue to the solution of this puzzle comes from observations showing that the increased apoptosis caused by V12 Ras is exacerbated by inhibiting PI3K activity. One possible explanation for this is that activated Ras triggers both a pro-apoptotic pathway and a separate PI 3-kinase-dependent anti-apoptotic pathway: in the absence of other factors, the former is dominant over the latter, such that the net effect of V12 Ras is to promote apoptosis. Active, GTP-bound Ras transduces signals through multiple intracellular targets. These include, amongst others, the p110 catalytic subunit of PI 3-kinase [36, 41], Raf (at the apex of the MAP kinase pathway [42–44]), and Ral.GDS, the exchange factor for Ral.GTPases [45–47]. We therefore investigated the individual contributions made by each of these effectors to cell survival using partial loss-of-function mutants located in the Ras effector loop that each activates only one of the above mentioned downstream pathways. These were developed by our ICRF colleague, Dr Julian Downward. Thus, V12 Ras interacts with all known Ras effectors, V12 S186 mutant interacts with none, V12 S35 andV12 E38 Ras both interact with Raf but not with PI3K or Ral.GDS, V12 C40 interacts with p110^{PI3K} but not Raf or Ral.GDS, and V12 G37 Ras interacts with Ral.GDS but not Raf or PI3K Figure 4). The various Ras effector mutants were expressed in Rat-1 fibroblasts and then assessed for their abilities to suppress apoptosis induced by c-Myc activation (Figure 5). p110^{PI3K}-specific V12 C40 Ras mutant protected cells from c-Myc as effectively as the constitutively active K227E mutant of p110

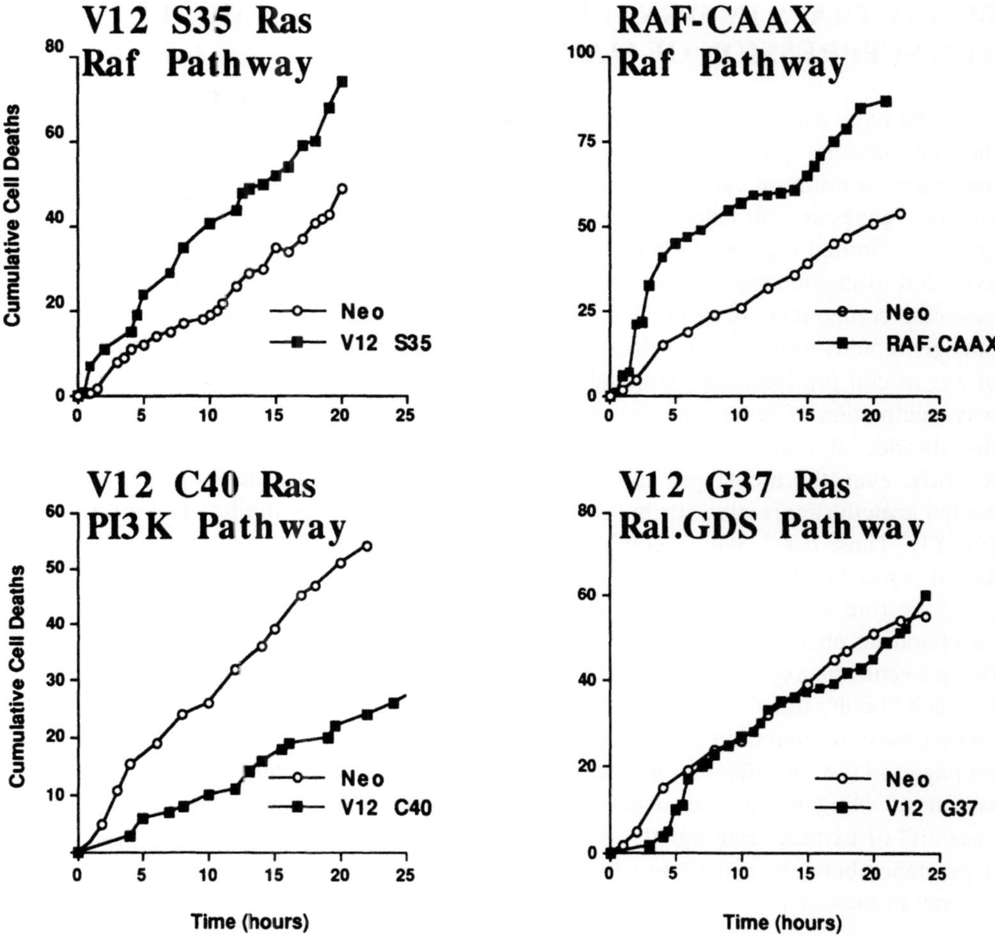

Figure 5. Effects of differing Ras effector mutant on c-Myc-induced apoptosis in Rat-1 fibroblasts. Rat-1 cells expressing OHT-dependent c-Myc were transfected either with control *neo* vector or with a vector driving constitutive expression of a Ras effector mutant as shown. Apoptotic cell deaths were monitored by time-lapse video microscopy as in Figure 2.

PI3K. In contrast, the RAL.GDS-specific V12 G37 Ras mutant conferred no protection whilst the Raf- specific S35 Ras mutant markedly promoted c-Myc-induced apoptosis. Apoptosis was also promoted by expression of a constitutively activated Raf-CAAX mutant whereas downstream inhibition of the Raf signalling pathway by the MAP kinase kinase inhibitor PD98059 afforded a modest suppression of apoptosis.

Taken together, the above data are most consistent with the notion that activation of the Raf pathway promotes apoptosis in fibroblasts whilst activation of the PI3K pathway suppresses it (Figure 4). We presume this indicates that, as with c-Myc, Ras does not make a decision as to which of its downstream effector pathways predominates in any situation: rather, *ras* enables multiple potential signalling pathways but the net outcome is determined by cross talk with other signals. Importantly, however, the net outcome of oncogenic *ras* activation in isolation is to trigger the dominant Raf-mediated death pathway. This provides a mechanism that would greatly restrict the probability of survival of any somatic cell that sustains a *ras* activating mutations.

BOOBY TRAPS IN SIGNALLING PATHWAYS — A MODEL FOR THE SUPPRESSION OF MALIGNANCY

We have suggested previously that the obligate dual action of c-Myc in promoting the contradictory processes of proliferation and programmed cell death represents an important restraint on carcinogenic progression [48]. This is because mutations that activate *myc* genes are intrinsically unstable and result in suicide of any affected cell that outgrows its limited supply of survival factors. We believe that this hypothesis can now be extended to include the oncogene *ras*. Our data indicate that even though *ras* may be an essential component of survival signalling, its dominant pro-apoptotic effect make its oncogenic activation largely self-limiting. This situation is reminiscent of the dual effect of *ras* in cell proliferation. Although *ras* is involved in most mitogenic signalling pathways, activation of *ras* in isolation is potently cytostatic in normal cells [49, 50]. Thus, in the absence of other factors, activation of *ras* is both cytostatic and cytotoxic. More recently, even the archetypal anti-apoptotic oncogene *bcl*-2 has demonstrated an unexpected growth suppressive attribute—deregulated *bcl*-2 potently inhibits cell proliferation [51–53]. Thus, *bcl*-2, whilst conferring resistance to apoptosis, also acts to curtail the clonal expansion of cells in which it becomes activated.

We find it intriguing that the *independent* oncogenic activation of each of the three co operating oncogenes, *myc*, *ras* or *bcl*-2, effectively triggers a "fail-safe" mechanism that prevents the expansion of the affected cell. We hypothesise that evolution has selected for such "booby traps" in metazoan signalling pathways such that activation of each separate pathway by mutation ends up as a literal "dead end." At the same time, however, the propagation (i.e. proliferation *and* survival) of somatic cells is easily achieved by the activation of obligatorily inter-dependent pathways activated in concert by the correct "gestalt" of extracellular signals. An understanding of the molecular basis of such interdependence between signalling pathways may well suggest novel routes for the effective therapy of tumours.

REFERENCES

1. Evan, G., Wyllie, A., Gilbert, C., Littlewood, T., Land, H., Brooks, M., Waters, C., Penn, L. and Hancock, D., 1992, Induction of apoptosis in fibroblasts by c-*myc* protein, Cell. 63: 119–125.
2. Eilers, M., Schirm, S. and Bishop, J.M., 1991, The MYC protein activates transcription of the alpha-prothymosin gene, EMBO J. 10: 133–141.
3. Littlewood, T. and Evan, G., 1995, Transcription factors 2: helix-loop-helix proteins, Protein Profile. 2. ed Sheterline, P, Acadamic Press, London.
4. Kato, G.J., Barrett, J., Villa, G.M. and Dang, C.V., 1990, An amino-terminal c-myc domain required for neoplastic transformation activates transcription, Mol Cell Biol. 10: 5914–5920.
5. Amati, B., Dalton, S., Brooks, M., Littlewood, T., Evan, G. and Land, H., 1992, Transcriptional activation by c-Myc oncoprotein in yeast requires interaction with Max, Nature. 359: 423–426.
6. Kretzner, L., Blackwood, E. and Eisenman, R., 1992, Myc and Max possess distinct transcriptional activities, Nature. 359: 426–429.
7. Amati, B., Brooks, M., Levy, N., Littlewood, T., Evan, G. and Land, H., 1993, Oncogenic activity of the c-Myc protein requires dimerisation with Max, Cell. 72: 233–245.
8. Amati, B., Littlewood, T., Evan, G. and Land, H., 1994, The c-Myc protein induces cell cycle progression and apoptosis through dimerisation with Max, EMBO J. 12: 5083–5087.
9. Blackwood, E.M. and Eisenman, R.N., 1991, Max: a helix-loop-helix zipper protein that forms a sequence-specific DNA-binding complex with Myc, Science. 251: 1211–1217.
10. Littlewood, T., Amati, B., Land, H. and Evan, G., 1992, Max and c-Myc/Max DNA binding activities in cell extracts., Oncogene. 7: 1783–1792.

11. Ayer, D.E., Kretzner, L. and Eisenman, R.N., 1993, Mad: a heterodimeric partner for Max that antagonizes Myc transcriptional activity, Cell. 72: 211–222.

12. Benvenisty, N., Leder, A., Kuo, A. and Leder, P., 1992, An embryonically expressed gene is a target for c-Myc regulation via the c-Myc-binding sequence, Gene Devel. 6: 2513–2523.

13. Bello Fernandez, C., Packham, G. and Cleveland, J.L., 1993, The ornithine decarboxylase gene is a transcriptional target of c-Myc, Proc Natl Acad Sci U S A. 90: 7804–7808.

14. Haupt, Y., Alexander, W.S., Barri, G., Klinken, S.P. and Adams, J.M., 1991, Novel zinc finger gene implicated as myc collaborator by retrovirally accelerated lymphomagenesis in E mu-myc transgenic mice, Cell. 65: 753–763.

15. van Lohuizen, M., Verbeek, S., Scheijen, B., Wientjens, E., van der Gulden, H. and Berns, A., 1991, Identification of cooperating oncogenes in E mu-myc transgenic mice by provirus tagging, Cell. 65: 737–752.

16. Askew, D., Ashmun, R., Simmons, B. and Cleveland, J., 1991, Constitutive c-*myc* expression in IL-3-dependent myeloid cell line suppresses cycle arrest and accelerates apoptosis, Oncogene. 6: 1915–1922.

17. White, E., Cipriani, R., Sabbatini, P. and Denton, A., 1991, Adenovirus E1B 19-kilodalton protein overcomes the cytotoxicity of E1A proteins, J Virol. 65: 2968–2978.

18. Shan, B. and Lee, W.H., 1994, Deregulated expression of E2F-1 induces S-phase entry and leads to apoptosis, Mol Cell Biol. 14: 8166–8173.

19. Qin, X., Livingston, D., Kaelin, W. and Adams, P., 1994, Deregulated transcription factor E2F-1 expression leads to S-phase entry and p53-mediated apoptosis., Proc Natl Acad Sci U S A. 91: 10918–10922.

20. Wu, X. and Levine, A.J., 1994, p53 and E2F-1 cooperate to mediate apoptosis, Proc Natl Acad Sci U S A. 91: 3602–3606.

21. Galaktionov, K., Chen, X. and Beach, D., 1996, Cdc25 cell-cycle phosphatase as a target of c-Myc, 382: 511–517.

22. Kauffmann-Zeh, A., Rodriguez-Viciana, P., Ulrich, E., Gilbert, C., Coffer, P. and Evan, G., 1997, Suppression of c-Myc-induced apoptosis by Ras signalling through PI 3-kinase and PKB, Nature. In Press.

23. Krajewski, S., Tanaka, S., Takayama, S., Schibler, M., Fenton, W., and Reed, JC., 1993, Investigation of the subcellular-distribution of the Bcl-2 oncoprotein - residence in the nuclear-envelope, endoplasmic-reticulum, and outer mitochondrial-membranes, Cancer Res. 53: 4701–4714.

24. Nakai, M., Takeda, A., Cleary, M.L. and Endo, T., 1993, The *bcl*-2 protein is inserted into the outer-membrane but not into the inner membrane of rat-liver mitochondria *in vitro*, Biochem Biophys Res Comms. 196: 233–239.

25. Hockenbery, D.M., Zutter, M., Hickey, W., Nahm, M. and Korsmeyer, S.J., 1991, BCL2 protein is topographically restricted in tissues characterized by apoptotic cell death, Proc Natl Acad Sci U S A. 88: 6961–6965.

26. Adams, J.M. and Cory, S., 1991, Transgenic models for haemopoietic malignancies, Biochim Biophys Acta. 1072: 9–31.

27. Fanidi, A., Harrington, E. and Evan, G., 1992, Cooperative interaction between c-*myc* and *bcl*-2 proto-oncogenes, Nature. 359: 554–556.

28. Bissonnette, R., Echeverri, F., Mahboubi, A. and Green, D., 1992, Apoptotic cell death induced by c-*myc* is inhibited by *bcl*-2, Nature. 359: 552–554.

29. Wagner, A.J., Small, M.B. and Hay, N., 1993, Myc-mediated apoptosis is blocked by ectopic expression of *bcl*-2, Mol Cell Biol. 13: 2432–2440.

30. Möröy, T. and Zörnig, M., 1996, Regulators of life and death: the *bcl*-2 gene family, Cell Physiol Bichem. 6: 312–336.

31. LeRoith, D., Baserga, R., Helman, L. and Roberts, C.T. Jr., 1995, Insulin-like growth factors and cancer [see comments], Ann Intern Med. 122: 54–59.

32. Rodriguez, G., T., Collins, M.K., Garcia, I. and A., L.-R., 1992, Insulin-like growth factor-I inhibits apoptosis in IL-3-dependent hemopoietic cells, J Immunol. 149: 535–540.

33. Harrington, E., Fanidi, A., Bennett, M. and Evan, G., 1994, Modulation of Myc-induced apoptosis by specific cytokines, EMBO J. 13: 3286–3295.

34. Barres, B.A., Hart, I.K., Coles, H.S., Burne, J.F., Voyvodic, J.T., Richardson, W.D. and Raff, M.C., 1992, Cell death in the oligodendrocyte lineage, J Neurobiol. 23: 1221–1230.

35. Yao, R. and Cooper, G.M., 1995, Requirement for phosphatidylinositol-3 kinase in the prevention of apoptosis by nerve growth factor, Science. 267: 2003–2006.

36. Rodriguez Viciana, P., Warne, P.H., Dhand, R., Vanhaesebroeck, B., Gout, I., Fry, M.J., Waterfield, M.D. and Downward, J., 1994, Phosphatidylinositol-3-OH kinase as a direct target of Ras, Nature. 370: 527–532.

37. Burgering, B.M. and Coffer, P.J., 1995, Protein kinase B (c-Akt) in phosphatidylinositol-3-OH kinase signal transduction, Nature. 376: 599–602.

38. Franke, T.F., Tartof, K.D. and Tsichlis, P.N., 1994, The SH2-like Akt homology (AH) domain of c-*akt* is present in multiple copies in the genome of vertebrate and invertebrate eucaryotes. Cloning and characterization of the Drosophila melanogaster c-*akt* homolog Dakt1, Oncogene. 9: 141–148.

39. Chung, J., Grammer, T.C., Lemon, K.P., Kazlauskas, A. and Blenis, J., 1994, PDGF- and insulin-dependent pp70^{S6k} activation mediated by phosphatidylinositol-3-OH kinase, Nature. 370: 71–75.

40. Chung, J., Kuo, C.J., Crabtree, G.R. and Blenis, J., 1992, Rapamycin-FKBP specifically blocks growth-dependent activation of and signalling by the 70 kd S6 protein kinases, Cell. 69: 1227–1236.

41. Rodriguez Viciana, P., Marte, B.M., Warne, P.H. and Downward, J., 1996, Phosphatidylinositol 3' kinase: one of the effectors of Ras, Philos Trans R Soc Lond B Biol Sci. 351: 225–231.

42. Warne, P.H., Viciana, P.R. and Downward, J., 1993, Direct interaction of Ras and the amino-terminal region of Raf-1 *in vitro*, Nature. 364: 352–355.

43. Zhang, X.F., Settleman, J., Kyriakis, J.M., Takeuchi Suzuki, E., Elledge, S.J., Marshall, M.S., Bruder, J.T., Rapp, U.R. and Avruch, J., 1993, Normal and oncogenic p21ras proteins bind to the amino-terminal regulatory domain of c-Raf-1, Nature. 364: 308–313.

44. Vojtek, A.B., Hollenberg, S.M. and Cooper, J.A., 1993, Mammalian Ras interacts directly with the serine/threonine kinase Raf, Cell. 74: 205–214.

45. Albright, C.F., Giddings, B.W., Liu, J., Vito, M. and Weinberg, R.A., 1993, Characterization of a guanine nucleotide dissociation stimulator for a ras-related GTPase, EMBO J. 12: 339–347.

46. Kikuchi, A., Demo, S.D., Ye, Z.H., Chen, Y.W. and Williams, L.T., 1994, ralGDS family members interact with the effector loop of Ras p21, Mol Cell Biol. 14: 7483–7491.

47. Spaargaren, M. and Bischoff, J.R., 1994, Identification of the guanine nucleotide dissociation stimulator for Ral as a putative effector molecule of R-ras, H-ras, K-ras, and Rap, Proc Natl Acad Sci U S A. 91: 12609–12613.

48. Harrington, E., Fanidi, A. and Evan, G., 1994, Oncogenes and cell death, Curr. Opin. Genet. Dev. 4: 120–129.

49. Ridley, A.J., Paterson, H.F., Noble, M. and Land, H., 1988, Ras-mediated cell cycle arrest is altered by nuclear oncogenes to induce Schwann cell transformation, EMBO J. 7: 1635–1645.

50. Hirakawa, T. and Ruley, H.E., 1988, Rescue of cells from *ras* oncogene-induced growth arrest by a second, complementing, oncogene, Proc Natl Acad Sci USA. 85: 1519–1523.

51. Linette, G.P., Li, Y., Roth, K. and Korsmeyer, S.J., 1996, Cross talk between cell death and cell cycle progression: BCL-2 regulates NFAT-mediated activation, Proc Natl Acad Sci U S A. 93: 9545–95.

52. O'Reilly, L., Huang, D. and Strasser, A., 1996, The cell death inhibitor Bcl-2 and its hpmologues influence control of cell cycle entry, EMBO J. 15: 6979–6990.

53. Brady, H., Gil-Gómez, G., Kirberg, J. and Berns, A., 1996, Baxα perturbs T cell development and affects cell cycle entry of T cells, EMBO J. 15: 6991–7001.

DISCUSSION

Bernards: In your scheme you draw *myc* as inducing a number of target genes that we do not know, and then from that sort of *myc* target gene you have two arrows emerging, one going to proliferation, one going to death. The question, though, is whether you need the proliferative response in order to get death. Could you have all of the *myc* target genes activated but block the proliferative response of *myc* by, for instance, a strong CDK inhibitor co-expressing, and do you then still get apoptosis or not?

Evan: I think that is a very important question. There are three points I want to make here: one is that I am not saying that you can only enter apoptosis through a mitogenic program. There are many cells which are permanently quiescent; in fact, all cells in the body, even those that are post mitotic, can undergo apoptosis. You do not need to switch on the mitogenic machinery. What I am saying, however, is that if you do switch on the mitogenic machinery in some way that is entirely unclear, you also switch on the tendency to undergo apoptosis. Now those two arrows are not pathways, they are outcomes, they are logical outcomes. I do not know how it is that *myc* does this, I do not know whether it is downstream of myc, I do not know what the physical connection between the two processes is. There are a number of studies going on at the moment where people have tried to block things that they think are downstream of myc to stop cell cycle progression, to see if you still get apoptosis; I think the jury is still out. Part of the problem is that everything seems to be tritateable against everything else, so my understanding is that if you express enough *myc* you can override any CDK inhibitor.

Bernards: But the converse is also true, you can express levels of inhibitor that cannot be overcome by any level of myc, right?

Evan: Well, I do not know that.

Bernards: At low levels of p27 you override with *myc,* and at high levels of p27 you do not override. So you could do the experiment under prohibitive levels.

Evan: Yes, so that is the open question at the moment. The question is whether or not these do gate each other in some irrevocable way, or whether it is all tritateable one against the other, and that is not clear. That would need to be sorted out before you could do that experiment.

Hartwell: We learned some time ago that the nuclease Mu2 is not necessary for cell death. Now, we have just learnt that the ICE proteases are not necessary for cell death, so this really opens up. again, the issue of whether lower organisms like yeast apoptose. At this point in time the only test would be perhaps BCL2 BAKs effects. Do you want to comment at all on what you find there?

Evan: It becomes very difficult to know, now, what you mean by death. That is why I have tended to back off a little bit and talk about what I see is the biological logic behind this. The biological logic is how you ensure that certain cells that may be of risk to you, you cannot propagate indefinitely.

Evan: Even the clonogenicity is a vexed issue because it is not clear to what extent cells that have been triggered to die, that are being protected by something like Bcl-2 and that survive long-term in culture can actually go on to propagate. That really is not shown in most studies that I have looked at, and the studies that we have done, the clonogenicity has disappeared, so the cells are alive, which may be useful from the point of view of the somatic tissue they can provide support functions in the soma, but they are genetically dead if you hit them hard enough. In terms of yeast, all we know is that killer members of the Bcl-2 family really kill yeast. That is, the chromatin becomes degraded, the nuclear membrane becomes broken up and they vacuolate and they show many of the hallmarks of apoptosis in mammalian cells, but of course, not all. I guess what would need to be sorted out is whether there is a specific protease that mediates that, and that really is not clear.

Livingston: From the Z VAD experiment, can one rule out the possibility that the same ICE product or products work in two ways?

Evan: The answer is that we know almost nothing about the intracellular pharmacokinetics and pharmacology of Z VAD. So it is possible that there is an ICE-like protease of different specificity that is involved in membrane blebbing or with a different sensitivity to the levels of the inhibitor. But it also remains possible, that blebbing is not driven by

ICE proteases at all, and if you think back to the nematode we are still looking for a function which is the said for function in the nematode.

Livingston: How was death induced in the Z VAD experiments?

Evan: Myc, E1A, BAK, BAKs, serum depravation, DNA damage.

Livingston: How about activating pro-ICE?

Evan: There is a slight complexity here. Our initial impressions, and these are hard experiments to do because the pro-ICE and the active-ICE, which is what you really want to know, are very difficult to do because everything has got to be under very tight induceable promoters or regulateable switches. But our feeling is that there will be triggers of cell death which do not speak to the blebbing machinery and one example of this would be if you switch on an ICE protease in a cell atopically and the hint would be that that is the way in which TNF and FAS induce apoptosis. TNF and FAS directly recruit an ICE protease to the receptor in a ligand dependent fashion. Now we are trying to do these experiments and we would predict that in those situations you would not get blebbing in situations where you hit directly the ICE proteases.

Pagano: Do you think that *myc* is involved in apoptosis? I mean, when you overexpress *myc*, as when you overexpress CDC-25, E1A or E2F in some conditions you induce apoptosis. Do you have any evidence that these genes have a physiological role in inducing apoptosis in the cells of our body or not?

Evan: O.K. Although we can provide evidence, which I think is important enough, that the deregulated expression of these genes which happens in so many tumors promotes apoptosis, is there any evidence that this happens in a normal day-to-day cell that switches on myc as part of its normal proliferative machinery? All I can say is that there is no evidence that that is not the case. We know that physiological levels of myc under ectopic control are sufficient to induce apoptosis if the cell is not being bathed in large excesses of anti-apoptotic signaling factors. So the question really then is, what is the availability of anti-apoptotic factors in somatic tissues? So, if myc comes on in a cell, it is going to die unless there are anti-apoptotic factors around which suppress that. How available are they? We are attempting to answer that question by using a switchable transgenic strategy by making a myc protein which has an activatable switch on its C terminus. Because it is an activatable protein, the expression of the gene encoding that protein could be targeted using standard transgenic technology. I cannot tell you the answer yet because we have only got the lines through now.

Livingston: Yes and in keeping with that there are two recent reports, one from Ed Harlow's laboratory and one from Mike Greenberg's and our own which say two things. The first is that an animal devoid of the E2F1 gene for its entire life develops late onset neoplasms. Secondly in those same animals there is a clear cut apoptotic defect in those cells in which it has been sought, i.e., lymphocytes. Put the two observations together, and one can suggest that E2F1 has a tumor suppressor function which could be mediated through an apoptosis-promoting effect.

Evan: I fully agree with that. I think these are really ying-yang type things, that all the things that promote proliferation have also the capacity to induce apoptosis. It is so certainly in metazone cells, certainly in vertebrate cells. Therefore the amount of proliferation and propagation of these cells you can get is determined by the factors that suppress cell death and in the main we think that those are soluble survival factors which are in short supply in somatic tissues. So there is a ceiling that prevents clonal expansion which is the availability of survival factors.

Nasmyth: Where is the pro ICE in the cell?

Evan: My understanding is that the ICE proteases have favored cell compartments, I think, ICH1 is found in the nucleus, others are found in the plasma membrane and in the cytoplasm. I do not think a clear picture is really emerging in the same way that a clear picture is not emerging as to whether these are in hierarchies relative to one another. It is not clear what their hierarchical relationship is. It is not clear whether they take part in parallel or in network type pathways. There is one thing that I will say that is very important, it is not what you asked, but it is very important. Nobody has ever found a tumor cell that cannot undergo apoptosis. When it undergoes apoptosis it does so with the same kinetics as a sensitive cell. That is, when that individual cell starts to bleb, it has gone in twenty minutes. This argues very strongly that the basal apoptotic machinery is not lost in cancer, even while there is clear evidence that suppression of apoptosis is very powerfully selected for. So it has either got to be redundant, or the apoptotic basal machinery is wired into essential functions of the cell that do other things, or, as I am suggesting, even if you lost it, it would not keep that cell alive because you could still die by the sort of cytoplasmic death. I think it is probably all of those three. It means that the program is always there and that capacity for self destruction exists in every tumor. It would be active were it not for the mutations that were preventing it from being active.

Nasmyth: So there is no indication whether the ECL2 members and the pro ICEs would find themselves in the same compartments.

Evan: There is nothing at the moment, apart from the TNF FAS signal transduction pathway, whether there is a direct physical and logical connection. There is absolutely nothing to link survival factors with BCL2 family members with pro ICE, with myc, with p53. The only thing I would say is what I said on the first day. The Drosophila DNA damage induced cell death appears to be in great part, mediated through reaper, and reaper is a death domain protein which is homologous to the killer domain of TNF receptor in FAS. It is a very ancient pathway for transducing DNA damage and other physical damage into the apoptotic program. The FAS receptor in response to ligand transmembrane has a death domain and this death domain is basically an oligomerization domain which recruits either directly in the case of FAS CD95 or indirectly in the case of TNF, another death domain protein which is called FADD, which has already been published, identified, and this death domain protein has a death effector domain and this death effector domain is another oligomerization domain that recruits the pro domain of a novo ICE protease. You get into an infinite regress of proteases which cleave other proteases and so on. The ICE proteases come in three families on the basis of sequence, which I think has misled everyone. But they come in two structural families, there are those with huge prodomains which get cleaved into the active ICE protease p20 and p10. There are those with almost no prodomain.

Pagano: I would like to go back to E2F and myc. There is, I think, a basic difference between E2F and myc and E2F and other genes able to induce apoptosis. Myc,

CDC25, cylin D1 and cylin E are all overexpressed in tumors. There are numerous reports showing that all these guys are overexpressed. E2F is very rarely overexpressed in tumors. So, I can understand the role of E2F in apoptosis in physiological conditions, still I am not completely convinced by myc or cylin D1 roles in apoptosis.

Evan: One question that I really do not know the answer to is who is limiting in a cell? One thing I know for certain is that myc is limiting because even in a fully logphased cell there is only about 1200 molecules of the protein in a cell. You know it is there, you know if you block it with a dominant-negative or an antisense, the cell stops but you cannot really see it. So myc is certainly limiting; I do not know if E2Fs are limiting.

Livingston: There are plenty of people here who have worked on this: René Bernards, Willie Krek. I think the consensus would be that cells are not replete with many molecules of any of the E2Fs.

Bernards: Certainly not compared to the pocket proteins.

Evan: But then the question is, is there something importantly, qualitatively important about what E2F does versus things which we presume are upstream or sidestream. Or is it just an accident of the way that the E2F system is configured, seeing that they do overlapping things in a slightly different way. Such that if you were to overexpress, it would not give you a malignant phenotype. You could imagine that E2F was slightly more efficient at triggering apoptosis versus proliferation under factor deprivation conditions. It would never appear as an oncogene. I think myc barely makes it as an oncogene and, in fact, all the data throughout the years have told you this. You can transgenically express myc or any other oncogene in every cell in the tissue of an animal and you get a clonal tumor after several months. Well that is not exactly a very efficient oncogenic process.

Bernards: Just to comment on that, in my laboratory we have made transgenic animals that overexpress E2F in T cells and you get about ten fold endogenous levels of E2F. No apparent pathology was seen in six months.

Draetta: Why not take the paradigm of yeast cell cycle control versus checkpoint control? Elements that control the cell cycle in yeast are required at each cell cycle. On top of that control there are gene functions that are normally not needed as that that come into action at specific points. Whenever you talk about apoptosis, you talk about functions that are regulating apoptosis, about executors. Yet I feel that there should be a third level which is the level of the sensors that you still miss. You can induce apoptosis with myc and by deregulating almost everything. Myc could induce things that IGF1 could also induce under different conditions and probably if you study the different cellular context you would have opposing effects by other things. There has to be some functions that sends this whole process and transduce it to the apoptotic machinery. I still sense that this is not evident from any of the studies except in the specific case of FAS but that is a very linear pathway.

Evan: One thing I would say what we have attempted to do. The original idea was to turn myc on with serum. When you are driving it with myc, the cell gets asynchronous; that is the reason they die. The most extreme view that derives from that is that when you have serum or IGF1 around, you modify what myc does. What we have attempted to show is that all the evidence seems to suggest that IGF1 has no effect on what myc does. It acts way downstream of everything that myc ever did, that p53 ever did, anything that cell cycle ever did to just prevent or suppress the activation of the ICE protease family.

Draetta: Do you actually see the delayed process of ICE?

Evan: Yes you do. At the moment the simplest hypothesis is that what IGF1 does, is like BCL2, that says do not activate your ICE proteases. There is no evidence of any back talk or cross talk modifying what myc does. So if you follow that line of logic through, myc does what it does, whether or not there is IGF1 there. That is the immediate down-stream programs that myc elicits are the same in the cell whether IGF1 is there or whether it is not there. It is just that if you have got IGF1 there, eventually you just choke down the cell death program. You can explain everything by that simple argument. There is no reason to believe that myc does a certain type of thing in one circumstance and a different type of thing in another circumstance. You could explain it all by saying whenever myc comes on, the potential to undergo apoptosis also comes on, in some way. We do not know how that is coupled. If you have got something that blocks apoptosis you will go off and propagate. If you do not have anything that blocks apoptosis, the chances are you are going to die. So that is why we have tried to find out where it is that IGF1 is working. All the evidence points that it is working as far down at the trigger as anybody would ever have guessed. It certainly does not modify, as far as we can tell, the transcriptional pro-gram that myc implements. There is no evidence for that or indeed anything that myc does. Another very important point which again is part of what you are saying is, does the death arise because of an internal conflict in the cell, or does the cell death arise because of perfectly normal cell that just requires more signals from the outside to stay alive? IGF1 does not react as to what it is that is promoting death. The only difference that you will see is if you have a very potent and unremitting death signal such as continuous DNA damage, IGF1 will give you delay, but the cell will eventually die. The real question from the point of view of propagation of DNA damage into a tumor is the balance. To what extent IGF1 will enable enough cells with DNA damage to survive long-term and to propagate, that really is not clear. The same argument stands for the whole of suppression of apoptosis. BCL2 is a very effective suppressor of immediate drug induced toxicity but it does not have very much effect on clonogenicity. In a large tumor a subtle change in the clonogenic potential, just a slight increase in the survival of a few cells with DNA damage, could be enough to kill the patient.

13

ANTITUMOR DRUGS AND YEAST CELL CYCLE CHECKPOINTS

Martin Weinberger,[1] Lisa Black,[1] Terry A. Beerman,[2] Joel A. Huberman,[1] and William C. Burhans[1*]

[1]Department of Molecular and Cellular Biology
[2]Department of Experimental Therapeutics
Roswell Park Cancer Institute
Buffalo, New York 14263

DNA replication is inhibited when cells are subjected to DNA damage during the S phase of the eukaryotic cell cycle (reviewed in Murnane, 1995, Kaufmann, 1995). The dose-dependent magnitude of this inhibition is biphasic in nature—an initial steep component occurs at low levels of damage, followed by a shallower component at higher levels. Analysis of cellular DNA pulse-labeled shortly after inducing DNA damage suggests that these two components correspond to inhibitory effects on two fundamentally different processes involved in DNA replication–initiation of DNA replication at origins of replication, and subsequent elongation of nascent chains at replication forks.

Very little is known about the molecular mechanisms which underly the S phase inhibitory effects of DNA damage in eukaryotic cells. Earlier models suggested that the inhibitory effect on initiation of DNA replication induced by ionizing radiation might occur in *cis* as a result of alterations in chromatin structure produced by radiation-induced strand breaks (Povirk, 1977). The elongation arrest observed at high doses of UV radiation also was thought to occur in *cis* when lesions formed a physical block to replication fork progression (Berger and Edenberg, 1986). However, the results of more recent experiments indicated that these inhibitory effects can occur in *trans* (Wang et al., 1996 and references therein). In some cases, they occur as part of an intra-S-phase checkpoint response which presumably protects cells and organisms from the deleterious consequences of DNA damage. For instance, it was recently demonstrated that the slow progression through S phase of *S. cerevisiae* cells subjected to ultraviolet radiation or treated with the alkylating agent methylmethane sulfonate (MMS) is mediated in part by the products of the *MEC1* and *RAD53* genes (Paulovich and Hartwell, 1995). An intra-S-phase DNA damage checkpoint also occurs in mammals - mutations in the *ATM* gene, such as those which occur in the cells of patients suffering from the hereditary disorder ataxia telangectasia,

---

* To whom all correspondence should be sent.

are associated with a reduced inhibitory effect on initiation and elongation of nascent DNA chains in response to DNA damage (Painter and Young, 1980). Structural similarities between the *MEC1* gene and the mammalian *ATM* gene have led to the suggestion that the intra-S-phase checkpoint response is conserved from yeast to mammals (Zakian, 1995). This response is distinguishable from the more characterized DNA damage checkpoint response that blocks entry into S phase from the G1 phase of the cell cycle in that it is triggered by damage which occurs *after* cells have entered S phase.

Further characterization of the inhibitory effects of DNA damage on DNA replication has been hindered by a number of factors. First, previous studies have distinguished between inhibitory effects on initiation from those on elongation by indirectly analyzing nascent DNA intrinsically labeled with DNA precursor molecules, such as tritiated thymidine and/or bromodeoxyuridine, on sucrose gradients or by fiber autoradiography. These techniques are difficult or impossible to apply to the analysis of DNA replication in the experimental systems that are most useful as model eukaryotic replicons, those of yeast and small viral replicons such as simian virus 40 (SV40). Consequently, there are few reports in the literature which distinguish between DNA damage-induced initiation and elongation effects on SV40 DNA replication, for instance, and none that identify specific effects on either process in yeast. Second, experiments that employ indirect labeling techniques can be difficult to interpret because of a number of complicating factors, such as differences in rate of replication fork movement, effects of the size of intracellular precursor pools on radiolabel incorporation, radiolabel incorporation due to DNA repair, and potential inhibitory effects on DNA replication related to radioisotopic labeling of DNA. Finally, although much is known about how nascent chain elongation occurs in eukaryotic cells, much less is known about the mechanisms by which initiation of DNA replication occurs, especially in mammals.

We recently developed an assay that clearly and simply distinguishes between initiation- and elongation-specific inhibitory effects independently of an intrinsic label. The assay employs a neutral-neutral two-dimensional gel electrophoresis technique for analyzing DNA replication. This technique cleanly separates replicating from nonreplicating DNA on the basis of the unique structure of replicating DNA and the correspondingly unique migration characteristics of this DNA on agarose gels (Brewer and Fangman, 1987). In this assay, initiation-inhibitory effects are detected as a decrease in the number of replication intermediates ("RIs") as previously formed RIs mature in the absence of new initiation events. Elongation inhibitory effects are detected as an accumulation of RIs. The assay is particularly suitable for analyzing inhibitory effects on DNA replication in small viral genomes, such as SV40, and in organisms in which nascent DNA cannot be intrinsically labeled in a facile manner, such as yeast.

We employed this assay to study the inhibitory effects on SV40 DNA replication of two potent experimental antitumor agents which damage DNA, adozelesin and C-1027, (Cobuzzi et al., 1996; McHugh et al., 1997). Adozelesin is a minor groove binding drug which alkylates the N3 position of adenine (Boger and Johnson, 1996), and C-1027 is an enediyne which induces DNA strand scissions (Nicolau et al., 1993). The results of this analysis indicated that these drugs induce specific inhibitory effects on both the initiation and elongation of nascent SV40 DNA chains, similar to the effects of other DNA damaging agents on cellular DNA replication detected by other techniques. In both cases, the inhibitory effects on initiation of SV40 DNA replication were observed in association with levels of DNA damage that correspond to far less than one lesion per SV40 genome, indicating that they occur in *trans*. This is consistent with the possibility that these effects are part of an intra-S-phase checkpoint response to DNA damage.

To further explore the potential relationship between an intra-S-phase checkpoint response and the inhibitory effect of these compounds on DNA replication, we analyzed the effects of adozelesin on DNA replication in the budding yeast *S. cerevisiae*. Treatment of *S. cerevisiae* cultures with adozelesin caused a cell division arrest as large-budded cells, and FACS analysis of arrested cell populations indicated they were blocked within S phase, similar to control cultures treated with the DNA replication inhibitor hydroxyurea. However, adozelesin's inhibitory effect on *S. cerevisiae* DNA replication was abolished in isogenic strains in which the *MEC1* or *RAD53* gene was disrupted. Therefore, adozelesin-induced inhibition of *S. cerevisiae* DNA replication occurs as part of the *MEC1/RAD53*-dependent intra-S-phase checkpoint response to DNA damage which which was recently described in this organism.

2D gel analysis of replicating DNA isolated from adozelesin-treated *S. cerevisiae* cells indicated that, similar to its effects on SV40 DNA replication, adozelesin induced a rapid decrease in the numbers of RIs, but not when the cells were also treated with an elongation inhibitor to block their maturation. Thus, the decrease in RIs observed in *S. cerevisiae* cells treated with adozelesin alone occurs because RIs mature in the absence of new initiation events. However, this inhibitory effect may be transient—RIs rapidly accumulate once again shortly after the disappear. A similar inhibitory effect on initiation occurred in *mec1* and *rad53* mutant strains of *S. cerevisiae*, although the magnitude of this effect may be reduced compared to wild type cells. The reduced inhibitory effect on initiation in *mec1* and *rad53* strains suggests that the intra-S-phase checkpoint mediated by the *MEC1* and *RAD53* genes involves an inhibitory effect on initiation of DNA replication, similar to the initiation inhibitory effects mediated by ATM in mammals. However, the RIs which eventually accumulated in adozelesin-treated wild type strains after prolonged treatment persisted for a period of time much longer than the length of S-phase in untreated cells, indicating that adozelesin also blocks *S. cerevisiae* DNA replication at the level of elongation of nascent DNA chains. Therefore, the *S. cerevisiae* intra-S-phase checkpoint response may also involve an arrest of DNA synthesis at the level of nascent chain elongation.

A number of proteins directly involved in initiation of DNA replication in *S. cerevisiae* have been identified in recent years—these include the subunits of the Origin Recognition Complex (ORC), which interacts with sequences at DNA replication origins, as well as other proteins which play various roles in regulating initiation of DNA replication (reviewed in Diffley, 1995). One possibility is that drug-induced alterations in the function of some of these proteins are responsible for the initiation inhibitory effect on *S. cerevisiae* DNA replication triggered by adozelesin. To address this question, we screened a number of strains of *S. cerevisiae* containing temperature-sensitive mutations in proteins involved in DNA replication for adozelesin-induced changes in viability and/or growth. The mutant proteins produced by these strains are partially defective in function when the strains are incubated at a semi-permissive temperature. In the absence of drug treatment, the defective function of these proteins does not significantly alter the growth or viability of mutant strains compared to untreated isogenic wild type strains. However, when the mutant strains are also treated with adozelesin, additional drug-induced alterations in the function of specific mutant proteins affected by the drug treatment result in slower growth and/or loss of viability compared to adozelesin-treated isogenic wild type strains and strains containing mutations in genes whose products are not affected by the drug. In principle, this assay is similar to a synthetic lethality assay of strains containing two conditional mutant genes (Guarente, 1993). A functional interaction between the two mutant proteins is indicated in the synthetic lethality assay by the lethal phenotype of the double mutant under conditions in which each of the single mutant strains is viable. In our screen for drug effects, effectors of the response triggered by the drug which function upstream or downstream of the mutant protein play a role analogous to that of the second mutation in the synthetic lethality screen.

In preliminary experiments, adozelesin treatment caused a 1–2 log decrease in the viability of *S. cerevisiae* strains containing temperature-sensitive mutations in the *ORC2* gene compared to adozelesin-treated isogenic wild type control strains and several other mutant strains. In addition, strains containing mutations in the *ORC5* and *CDC28* genes exhibited reduced growth kinetics as a result of adozelesin treatment. All three of these genes produce proteins which are involved in various aspects of initiation of DNA replication. Thus, it is likely that adozelesin's initiation-specific inhibitory effect on DNA replication in wild type *S. cerevisiae* cells is related to adozelesin-induced alterations in the function of these proteins. No significant effect on viability or growth compared to isogenic wild type strains was observed in the seven additional mutant strains—*cdc2, cdc6, cdc7, cdc8, dbf4, cdc14,* and *cdc45*—which were examined in the screen. Although several of these strains also produce mutant proteins involved in initiation of DNA replication, others contain mutations in genes involved in other aspects of DNA metabolism. Thus, adozelesin's effect on the viability and/or growth of strains containing mutations in *ORC2, ORC5* and *CDC28* appears to be related to a specific drug-induced response which impinges directly or indirectly on the function of the proteins encoded by these genes rather than the result of a nonspecific cytotoxic effect on a number of different cellular processes.

In summary, these data represent the first evidence for DNA damage-induced initiation- and elongation-specific inhibitory effects on DNA replication in yeast, similar to the replication-inhibitory effects of DNA damage in mammals. They also suggest that the initiation and elongation-inhibitory effects correspond to the *MEC1/RAD53*-mediated intra-S-phase checkpoint response recently identified in *S. cerevisiae*. The initiation-inhibitory response appears to directly impinge on the function of proteins involved in initiation of DNA replication.

The ability to distinguish between initiation- and elongation-inhibitory effects induced by DNA damage in yeast should facilitate efforts to understand the nature of the similar effects observed in mammals, including their role in the intra-S-phase DNA damage checkpoint response. Whether or not this response is related to the potent antitumor activity inhibitory of adozelesin and other DNA damaging drugs remains unclear. If the cytotoxic effects of these drugs are, in fact, related to this response, elements of this response might provide new targets for the development of effective anticancer drugs.

REFERENCES

Berger, C. A., and Edenberg, H. J. (1986). Pyrimidine dimers block simian virus 40 replication forks. Mol. Cell. Biol. 6:3443–3450.

Boger, D. L., and Johnson, D. S. (1996). CC-1065 and the Duocarmycins: understanding their biological function through mechanistic studies. Angewandte Chemie 35:1438–1474.

Brewer, B. J., and Fangman, W. L. (1987). The localization of replication origins on ARS plasmids in *S. cerevisiae*. Cell 51:463–471.

Cobuzzi, R. J., Jr., Burhans, W. C., and Beerman, T. A. (1996). Inhibition of initiation of simian virus 40 DNA replication in infected BSC-1 cells by the DNA alkylating drug adozelesin. J. Biol. Chem. 271:19852–19859.

Diffley, J. F. X. (1995). The initiation of DNA replication in the budding yeast cell division cycle. Yeast 11:1651–1670.

Guarente, L. (1993) Synthetic enhancement in gene interaction: a genetic tool come of age *Trends in Gen.* 9, 362–366.

Kaufmann, W. K. (1995). Cell cycle checkpoints and DNA repair preserve the stability of the human genome. Cancer Metastasis Rev. 14:31–41.

McHugh, M.M., Beerman, T.A., and Burhans, W.C. (1996) The DNA-damaging enediyne C-1027 inhibits initiation of intracellular SV40 DNA replication in *trans Biochemistry, in press.*

Murnane, J. P. (1995). Cell cycle regulation in response to DNA damage in mammalian cells: a historical perspective. Cancer Metast. Revs. 14:17–19.

Nicolaou, K. C., Smith, A. L., & Yue, E. W. (1993) Proc. Natl. Acad. Sci. USA 90, 5881–8.

Painter, R. B., and Young, B. R. (1980). Radiosensitivity in ataxia-telangiectasia: a new explanation. Proc. Natl. Acad. Sci. USA 77:7315–7317.

Paulovich, A. G., and Hartwell, L. H. (1995). A checkpoint regulates the rate of progression through S phase in S. cerevisiae in response to DNA damage. Cell 82:841–847.

Povirk, L. F. (1977). Localization of inhibition of replicon initiation to damaged regions of DNA. J. Mol. Biol. 114:141–151.

Wang, Y., Huq, M. S., Cheng, X., and Iliakis, G. (1996). Evidence for activities inhibiting in trans initiation of DNA replication in extracts prepared from irradiated cells. Radiat. Res. 145:408–418.

Zakian, V. A. (1995). *ATM*-related genes: What do they tell us about functions of the human gene? Cell 82:685–687.

DISCUSSION

Hartwell: How do you interpret the increased death of the *orc2* mutant strain following treatment with adozelesin?

Burhans: What I think is happening is that at the semipermissive temperature, initiation function is partially compromised by the mutation.These cells are sick compared to the congenic wild type cells, even in the absence of drug treatment. Presumably, the decrease in viability that occurs when we treat them with adozelesin sort of kicks them over the edge by triggering an inhibitory effect that further decreases the function of orc2p. Our results don't prove this, because the possibility remains that, when we treat the *orc2* mutant strain with drug, we are simply making sick cells even sicker in a nonspecific way. To address this question, we are using other types of assays to look for differential effects in the *orc2* mutant strain compared to wild type cells.

Wahl: You went from inhibition of SV40 DNA replication to yeast. But you did not say anything about drugs and cellular replication origins.

Burhans: We do not know yet whether or not there is a similar effect at cellular origins in mammals. The only way to address this question is by 2D gel analysis. I think the 2D gels provided definitive information in the case of SV40, but as you know, they are very difficult to perform on mammalian DNA intermediates.

Wahl: Well, I was actually going to go one step back from that, because I know how hard that experiment is. What is the effect of this drug on S phase progression in normal and AT cells—do AT cells undergo a rapid S phase, just as *mec1* cells do?

Burhans: The drug very effectively blocks cellular DNA replication in mammalian cells at the same concentrations that it blocks SV40 DNA replication. The inhibitory effect on cellular DNA replication in AT cells is, as far as we can tell, identical to that in normal cells. This is also the case for a matched pair of p53 plus and p53 minus Li-Fraumeni fibroblasts we obtained from George Stark.

Wahl: So then it behaves differently than *mec1*, is that what I am hearing?

Burhans: Yes. Apparently, at least in the assays we've done, there is not very much of a different effect in AT cells compared to normal cells, and this is different from the comparison between *mec1* and wild type cells in *S. cerevisiae*. But I think it should be kept in mind that the effect of DNA damage on DNA replication observed in AT cells by other laboratories can be very subtle—for instance, maybe a 50% reduction in the inhibitory effect on DNA replication compared to normal controls. So it is not yet clear to me that there isn't a different effect, but because of its subtle nature, we missed it, or perhaps it is just missing in the strains of AT cells we are using in our experiments.

Nasmyth: So the synthetic lethality with your mutants might suggest that the damage response is going for ORC/CDC6/MCM-type complexes with regard to an effect on initiation. Yet, the fact that you get, I mean, one suspects that these complexes are bypassed in SV40, and yet SV40 DNA replication is inhibited as well. So one might suspect that it is affecting a step subsequent to the steps involving these complexes.

Burhans: Yes, I would agree.

Nasmyth: Did you test what happened in the *mec1* and *rad53* mutants? Did they respond in the same way? Did you look at what the effect on viability was in these strains?

Burhans: Yes. While the viability of congenic wild type strains was reduced by about 20%, the viability of the *mec1* and *rad53* strains was reduced by an additional four logs.

Nasmyth: So they die very rapidly.

Burhans: Yes. At one point we were looking for divided nuclei as another indication of an initiation-specific effect, since this is a phenotype of some initiation mutants. In fact, we did not see divided nuclei in the wild type cells. However, it became apparent we were seeing lots of divided nuclei in the *mec1* strain after treatment with the drug. But this may not be related just to an intra-S-phase checkpoint. *Mec1* cells are also deficient in the G2 checkpoint induced by DNA damage, and this would also result in elevated numbers of divided nuclei as the cells go through mitotic catastrophe. So the reason why the *Mec1* cells die more efficiently is probably related to the absence of the G2 checkpoint as well as the intra-S-phase checkpoint.

NEOPLASTIC PROGRESSION IN BARRETT'S ESOPHAGUS

Michael T. Barrett,[1] Carissa A. Sanchez,[1] Patricia C. Galipeau,[1] Katayoun Neshat,[2] David S. Cowan,[1] Douglas S. Levine,[2] and Brian J. Reid[1,2*]

[1]Program in Cancer Biology
Public Health Sciences
Fred Hutchinson Cancer Research Center
Seattle, Washington 98104
[2]Division of Gastroenterology 356424
Department of Medicine
University of Washington Medical Center
Seattle, Washington 98195

In 1976, Nowell hypothesized that cancer develops as a consequence of an acquired genomic instability that predisposes to the development of abnormal clones of cells with accumulated genetic errors (Nowell, 1976). Some clones gain selective proliferative advantages, and eventually a subclone evolves that has acquired the capacity for invasion, becoming an early carcinoma. There is now substantial evidence that human cancers develop in association with a process of genetic instability and clonal evolution that leads to the accumulation of genetic errors (Vogelstein, et al., 1988; Sidransky, et al., 1992; Raskind, et al., 1992; Huang, et al., 1992). However, few human model systems have been developed in which it is possible to investigate the sequence of events that leads to the development of a carcinoma *in vivo*. Many studies have focused on the genetic abnormalities that are present in advanced carcinomas in surgical specimens. Although the study of advanced cancers is useful for identifying genetic abnormalities that have accumulated in the cancer, the ability of this approach to determine the order in which genetic and other abnormalities develop during earlier stages of neoplastic progression is limited.

Barrett's esophagus is a condition in which the normal squamous epithelium of the esophagus is replaced by a metaplastic columnar epithelium (Phillips and Wong, 1991). It

* Address Correspondence to: Brian J. Reid, MD, PhD, Program in Cancer Biology, Division of Public Health Sciences, Fred Hutchinson Cancer Research Center, Seattle, Washington 98104. Telephone: (206) 667-2875; FAX: (206) 667-5815.

develops as a complication in approximately 10–20% of patients with chronic gastroesophageal reflux and predisposes to the development of adenocarcinomas of the esophagus and gastroesophageal junction (Hamilton, et al., 1988; Phillips and Wong, 1991; Reid, 1991). In the past two decades, the incidence of these Barrett's-associated cancers increased more rapidly than that of any other cancer in the United States (Blot, et al., 1991). Unfortunately, Barrett's adenocarcinomas are rarely detected in time for cure, and 93% of patients who develop an esophageal adenocarcinoma will eventually die of their disease (Miller, et al., 1989).

Barrett's esophagus is a unique model system in which to investigate intermediate events of human epithelial neoplasia (Neshat et al., 1995). Because patients with Barrett's esophagus typically have symptoms of gastroesophageal reflux, such as heartburn or indigestion, they frequently seek medical attention before they develop cancer (Phillips and Wong, 1991). The Barrett's epithelium can be safely visualized and biopsied during upper gastrointestinal endoscopy (Levine and Reid, 1992). At the present time, total removal of Barrett's epithelium requires esophagectomy, a procedure with substantial morbidity and mortality (Muller, et al., 1990). However, a systematic protocol of endoscopic biopsies can detect cancers arising in Barrett's esophagus when they are early and curable (Reid, et al., 1988a; Levine, et al., 1993). Therefore, the standard of care for many patients includes periodic endoscopic biopsy surveillance for the early detection of cancer (Spechler, 1987; Levine and Reid, 1992). Thus, intermediate events in neoplastic progression can be evaluated by serial biopsies of the same patient over time and related to progression (Reid, et al., 1992). Furthermore, in addition to cancer, esophagectomy specimens very frequently contain the surrounding premalignant epithelium in which the cancer arose, permitting the study of multiple stages of neoplastic progression in a single esophagectomy specimen (Reid, et al., 1988a; Rabinovitch, et al., 1988).

GENETIC AND CELL CYCLE ABNORMALITIES IN BARRETT'S ADENOCARCINOMAS

Barrett's adenocarcinoma, like many other human malignancies, has a high prevalence of genetic and cell cycle abnormalities (Table 1) (Reid, et al., 1987; Blount, et al., 1991; Boynton, et al., 1992; Reid, et al., 1993; Neshat, et al., 1995, Barrett, et al., 1996a). None of these abnormalities are present in control biopsies from columnar epithelium of

Table 1. Barrett's adenocarcinomas –
prevalence of abnormalities

Abnormality	Prevalence (%)
↑G1 fractions	15/17 (88%)
↑S phase fractions	24/28 (86%)
↑4N fractions	18/28 (64%)
Aneuploid	25/28 (89%)
17p allelic loss	34/36 (94%)
p53 mutations	14/16 (88%)
5q allelic loss	20/24 (83%)
9p allelic loss	19/26 (73%)
13q allelic loss	10/20 (50%)
18q allelic loss	10/21 (48%)

the upper gastrointestinal tract or other constitutive tissues; they all develop as somatic events in Barrett's metaplastic epithelium during the progression to cancer.

ANEUPLOIDY IN BARRETT'S ESOPHAGUS

The prevalence of aneuploid cell populations increases with increasing histological risk of malignancy in Barrett's esophagus (Reid, et al., 1987; McKinley, et al., 1987; Fennerty, et al., 1989; Robaszkiewicz, et al., 1991; Reid, 1991). In the largest series of consecutive patients, aneuploid cell populations were not detected in biopsies from 44 patients who had gastroesophageal reflux disease without Barrett's metaplastic epithelium. However, aneuploid cell populations were found in biopsies from 3 of 70 patients with metaplasia (4%), 2 of 32 patients whose biopsies were in the indefinite for dysplasia or low-grade dysplasia range (6%), 5 of 8 patients with high-grade dysplasia (63%), and 25 of 28 cancers (89%) (Reid, 1991).

We investigated the distribution of aneuploid cell populations in Barrett's esophagus by taking endoscopic biopsy specimens at different levels of the metaplastic epithelium and by sampling the mucosa of esophagectomy specimens in a grid array (Reid, et al., 1987; Rabinovitch, et al., 1988). Many aneuploid cell populations are localized to a single region of the esophageal mucosa, but some spread to involve large areas of the esophagus. For example, one patient had the same 2.2N aneuploid cell population at each level of a ten centimeter length of metaplastic epithelium. Control biopsies from gastric and squamous mucosa were diploid, indicating that this aneuploid cell population developed in the metaplastic epithelium as the result of a somatic genetic event. Cytogenetic analysis of endoscopic biopsies from this patient confirmed that the aneuploid cell population contained clonal karyotypic abnormalities that were found at multiple levels of the metaplastic epithelium (Raskind, et al., 1992). These results suggest that abnormal clones of cells can spread by a process of cell division to involve large regions of esophageal mucosa in persons with Barrett's esophagus (Rabinovitch, et al., 1988; Reid, et al., 1992; Raskind, et al., 1992).

Some patients have multiple aneuploid cell populations in their endoscopic biopsies or in their esophagectomy specimens. We found multiple (2–14) aneuploid cell populations in 12 of 14 patients (86%) who had high-grade dysplasia, adenocarcinoma, or both in Barrett's esophagus (Rabinovitch, et al., 1988). Different aneuploid cell populations occupied defined, but sometimes overlapping, spatial distributions in the Barrett's epithelium, suggesting that they represented abnormal clones of cells that had expanded to involve variable regions of esophageal mucosa. In seven patients, we investigated the relationship between multiple aneuploid cell populations that were present in premalignant epithelium and the ploidy of the invasive carcinoma that developed within the multiple aneuploidies (Blount, et al., 1990). In six of the seven patients, the cancer contained only one of the multiple aneuploid cell populations present in the premalignant epithelium. The seventh and largest cancer contained five different ploidies in the cancer itself.

Our results indicate that endoscopic biopsies from most patients with Barrett's esophagus are diploid, but some patients develop abnormal clones of cells with large changes in DNA content or ploidy. Some of these clones can spread to involve large regions of esophageal mucosa. With continued genetic instability, multiple aneuploid cell populations can evolve in the premalignant epithelium, and one of the aneuploid cell populations may acquire the capacity for invasion, becoming an early carcinoma. If the cancer is not resected at this stage, continued instability can lead to multiple aneuploid cell populations within the cancer itself.

THE CELL CYCLE IN BARRETT'S ESOPHAGUS

Cell cycle checkpoints cause arrest at specific stages of the cell cycle in response to genotoxic injury or a failure to complete previous events of the cell cycle (Weinert and Hartwell, 1988; Hartwell and Weinert, 1989; Dasso and Newport, 1990; Enoch and Nurse, 1990; Brown, et al., 1991). Previous studies have shown that some, but not all, patients with Barrett's esophagus have increased proliferative fractions in the metaplastic epithelium (Herbst, et al., 1978; Pellish et al., 1980; Reid, et al., 1987; Gray, et al., 1992). However, it has been difficult to assess the transitions from G_0 to G_1 to S phase during neoplastic progression *in vivo* because most assays of proliferation cannot simultaneously measure all three intervals in human biopsies. Therefore, we used a multiparameter flow cytometric assay that simultaneously measures Ki67 (a proliferation associated nuclear antigen present in cells in late G_1, S, G_2, and mitosis, but not in G_0) and DNA content to assess proliferation in biopsies from patients with Barrett's esophagus (Reid, et al., 1993). Biopsies from control columnar epithelium of the upper gastrointestinal tract (fundic gland mucosa and cardiac gland mucosa) had low G_1 and S phase fractions, suggesting that the cells were predominantly in G_0. Increased Ki67-positive G_1 fractions were found in Barrett's metaplastic epithelium at an early stage of neoplastic progression, but S phase fractions typically remained normal, suggesting that regulatory mechanisms at the G_1/S phase transition prevented uncontrolled progression of G_1 cells into S phase even if G_1 fractions increased to very high levels. At later stages of neoplastic progression, increased S phase fractions developed, usually in association with aneuploidy, high-grade dysplasia, or carcinoma. Only 19 of 73 diploid biopsies (26%) from Barrett's esophagus at all histological stages of progression had increased S phase fractions compared with 21 of 22 aneuploid cell populations (95%) (p<0.001).

Some patients also develop elevated 4N (G_2/tetraploid) fractions, and the prevalence of elevated 4N fractions increases with increasing histological risk of cancer (Reid, et al., 1987; Reid, 1991). Increased 4N fractions were found in none of 44 patients who had gastroesophageal reflux disease without Barrett's metaplasitic epithelium and none of 70 patients who had metaplasia without dysplasia, but they were detected in 7 of 32 patients with indefinite/low-grade dysplasia (22%), 7 of 8 patients with high-grade dysplasia (88%), and 18 of 28 patients with cancer (64%) (Reid, 1991).

ANEUPLOIDY AND INCREASED 4N FRACTIONS AS PREDICTORS OF PROGRESSION

We prospectively evaluated 62 patients for a mean of 34 months (Reid, et al., 1992). Of these 62, 13 had either aneuploidy or increased 4N fractions as their initial flow cytometric abnormality. Of these 13, nine progressed to develop high-grade dysplasia or cancer that was not present at the initial endoscopy. None of 49 patients without aneuploidy or increased 4N fractions progressed to high-grade dysplasia or cancer (p<0.0001). The temporal course of progression from aneuploidy or increased 4N fractions to cancer was variable, ranging from 18 to 84 months. Patients whose biopsies were diploid and had normal 4N fractions did not progress to high-grade dysplasia and cancer during the time course of this study, but 7 of the 49 patients subsequently progressed to develop aneuploid cell populations or increased 4N fractions during prospective follow-up.

In a subsequent study, we investigated the hypothesis that patients whose biopsies had increased 4N fractions were at increased risk for progression to aneuploidy (Galipeau, et al., 1996). In this study, we prospectively evaluated 90 patients, whose biopsies were initially diploid, to determine whether increased 4N fractions predict subsequent progression to aneuploidy. Eight of 75 patients (11%) with no 4N abnormalities progressed to aneuploidy over a mean follow-up of 52 months. In contrast, 11/15 patients (73%) with 4N abnormalities subsequently developed aneuploid populations during a mean follow-up of 17.2 months (p<0.0001).

ORDERING EVENTS OF NEOPLASTIC PROGRESSION IN BARRETT'S ESOPHAGUS

The formalism for investigating the order of genetic events in a multistep pathway was developed as part of the early investigations of the yeast cell cycle. Basically, any two events, A and B, of a multistep pathway can have one of the following four relationships (Figure 1): 1) A can consistently precede B; for example, A could cause B, or A could create a condition that is permissive for B to occur; 2) B can consistently precede A, which rules out the possibility that A causes B in those cases; 3) A and B can be detected simultaneously, for example, A and B may affect the same step (interdependent events) or 4) A and B can occur in a random order (independent events) (Hereford and Hartwell, 1974). We have used this formalism to investigate the order in which two events occur relative to each other in patients with Barrett's esophagus by evaluating serial endoscopic biopsies and "mapped" biopsies from esophagectomy specimens.

ALLELOTYPE OF ESOPHAGEAL ADENOCARCINOMA

Different types of cancers are characterized by high frequencies of loss of heterozygosity (LOH) at specific sites in the genome. These include colorectal carcinoma (5q,17p,18q) (Vogelstein, et al., 1989), osteosarcoma (3q,13q,17p,18q) (Yamaguchi, et al.,

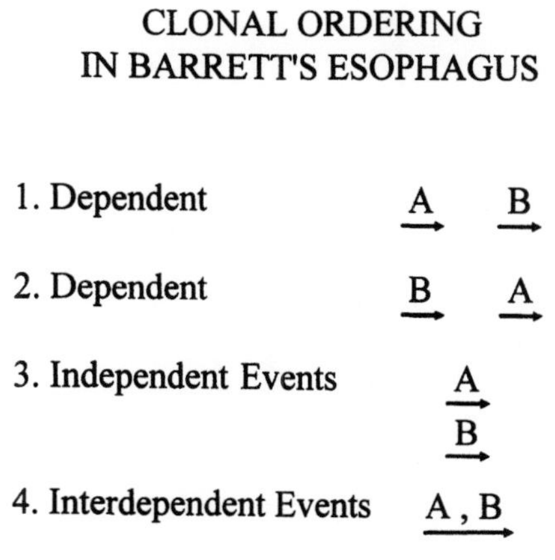

CLONAL ORDERING
IN BARRETT'S ESOPHAGUS

1. Dependent   A →   B →

2. Dependent   B →   A →

3. Independent Events   A →
      B →

4. Interdependent Events   A , B →

Figure 1. Four possible relationships between two events, A and B, of a multistep pathway.

1992), breast carcinoma (13q,16q,17p) (Sato, et al., 1990), renal cell carcinoma (3p,5q,6q,10q,11q,17p,19p) (Thrash-Bingham, et al., 1995) and endometrial carcinoma (6p,10q,12p,15q,17p) (Peiffer, et al., 1995). Regions of high frequency LOH have been shown to include specific tumor suppressor genes that are involved in neoplastic progression of several cancers.

However, some human cancers of the upper gastrointestinal tract have an excess of genetically normal stromal cells that make LOH investigations difficult (Boynton, et al., 1991; Hirohashi and Sugimara, 1991). Previously, flow cytometric cell sorting allowed us to investigate LOH of a single locus in small highly purified endoscopic biopsies (Blount, et al., 1993, 1994; Neshat, et al., 1995). However, in order to investigate multiple genetic loci in a single flow-sorted sample, we adapted a PCR-based technique of whole genome amplification to the study of genetic lesions in somatic cells (Barrett, et al., 1995).

DNA from flow-sorted aneuploid and matching normal cells from 22 different patients with esophageal adenocarcinoma (n=20) or high-grade dysplasia (n=2) were amplified by whole genome amplification, then screened for LOH using highly polymorphic simple sequence repeat markers in PCR assays for each autosomal chromosome arm, with the exception of the short arms of the acrocentric chromosomes 13, 14, 15, and 22 (Barrett, et al., 1996a). The frequency of LOH at different loci varied from 0% (20q, 0 of 8; 21p 0 of 10) to 100% (17p, 20 of 20) (Figure 2). The highly significant rates of LOH observed in the allelotype analysis at 17p (100%, p< 0.0001) and 5q (80% , p<0.0001) were consistent with our previous reports (Huang, et al., 1992; Neshat, et al., 1995). After exclusion of 17p and 5q, the frequency of LOH at 9p (64%) in our initial study was found to be significantly higher than expected for the largest of 38 binomial observations (p=0.02). After exclusion of 17p, 5q and 9p the estimated background frequency of allelic loss was 0.23 with a 95% confidence interval of 0.20 to 0.27. Chromosome arms most likely to contain tumor suppressor genes involved in esophageal adenocarcinomas were identified by dividing the arms into three groups according to the probability that the

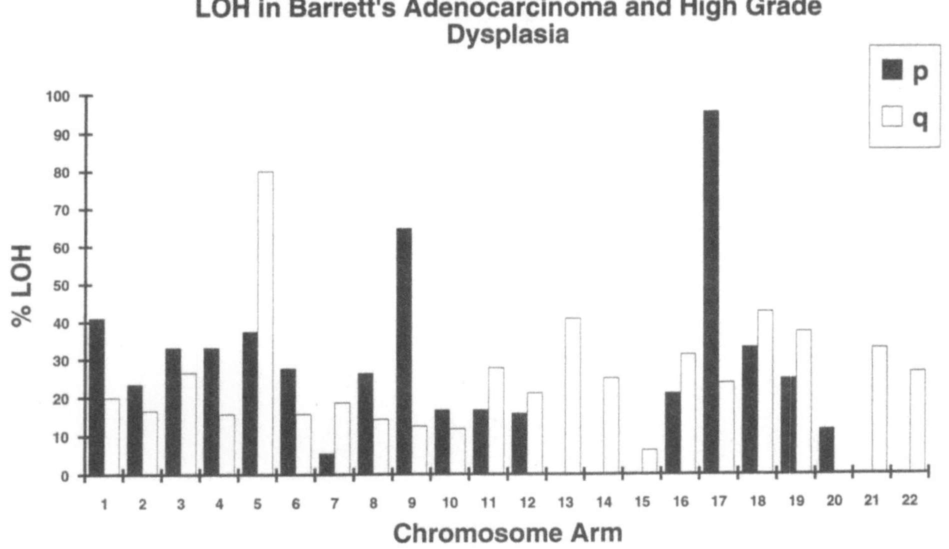

Figure 2. Allelotype of Barrett's adenocarcinoma.

observed frequency could be explained by random fluctuations about the background LOH frequency. Based on a binomial distribution with mean 0.23, observed LOH frequencies having probabilities less than 0.05 of occuring at random were 17p (100%), 5q (80%), 9p (64%) and 13q (43%) (i.e. $p<0.05$). Observed LOH frequencies with $0.05<p<0.10$ were 18q (43%) and 1p (41%). Subsequent analyses of aneuploid cell populations in a larger set of patients has shown that the prevalence of loss at 13q (14/30; 47%) and 18q (17/36; 47%) are consistently observed with a $p<0.05$, while LOH at 1p (9/33; 28%) has $p>0.10$ of occuring at random.

There were 126 examples of LOH in which the patient was informative for both the p and the q arms of the chromosome showing LOH. In 86 of these cases (68%), LOH involved only a single arm, indicating that the majority of LOH events detected in this study were subchromosomal, such as might result from mitotic recombination or interstitial deletions, rather than nondisjunction of an entire chromosome.

The fractional allelic loss (FAL) (i.e. the ratio of chromosomal arms showing allelic loss to the number of informative arms in each patient) ranged from 0.07 to 0.5 with a median of 0.28. There was a significant trend for FAL to increase with depth of invasion ($p=0.001$). The mean FAL was 0.095 in high-grade dysplasia without invasion ($n=2$), 0.264 in early adenocarcinomas (intramucosal and submucosal, $n=10$), and 0.343 in advanced adenocarcinomas in which the level of invasion extended at least to the muscularis propria ($n=10$). Previous allelotype studies have suggested that FAL is an indicator of aneuploidy (Vogelstein, et al. 1989). However increasing ploidy, as assessed by increasing DNA content, in patients with Barrett's esophagus did not correlate with increasing allelic loss events. This suggests that allelic losses and changes in ploidy are two distinct consequences of the genetic instability that arises during neoplastic progression.

p53 IN BARRETT'S ESOPHAGUS

p53 mutations and 17p allelic losses are among the most common abnormalities found in human cancers (Hollstein, et al., 1991: Harris and Hollstein, 1993). In addition, inactivation of p53 by 17p allelic loss and p53 mutation has been reported in premalignant conditions of the colon, lung, and squamous esophagus, among others (Vogelstein, et al., 1988; Baker, et al., 1990; Bennett, et al., 1992; Sozzi, et al., 1992). Our results, and those of others, suggest that 17p allelic losses, p53 mutations, and p53 overexpression, develop as relatively early events in the progression to cancer in Barrett's esophagus (Casson, et al., 1991; Ramel, et al., 1992; Younes, et al., 1993; Blount, et al., 1994; Neshat, et al., 1994). Therefore, we have used 17p allelic losses and p53 mutations to assess the relationships among inactivation of p53 and other events of neoplastic progression in Barrett's esophagus.

THE ORDER OF 17p (p53) ALLELIC LOSSES AND CANCER IN BARRETT'S ESOPHAGUS

17p allelic losses were found in 34 of 36 Barrett's adenocarcinomas (94%). The order in which 17p allelic losses and cancer developed could not be determined in many cases because the cancer had overgrown its precursors. In 14 patients in whom the order could be determined, 17p allelic losses were detected in premalignant epithelium in 12 cases (86%), cancer and the 17p allelic loss were detected simultaneously in 1 case (7%), and cancer preceded the 17p allelic loss in one case (7%) (Neshat at al., 1995). We also investigated p53 mutations in Barrett's adenocarcinomas and premalignant epithelium. p53 mutations were detected in 14 of 16 cancers (88%) and in 6 of 9 patients with high-grade dysplasia (67%) (Neshat et al., 1994). In three cases, the same p53 mutation was found in both premalignant epithelium and cancer, indicating that both p53 mutations and 17p allelic losses develop before cancer in Barrett's esophagus.

THE ORDER OF 17p (p53) ALLELIC LOSSES AND ANEUPLOIDY IN BARRETT'S ESOPHAGUS

17p allelic losses were found in the aneuploid cell populations of 41 of 44 patients (93%) who had aneuploidy in Barrett's metaplastic epithelium. Because both aneuploidy and 17p allelic losses can be detected before cancer in Barrett's esophagus, we investigated the order in which the two events occurred relative to each other in the premalignant epithelium of 22 patients (Blount et al., 1994; Neshat et al., 1995). 17p allelic losses were detected in diploid cells in the premalignant epithelium of 21 of the 22 patients (95%). In the remaining patient, the 17p allelic loss and aneuploidy were detected simultaneously (p<0.00001). We sequenced the p53 gene in the diploid and aneuploid cell populations of three patients to confirm that the 17p allelic losses were associated with p53 mutations. In all three patients, diploid and aneuploid cell populations contained the same p53 mutation.

THE ORDER OF 17p (p53) ALLELIC LOSSES AND INCREASED 4N (G2/TETRAPLOID) FRACTIONS IN BARRETT'S ESOPHAGUS

Our results indicate that 17p (p53) allelic losses precede the development of aneuploidy and that increased 4N fractions predict progression to aneuploidy in Barrett's esophagus. Therefore, we investigated the relationship between 17p (p53) allelic losses and 4N fractions in 105 diploid biopsies from 34 patients to determine the relationship between inactivation of p53 and the development of 4N abnormalities in Barrett's epithelium (Galipeau, et al., 1996). The mean 4N (G_2/tetraploid) fraction was 2.7% in 66 biopsies that retained both 17p alleles, compared to 9.5% in 39 biopsies with 17p allelic loss (p=0.0001). In five patients, the remaining p53 allele was sequenced; all five had mutations in the remaining allele. In 10 patients, we investigated the change in 4N (G_2/tetraploid) fractions with acquisition of a 17p allelic loss. In every case, the mean 4N (G_2/tetraploid) fraction increased in association with development of a 17p allelic loss, rising from a total mean 4N (G_2/tetraploid) fraction of 3.3% in biopsies with two 17p alleles to 9.3% in biopsies with a 17p allelic loss. In every case, the same 17p allele was lost in 2N, 4N and, when available, aneuploid cell populations. In one case, we showed by DNA sequencing that the remaining 17p allele had the same mutation in 2N, 4N, and aneuploid cells. Therefore, inactivation of the p53 gene with loss of its checkpoint functions is interdependent with the development of an abnormal 4N intermediate and precedes the development of aneuploidy in Barrett's esophagus. These results are consistent with data indicating that increased 4N (G_2/tetraploid) cell populations identify a subset of patients with an increased risk of progression to aneuploidy in Barrett's esophagus *in vivo*.

THE ORDER OF 17p ALLELIC LOSSES AND 5q ALLELIC LOSSES IN BARRETT'S ESOPHAGUS

Chromosome 5q is the second most frequent site of allelic loss in Barrett's adenocarcinomas (Table 1). We investigated the order in which 17p allelic losses and 5q allelic losses occurred in 38 aneuploid cell populations from 14 patients with Barrett's esophagus (Blount, et al., 1993). 17p allelic losses developed before 5q allelic losses in 7 patients (50%), the two allelic losses were detected simultaneously in 4 patients (29%), and 17p allelic losses occurred without 5q allelic losses in 3 patients (21%).

THE ORDER OF 5q ALLELIC LOSSES AND CANCER IN BARRETT'S ESOPHAGUS

The above results indicate that both 5q allelic losses and cancer typically occur after 17p allelic losses during neoplastic progression in Barrett's esophagus. Therefore, we investigated the order in which 5q allelic losses and cancer developed in 21 patients with Barrett's esophagus (Blount et al., 1993; Neshat et al., 1995). 5q allelic losses preceded cancer in 3 patients (14%); cancer preceded 5q allelic losses in 4 patients (19%); the two events were detected simultaneously in 8 patients (38%); cancer occurred without 5q allelic losses in 5 patients (24%); and 5q allelic losses occurred without cancer in 1 patient (5%). These results suggest that 5q allelic losses are independent of the events that cause invasion during neoplastic progression in Barrett's esophagus.

THE ORDER OF 9p ALLELIC LOSSES AND CANCER IN BARRETT'S ESOPHAGUS

9p allelic losses were found in the aneuploid populations from 31 of 41 informative patients with Barrett's esophagus (75%). 9p allelic losses were found in aneuploid cell populations from 19 of 26 cancers (73%). Premalignant tissue was available for analysis from seven of the nineteen cancers with 9p allelic loss. In all seven cases (100%), allelic loss of 9p21 was detected in the premalignant epithelium (p=0.016). In all cases, constitutive tissues had two 9p21 alleles indicating that the 9p21 allelic losses developed as somatic events in premalignant epithelium. Furthermore, 9p allelic losses were detected in aneuploid cell populations from twelve of fourteen patients (85%) who had aneuploidy in premalignant epithelium without cancer in their Barrett's segment. Therefore, 9p allelic losses were detected in the premalignant epithelium of 19 patients, including twelve who had not developed cancer, and another seven who had progressed. These results indicate that 9p allelic losses can develop as early events before invasion during neoplastic progression in Barrett's esophagus (Barrett et al., 1996b).

THE ORDER OF 9p ALLELIC LOSSES AND ANEUPLOIDY IN BARRETT'S ESOPHAGUS

Patients with both 9p allelic losses and aneuploidy in premalignant epithelium were used to determine the order in which these two events occured relative to each other during neoplastic progression (Barrett et al., 1996b). Sixty-four frozen biopsies that were suitable for immunofluorescent staining with Ki67, allowing purification of proliferating epithelial cells by Ki67/DNA content multiparameter flow cytometric cell sorting, were available from 15 of these patients. Allelic losses at 9p21 were detected in sorted Ki67-positive diploid G_1 epithelial cell populations in 13 of the 15 patients (87%) with aneuploidy (p = 0.0002). In the remaining two patients both events were detected simultaneously. However, not all Ki67-positive diploid G_1 fractions had 9p allelic loss; in eight of the 13 patients, some sorted biopsies had two 9p alleles.

A MODEL OF NEOPLASTIC PROGRESSION IN BARRETT'S ESOPHAGUS

These results suggest a model of neoplastic progression in Barrett's esophagus (Figure 3). In response to chronic reflux of gastric contents into the esophagus, approximately 10–20% of patients will develop Barrett's metaplastic epithelium. Increased G_1 fractions develop as an early event in the metaplastic epithelium, but S phase fractions are typically normal at this stage of neoplastic progression. Subsequently, the p53 gene may be inactivated in diploid cells by a combination of mutation and 17p allelic loss and inactivation of p53 is associated with the development of increased 4N (G_2/tetraploid) fractions. In addition, diploid cells may also develop allelic loss at the CDKN2 locus on 9p21 in the presence (23%) or absence of mutation in the remaining allele. These events precede the development of aneuploid cell populations, which typically have increased S phase fractions when they are detected. Cancer and 5q allelic losses develop later. Even though 5q allelic losses are detected in 83% of Barrett's adenocarcinomas, they appear to be independent of the events that lead to invasion in the sense that they can be detected before invasion, simultaneously with invasion, or after invasion.

The development of increased G_1 fractions may be a consequence of the injury associated with chronic gastroesophageal reflux and the need to replace the injured epithelium by cellular proliferation. At this stage of neoplastic progression, cell cycle controls still prevent unregulated entry of cells into S phase, and the increased G_1 fractions may be

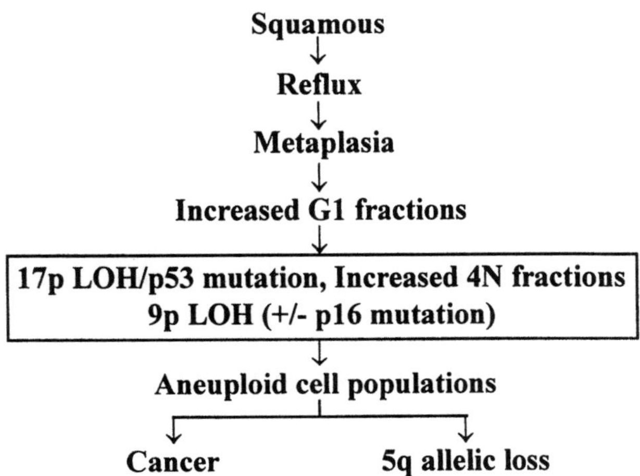

Figure 3. A model of neoplastic progression in Barrett's esophagus.

reversible. Inactivation of the p53 gene by mutation and allelic loss presumably eliminates its checkpoint functions because loss of p53 leads to a condition that is permissive for the subsequent evolution of aneuploid cell populations. The highly prevalent allelic losses at 9p21 ($\pm$ p16 mutations) may contribute to this process by disrupting a negative regulatory pathway that functions in G_1 to prevent entry into S phase. Increased 4N (G_2/tetraploid) fractions are found frequently in biopsies of patients in whom the p53 gene has been inactivated, suggesting that p53 deficiency either leads to tetraploidy in Barrett's esophagus or that cells may accumulate at the G_2 checkpoint as a backup mechanism for monitoring DNA damage when the p53 mediated G_1 checkpoint is lost. There is support for both possible mechanisms. We and others have shown that p53-null mouse embryonic fibroblasts develop tetraploidy spontaneously or after treatment with spindle inhibitors (Cross, et al., 1995; Deng, et al., 1995; Livingstone, et al., 1992). Our studies are also consistent with those that have shown that 17p allelic losses are correlated with aneuploidy in colorectal carcinomas and that p53 protein overexpression can be detected before aneuploidy during neoplastic progression in the colon (Offerhaus, et al., 1992; Carder, et al., 1993). However, it is also known that p53 is a component of a signal transduction pathway that responds to DNA damage by causing arrest in G_1 (Kastan, et al., 1992; Kuberitz, et al., 1992; Livingstone, et al., 1992; Yin, et al., 1992; Lu and Lane, 1993; El-Deiry, et al., 1993; Harper, et al., 1993; Xiong, et al., 1993). Inactivation of this checkpoint by loss of p53 function can allow cells to replicate unrepaired DNA, leading to double strand breaks that cause arrest at a subsequent checkpoint in G_2 (4N DNA content) (Kastan, et al., 1991). If the increased 4N cells that we observe in Barrett's esophagus are in G_2, then the G_2 checkpoint must either adapt, as it does in *Saccharomyces cerevisiae* (Sandell and Zakian, 1993), or be inactivated, because inactivation of p53 and the development of increased 4N fractions are followed by progression to aneuploidy, other allelic losses, and cancer *in vivo*.

SOMATIC GENETICS AND CANCER SURVEILLANCE IN BARRETT'S ESOPHAGUS

Periodic endoscopic biopsy surveillance is recommended for the early detection of cancer in Barrett's esophagus, but the histological indicators that are commonly used for cancer surveillance lack the precision necessary to accurately identify patients who are at increased risk of progression (Reid, et al., 1988b). Therefore, an understanding of the genetic and cell cycle abnormalities that lead to cancer in Barrett's esophagus has clinical implications for the management of the cancer risk in these patients. Only a small subset of Barrett's patients have increased 4N fractions or aneuploid cell populations in their Barrett's epithelium, but this subset has an increased risk of progression to high-grade dysplasia and cancer. Once high-risk patients are identified by flow cytometric or molecular tests, they can be placed in more frequent surveillance for the detection of adenocarcinoma at an early stage in which it is curable (Levine, et al., 1993). The majority of patients, who have not yet entered a pathway of genetic instability, may be followed by less frequent surveillance. Thus, genetic and cell cycle tests that can accurately distinguish patients who have an increased risk of progressing to cancer from those whose risk is low may improve the cure rate of patients who develop cancer while simultaneously decreasing health care costs by safely permitting less frequent surveillance of low-risk patients.

ACKNOWLEDGMENTS

We thank Susan Irvine, RN, and Christine Karlsen, for assistance in patient care, and Rodger C. Haggitt, MD, for histological interpretation. This research was supported by American Cancer Society Grant EDT-21E, NIH R01 CA 61202, and the Ryan Hill Research Foundation.

REFERENCES

Baker, S.J., Preisinger, A.C., Jessup, J.M., Paraskeva, C., Markowitz, S., Willson, J.K.V., Hamilton, S., and Vogelstein, B. 1990. p53 gene mutations occur in combination with 17p allelic deletions as late events in colorectal tumorigenesis. *Cancer Res.* **50:**7717–7722.

Barrett, M.T., Reid, B.J., and Joslyn, G. 1995. Genotypic analysis of multiple loci in somatic cells by whole genome amplification. *Nuc. Acids Res.* **23:**3368–3372.

Barrett, M.T., Galipeau, P., Sanchez, C., Emond, M.J., and Reid, B.J. 1996a. Determination of the frequency of loss of heterozygosity in esophageal adenocarcinoma by cell sorting, whole genome amplification and microsatellite polymorphisms. *Oncogene.* **12:**1873–1878.

Barrett, M.T., Sanchez, C.A., Galipeau, P.C., Neshat, K., Emond, M., Reid, B.J. 1996b. Allelic loss of 9p21 and mutation of the CDKN2/p16 gene develop as early lesions during neoplastic progression in Barrett's esophagus. *Oncogene,* in press.

Bennett, W.P., Hollstein, M.C., Metcalf, R.A., Welsh, J.A., He, A., Zhu, S., Kusters, I. Resau, J.H., Trump, B.F., Lane, D.P., and Harris, C.C. 1992. p53 mutation and protein accumulation during multistage human esophageal carcinogenesis. *Cancer Res.* **52:**6092–6097.

Blot, W.J,. Devesa, S.S., Kneller, R.W., and Fraumeni, Jr. J.F. 1991. Rising incidence of adenocarcinoma of the esophagus and gastric cardia. *JAMA.* **265:**1287–1289.

Blount, P.L., Rabinovitch, P.S., Haggitt, R.C. and Reid, B.J. 1990. Early Barrett's adenocarcinoma arises within a single aneuploid population. *Gastroenterology* **98:** A273.

Blount, P.L., Ramel, S., Raskind, W.H., Haggitt, R.C., Sanchez, C.A., Dean, P.J, Rabinovitch, P.S., and Reid. B.J. 1991. 17p allelic deletions and p53 protein overexpression in Barrett's adenocarcinoma. *Cancer Res.* **51:**5482–5486.

Blount, P.L., Meltzer, S.J., Yin, J., Huang, Y., Krasna, M.J., and Reid, B.J. 1993. Clonal ordering of 17p and 5q allelic losses in Barrett dysplasia and adenocarcinoma. *Proc. Natl. Acad. Sci. USA.* **90:**3221–3225.

Blount, P.L., Galipeau, P.C., Sanchez, C.A., Neshat, K., Levine, D.S., Yin, J., Suzuki, H., Abraham, J.M., Meltzer, S.J., and Reid, B.J. 1994. 17p allelic losses in diploid cells of patients with Barrett's esophagus who develop aneuploidy. *Cancer Res.* **54:**2292–2295.

Boynton, R.F., Huang, Y., Blount, P.L., Reid, B.J., Raskind, W.H., Haggitt, R.C., Newkirk, C., Resau, J.H., Yin, J., McDaniel, T., and Meltzer, S.J. 1991. Frequent loss of heterozygosity at the retinoblastoma locus in human esophageal cancers. *Cancer Res:* **51:**5766–5769.

Boynton, R.F., Blount, P.L., Yin, J., Brown, V.L., Huang, Y., Tong, Y., McDaniel, T., Newkirk, C., Resau, J.H., Raskind, W.H., Haggitt, R.C., Reid, B.J., and Meltzer, S.J. 1992. Loss of heterozygosity involving the *APC* and *MCC* genetic loci occurs in the majority of human esophageal cancers. *Proc. Natl. Acad. Sci. USA.* **89:**3385–3388.

Brown, M., Garvik, B., Hartwell, L., Kadyk, L., Seeley, T., and Weinert, T. 1991. Fidelity of mitotic chromosome transmission. In *Cold Spring Harbor Symposia on Quantitative Biology, Vol LVI.* pp. 359–365 Cold Spring Harbor Laboratory Press, Cold Spring Harbor, New York.

Carder, P, Wyllie, A.H., Purdie, C.A., Morris, R.G., White, S., Piris, J., and Bird, C.C. 1993. Stabilised p53 facilitates aneuploid colonal divergence in colorectal cancer. *Oncogene.* **8:**1397–1401.

Casson, A.G., Mukhopadhyay, T., Cleary, K.R., Ro, J.Y., Levin, B., and Roth, J.A. 1991. p53 gene mutations in Barrett's epithelium and esophageal cancer. *Cancer Res.* **51:**4495–4499.

Cross, S.M., Sanchez, C.A., Morgan, C.A., Schimke, M.K., Ramel, S., Izerda, R.L., Raskind, W.H., and Reid, B.J. 1995. A p53-dependent mouse spindle checkpoint. *Science* **267:**1353–1356.

Dasso, M., and Newport, J.R. 1990. Completion of DNA replication is monitored by a feedback system that controls the initiation of mitosis in vitro: Studies in xenopus. *Cell* **61:**811–823.

Deng, C., Zhang, P., Harper, J.W., Elledge, S.J., and Leder, P. 1995. Mice lacking p21$^{\text{CIP1/WAF1}}$ undergo normal development, but are defective in G1 checkpoint control. *Cell* **82:**675–684.

El-Deiry, W.S., Tokino, T., Velculescu, V.E., Levy, D.B., Parsons, R., Trent, J.M., Lin, D., Mercer, W.E., Kinzler, K.W., and Vogelstein, B. 1993. WAF1, a potential mediator of p53 tumor suppression. *Cell* **75**:817–825.

Enoch, T., and Nurse, P. 1990. Mutation of fission yeast cell cycle control genes abolishes dependence of mitosis on DNA replication. *Cell* **60**:665–673.

Fennerty, M.B., Sampliner, R.E., Way, D., Riddell, R., Steinbronn, K., and Garewal, H. 1989. Discordance between flow cytometric abnormalities and dysplasia in Barrett's esophagus. *Gastroenterology* **97**:815.

Galipeau, P.C., Cowan, D.S., Sanchez, C.A., Barrett, M.T., Emond, M.J., Levine, D.S., Rabinovitch, P.S., and Reid, B.J. 1996. 17p (p53) allelic losses, 4N (G$_2$/tetraploid) populations and progression to aneuploidy in Barrett's esophagus. *Proc. Natl. Acad. Sci. USA.* **93**:7081–7084.

Gray, M.R., Hall, P.A., Nash, J., Ansari, B., Lane, D.P., and Kingsnorth. A.N. 1992. Epithelial proliferation in Barrett's esophagus by proliferating cell nuclear antigen immunolocalization. *Gastroenterology* **103**:1769–1776.

Hamilton, S.R., Smith, R.R.L., and Cameron, J.L. 1988. Prevalence and characteristics of Barrett esophagus in patients with adenocarcinoma of the esophagus or esophagogastric junction. *Hum. Pathol.* **19**:942–948.

Harper, J.W., Adami, G.R., Wei, N., Keyomarsi, K., and Elledge, S.J. The p21 cdk-interacting protein Cip1 is a potent inhibitor of G1 cyclin-dependent kinases. *Cell* **75**:805–816, 1993.

Harris, C.C. and Hollstein, M. 1993. Clinical implications of the p53 tumor-suppressor gene. *N. Engl. J. Med.* **329**:1318–1327.

Hartwell, L.H., and Weinert, T.A. 1989. Checkpoints: Controls that ensure the order of cell cycle events. *Science* **246**:629–634.

Herbst, J.J., Berenson, M. M., McCloskey, D. W., and Wiser, W. C. 1978. Cell proliferation in esophageal columnar epithelium (Barrett's esophagus). *Gastroenterology* **75**:683–687.

Hereford, L.M. and Hartwell, L.H. 1974. Sequential gene function in the initiation of Saccharomyces cerevisiae DNA synthesis. *J. Mol. Biol.* **84**:445–461.

Hirohashi, S., and Sugimura, T. 1991. Genetic alterations in human gastric cancer. *Cancer Cells* **3**: 49–52.

Hollstein, M., Sidransky. D., Vogelstein, B., and Harris, C.C. 1991. p53 mutations in human cancers. *Science* **253**:49–53.

Huang, Y., Boynton, R.F., Blount, P.L., Silverstein, R.J., Yin, J., Tong, Y., McDaniel, T.K., Newkirk, C., Resau, J.H., Sridhara, R., Reid, B.J., and Meltzer, S.J. 1992. Loss of heterozygosity involves multiple tumor suppressor genes in human esophageal cancers. *Cancer Res.* **52**:6525–6530.

Kastan, M.B., Onyekwere, O., Sidransky, D., Vogelstein, B., and Craig, R.W. 1991. Participation of p53 protein in the cellular response to DNA damage. *Cancer Res.* **51**:6304–6311.

Kastan, M.B., Zhan, Q., El-Deiry, W.S., Carrier, F., Jacks, T., Walsh, W.V., Plunkett, B.S., Vogelstein, B., and Fornace, A.J. 1992. A mammalian cell cycle checkpoint pathway utilizing p53 and *GADD45* is defective in Ataxia-Telangiectasia. *Cell* **71**:587–597.

Kuerbitz, S.J., Plunkett, B.S., Walsh, W.V., and Kastan, M.B. 1992. Wild-type p53 is a cell cycle checkpoint determinant following irradiation. *Proc. Natl. Acad. Sci. USA.* **89**:7491–7495.

Levine, D.S., and Reid, B.J. 1992. Endoscopic diagnosis of esophageal neoplasms. Sivak, M., (eds). In *Gastrointestinal endoscopy clinics of North America* (eds. M. Sivak, H.W. Boyce and G.A. Boyce) pp. 395–413. W.B. Saunders, Philadelphia.

Levine, D.S., Haggitt, R.C, Blount, P.L, Rabinovitch, P.S, Rusch, V.W, and Reid, B.J. 1993. An endoscopic biopsy protocol can differentiate high-grade dysplasia from early adenocarcinoma in Barrett's esophagus. *Gastroenterology* **105**:40–50.

Livingstone, L.R., White, A., Sprouse, J., Livanos, E., Jacks, T., and Tlsty, T.D. 1992. Altered cell cycle arrest and gene amplification potential accompany loss of wild-type p53. *Cell* **70**:923–936.

Lu, X., and Lane, D.P. Differential induction of transcriptionally active p53 following UV or ionizing radiation: Defects in chromosome instability syndromes? 1993. *Cell* **75**:765–778.

McKinley, M.J., Budman, D.R., Grueneberg, D., Bronzo, R.L., Weissman, G.S., and Kahn, E. 1987. DNA content in Barrett's esophagus and esophageal malignancy. *Am J Gastroenterology* **82**: 1012–1015.

Miller, B.A., Ries, L.A.G., Hankey, B.F., Kosary, C.L., and Edwards, B.K. Cancer Statistics Review, 1973–1989. U.S. Department of Health and Human Services. NIH Publication No 92–2789.

Muller, J.M., Erasmi, H., Stelzner, M., Zieren, U., and Pichlmaier, H. 1990. Surgical therapy of oesophageal carcinoma. *Br. J. Surg.* **77**:845–857.

Neshat, K., Sanchez, C.A., Galipeau, P.C., Blount, P.L., Levine, D.S., Joslyn, G., and Reid, B.J. 1994. p53 mutations in Barrett's adenocarcinoma and high-grade dysplasia. *Gastroenterology* **106**:1589–1595.

Neshat, K., Sanchez, C.A., Galipeau, P.C., Cowan, D.S., Ramel, S., Levine, D.S., and Reid, B.J. 1995. Barrett's esophagus: a model of human neoplastic progression. IN: Symposium LIX: Molecular Genetics of Cancer. Cold Spring Harbor Sym. on Quant. Biol. LIX:577–583.

Nowell, P.C. 1976. The clonal evolution of tumor cell populations. *Science* **194**:23–28.

Offerhaus, G.J.A., De Geyter, E.P., Cornelisse, C.J., Tersmette, K.W.F., Floyd, J., Kern, S.E., Vogelstein, B., and Hamilton, S.R. 1992. The relationship of DNA aneuploidy to molecular genetic alterations in colorectal carcinoma. *American Gastroenterological Assoc.* **102**:1612–1619.

Peiffer, S.L., Herzog, T.J., Tribune, D.J., Mutch, D.G., Gersell, D.J., and Goodfellow, P.J. 1995. Allelic loss of sequences from the long arm of chromosome 10 and replication errors in endometrial cancers. *Cancer Res.***55**:1922–1926.

Pellish, L.J., Hermos, J.A., and Eastwood, G.L. 1980. Cell proliferation in three types of Barrett's epithelium. *Gut* **21**:26–31.

Phillips, R.W., and Wong, R.K.H. 1991. Barrett's esophagus: Natural history, incidence, etiology and complications. *Gastroenterol Clin North Am.* **20**:791–816.

Rabinovitch, P.S., Reid, B.J., Haggitt, R.C., Norwood, T.H., and Rubin, C.E. 1988. Progression to cancer in Barrett's esophagus is associated with genomic instability. *Lab Invest.* **60**:65–71.

Ramel, S., Reid, B.J., Sanchez, C.A., Blount, P.L., Levine, D.S., Neshat, K., Haggitt, R.C., Dean, P.J., Thor, K., and Rabinovitch, P.S. 1992. Evaluation of p53 protein expression in Barrett's esophagus by two-parameter flow cytometry. *Gastroenterology* **102**:1220–1228.

Raskind, W.H., Norwood, T., Levine, D.S., Haggitt, R.C., Rabinovitch, P.S., and Reid, B.J. 1992. Persistent clonal areas and clonal expansion in Barrett's esophagus. *Cancer Res.* **52**:2946–2950.

Reid, B.J., Haggitt, R.C., Rubin, C.E., and Rabinovitch, P.S. 1987. Barrett's esophagus: Correlation between flow cytometry and histology in detection of patients at risk for adenocarcinoma. *Gastroenterology* **93**:1–11.

Reid, B.J., Weinstein, W.M., Lewin, K.J., Haggitt, R.C., Van Deventer, G., DenBesten, L., and Rubin, C.E. 1988a. Endoscopic biopsy can detect high-grade dysplasia or early adenocarcinoma in Barrett's esophagus without grossly recognizable neoplastic lesions. *Gastroenterology* **94**:81–94.

Reid, B.J., Haggitt, R.C., Rubin, C.E., Roth, G., Surawicz, C.M., VanBelle, G., Lewin, K., Weinstein, W.M., Antonioli, D.A., Goldman, H., MacDonald, W., and Owen, D. 1988b. Observer variation in the diagnosis of dysplasia in Barrett's esophagus. *Hum Pathol.* **19**:166–178.

Reid, B.J. 1991. Barrett's esophagus and esophageal adenocarcinoma. *Gastroenterol Clin North Am.* **20**:817–834.

Reid, B.J., Blount, P.L., Rubin, C.E., Levine, D.S., Haggitt, R.C., and Rabinovitch, P.S. 1992. Flow cytometric and histologic progression to malignancy in Barrett's esophagus: prospective endoscopic surveillance of a cohort. *Gastroenterology* **102**:1212–1219.

Reid, B.J., Sanchez, C.A., Blount, P.L., and Levine, D.S. 1993. Barrett's esophagus: Cell cycle abnormalities in advancing stages of neoplastic progress. *Gastroenterology* **105**:119–129.

Robaszkiewicz, M., Hardy, E., Volant, A., Nousbaum, J.B., Cauvin, J.M., Calament, G., Robert, F.X., Saleun, J.P., and Gouerou, H. 1991. Analyse du contenu cellulaire en ADN par cytometrie en flux dans les endobrachyoesophages. *Gastroenterol Clin Biol.* **15**:703–710.

Sandell, L.L., and Zakian, V.A. 1993. Loss of a yeast telomere: arrest, recovery, and chromosome loss. *Cell* **75**:729–739.

Sato, T., Tanigami, A., Yamakawa, K., Akiyama, F., Kasumi, F., Sakamoto, G., and Nakamura, Y. 1990. Allelotype of breast cancer: cumulative allele losses promote tumor progression in primary breast cancer. *Cancer Res.*, **50**: 7184–7189.

Sidransky, D., Mikkelsen, T., Schwechheimer, K., Rosenblum, M.L., Cavanee, W., and Vogelstein, B. 1992. Clonal expansion of p53 mutant cells is associated with brain tumour progression. *Nature* **355**:846–847.

Sozzi, G., Miozzo, M., Donghi, R., Pilotti, S., Cariani, C.T., Pastorino, U., Della Porta, G., and Pierotti, M.A. 1992. Deletions of 17p and p53 mutations in preneoplastic lesions of the lung. *Cancer Res.* **52**:6079–6082.

Spechler, S.J. 1987. Endoscopic surveillance for patients with Barrett esophagus: Does the cancer risk justify the practice? *Ann Int Med.* **106**:902–904.

Thrash-Bingham, C.A., Greenberg, R.E., Howard, S., Bruzel, A., Bremer, M., Goll, A., Salazar, H., Freed, J.J., and Tartof, K.D. 1995. Comprehensive allelotyping of human renal cell carcinomas using microsatellite DNA probes. *Proc. Natl. Acad. Sci. USA.* **92**:2854–2858.

Vogelstein, B., Fearon, E.R., Hamilton, S.R., Kern, S.E., Preisinger, A.C., Leppert, M., Nakamura, Y., White, R., Smits, A.M.M., and Bos, J.L. 1988. Genetic alterations during colorectal tumor development. *N Eng J Med.* **319**:525–532.

Vogelstein, B., Fearon, E.R., Kern, S.E., Hamilton, S.R., Preisinger, A.C., Nakamura, Y., and White, R. 1989. Allelotype of colocrectal carcinomas. *Science*, **244**:207–211.

Weinert, T.A., and Hartwell, L.H. 1988. The *RAD9* gene controls the cell cycle response to DNA damage in *Saccharomyces cerevisiae*. *Science* **241**:317–322.

Xiong, Y., Hannon, G.J., Zhang, H., Casso, D., Kobayashi, R., and Beach, D. 1993. p21 is a universal inhibitor of cyclin kinases. *Nature* (Lond.), **366**:701–704.

Yamaguchi, T., Toguchida, J., Yamamuro, T., Kotoura, Y., Takada, N., Kawaguchi, N., Kaneko, Y., Nakamura, Y., Sasaki, M.S., and Ishizaki, K. 1992. Allelotype analysis in osteosarcomas: frequent allele loss on 3q,13q,17p, and 18q. *Cancer Res.*, **52**:2419–2423.

Yin, Y., Tainsky, M.A., Bischoff, F.Z., Strong, L.C., and Wahl, G.M. 1992. Wild-type p53 restores cell cycle control and inhibits gene amplification in cells with mutant p53 alleles. *Cell* **70**:937–948.

Younes, M., Lebovitz, R.M., Lechago, L.V., and Lechago, J. 1993. p53 protein accumulation in Barrett's metaplasia, dysplasia, and carcinoma: A follow-up study. *Gastroenterology* **105**:1637–1642.

DISCUSSION

Shay: Is there any evidence that changing the acid content of the stomach or preventing recurrent reflux or GERD will prevent the development of Barrett's?

Reid: There are no good randomized controls trials to indicate that if you control reflux at any state you will impact this progression. I think most of us believe that if you have not developed Barrett's and you are on effective treatment that you probably will not get Barrett's. But these patients require strong treatments. Once you develop Barrett's, then all bets are off. There are well documented cases in which patients have had surgery to prevent reflux and it has been effective; they have documented it by prolonged measurements of reflux in the esophagus and it is absent. And yet cancers have still developed. It is as though, at some point, this process develops a life of its own. All of us in this field tend to draw arrows in our models in one direction, namely, G1 leads to loss of p53 etc. In some cases this process is reversible in Barrett's esophagus and we have seen patients in whom the entire Barrett's segment has gone back to normal and we have other patients who have gone about 75% of the way in regressing. We do not understand what factors control that, and in fact that is one of the goals of our epidemiology study. We are not only trying to find out what the patients who progress do, we are also trying to find out what the patients who have regressed and get better do.

Shay: In reference to your point about the changes from 17LOH to p53 mutations to aneuploidy or tetraploidy or 4N, have you ever looked to see if one mutated p53 may result in endoduplication of the mutated p53? Maybe if you have one dominant negative mutant p53 inhibiting the wild type p53 that the cell that evolves out of that endoduplicates the mutated p53 and that may set off this whole wave of aneuploidy leading to eventual loss of the wild type. Can those types of studies be done in this kind of a model?

Reid: Two comments about that: You have to be very careful in a model like this when you order events that there is not a differential sensitivity in detection of the two events. That can lead to false positives. As an example of that you can detect a p53 mutation with levels of contamination of normal cells that would obscure a 17pLOH event. So you can get as an artifact an order when an order does not exist and that is the reason I would be worried about the first half of your question. The second half relates to what are these 4N abnormalities, and I think there are at least two possibilities: One is that what we are seeing, as a process of DNA damage, say that cells in G1 get a single strand break, without p53 that break gets replicated, you have a double strand break activating the G2 checkpoint accumulating with 4N. The other possibility is that there is loss of control leading to an endoreduplication event which means that the 4N would be tetraploid. That has been very difficult to answer *in vivo* and that is the reason that I leave them as 4N abnormalities. Some of the 4N cells go on to form a tetraploid-S, tetraploid-G2 suggesting that at least some of the 4N's are tetraploid cell populations.

Wahl: Your increase in Barrett's is strikingly reminiscent of the increase in fast food outlets in California. I am curious about the relationship between p16 and p53. Do you ever see p16 losses without p53, or is there a dependence on p53 first and then p16?

Reid: I did not show that data because we are still accumulating it. We do have a clear separation in some patients, and in those patients in whom we have a clear separation, it is always inactivation of p53 first, followed by 9p allelic loss and inactivation of p16. So in those cases in which we have been able to order them up until now it has been inactivation of p53 first. That work is still going on and it clearly shows that that order can occur in some patients. We are not sure it is an obligatory order until we do a larger number of patients. I do not know, everybody notices the fast food thing and the strongest risk factors for Barrett's cancers are being white, being male, being overweight, having a high-fat diet so there is the correlation.

Croce: In one of your slides showing loss of heterozygocity in Barrett's, you show seven out of seven loss of heterozygocity in 9p21, and then you said that you think that p53 mutation precedes p16. According to your data it would seem just the opposite because in all the cases of Barrett's you have loss of heterozygocity and 9p21 would seem to precede p53 mutation because, if I remember, you have a lower loss of heterozygocity at 17p in Barrett's. In addition, in a head and neck cancer that has been studied very carefully for loss of heterozygocity using very large markers for the possible critical region, a loss of heterozygocity at 9p21 is extremely common; in fact, it seems to be one of the earliest, if not the earliest, change. And that, in general, goes together with loss of heterozygocity 3p1–4.2, and very often it is difficult to decide whether p16 loss goes first or 3p1–4.2 goes first. Could you give us some indication of what you think about the sequence of events and in addition whether you have looked at loss of heterozygocity 3p1–4.2 in Barrett's esophagus?

Reid: We had seven patients who had cancer and had pre-malignant epithelium for which we could evaluate the order in which they occurred and in all seven the 9p allele losses developed before the cancer. I have not shown you this data, but if you then carry that back to an earlier stage and ask have all increased G1 fractions lost 9p--the answer is no. So the G1 fractions develop before the 9p allelic losses, the Barrett's epithelium develops before the 9p allelic loss. Both 17p and 9p develop at approximately the same stage of progression in the sense that they are after the increased G1 fractions but before aneuploidy. In those cases where we have been able to separate them, the 17p occurs first, the 9p subsequently. That does not rule out the possibility that they may be independent and will subsequently come up with other events in which the reverse order is true. I think it suggests that there is not an obligatory 9p then 17p order. We are aware of the data in head and neck. 3p is not a high region of allelic loss in Barrett's, it is one of those background ones that is near 23% background level.

Pinedo: You have a certain group of patients that come in with adenocarcinoma and no Barrett's—what percentage is this of the whole population of patients with adenocarcinoma at the moment? You have a selective group of patients of course with a lot of Barrett's but you must have also patients coming in with esophageal adenocarcinomas without a Barrett history.

Reid: Yes, a couple of comments that may get at your question. The first is that we have looked at patients who present with adenocarcinoma—how many of them have Barrett's? And in both of our series, if they are clearly esophageal adenocarcinomas the vast, vast majority, greater than 90% will have Barrett's in the surrounding epithelium. So most of those are arising in Barrett's. In our experience, and in Alan Cameron's experience also, carcinomas arising at the GE junction, the so-called cardiacarcinomas, if you look very carefully you will find evidence of Barrett's in about 75% of those cases. The cases that you do not find evidence of the metaplastic epithelium tend to be the larger tumors, and what people think is that the tumor actually overgrew a small short segment of Barrett's esophagus. The length of the Barrett's can vary.

Pinedo: And my second question is, with all this data you must have also cases where your hands are itching to do a preventive operation. What is your attitude, at this moment, with all this data being available?

Reid: Well, my attitude is that there is no substitute for your own esophagus. I guess I have had the bad fortune to recommend preventive operations and had the patient die and there was no cancer in the surgical specimen. When you do that you start to wonder how long they would have lived had you not been so aggressive. Our data at the moment, and this was just repeated by Stephen Sontag in Chicago, suggests that almost any combination of abnormalities that have been studied long enough to be measured do not identify a subset that is at high enough risk to do it prophylactically. You do identify subsets that are at high enough risk to engage in intensive surveillance for the early detection. Most people use high grade dysplasia. In our study, if you follow patients out for six years, less than half of those patients will develop cancer. In Stephen's study, about 75% of them remain cancer-free and in both studies you are able to detect the cancers when they were early and curable, thereby saving 50 to 75% of the patients. Now, theoretically, I think it is possible to come up with the combination of markers that may allow you to carry out the operation preventively and feel comfortable about that. And that requires taking the kind of data I have shown you here, which is ordering data, and turn it into what I would call predictor data.

Gatti: In your specimen you noticed many aneuploid cell population which appear to be *quasi* triploid. Do you know how the cells arise, what kind of mechanism is involved in the formation of those cells?

Reid: Do I know how they arise? No, I can tell you some data that we have. Our laboratory described p53 dependent spindle checkpoint in mice. As others have shown, p53 knockout mouse cells develop amplified numbers of centrosomes. The difference between our studies and George Vande Woude's data is that we find those amplified centrosomes only under conditions in which there has been increasing ploidy. George did not measure that. What we think is going on is not that p53 controls centrosome number directly but that when a cell undergoes an event whereby it re-enters S-phase without completing chromosome segregation in mitosis, it can also reduplicate the centrosomes leading to too many centrosomes per cell. We have shown that those can participate in multipolar mitosis and can lead to gross segregation of chromosome abnormalities in the cell which may give rise to a near triploid cell population like you are talking about. Furthermore, Sanchez has shown, not *in vivo* but in the *in vitro* cultured Barrett's epithelium, that you get the same process of multipolar mitosis and multicentrosomes.

Anderson: It seems like with Barrett's you have a long term exposure to a premalignant condition and it also seems like this condition produces pain in the patient. I was wondering whether your epidemiologists in Seattle have looked at the possibility that the patients are self-administering non-steroidal anti-inflammatories to deal with their pain, and that you may be getting some chemo-preventive effect analogous to what people are seeing with colorectal?

Reid: That is on our questionnaire. I do not know any data on that at the moment. There is now a national co-operative study to study the epidemiology of esophageal adenocarcinoma in a population based study. It is in their questionnaire and we do not know the final results of that yet.

Giaretti: I would like to add a comment on the previous question concerning the triploid tumor and also the increased tetraploidy. In other model systems of colorectal cancer we have seen that near triploid, by DNA content, is very low, it does not support the interpretation that results from losing chromosome from a tetraploid coming down. Also in neuroblastoma for example, you measure DNA index of 1.5 or 3N. These are not K-RAS mutated, while near diploid are 80% mutated. And preliminary data also suggests that in this model system p53 is not mutated in near triploid. So they look to have a different history, some mechanisms of triploidy appear to be different from that of tetraploidy or near diploidy. To demonstrate tetraploidy by sorting, I think it is probably not appropriate, or at least you may induce tetraploidization *in vitro* instead. Because first of all you have to sort and this cell may be changing. Another possibility is that you really increase the number of mitotic abnormal cell in mitosis when you see just a slight increase of 4N. Because mitotic cells may delay division. Then they could go to near diploidy as in the colorectal model system. And, in fact, I have seen, and we have also done, Barrett's esophagus many of the aneuploid cases are near diploid. So one possible interpretation is not just this jump from diploid to tetraploid but just delaying abnormal mitosis delaying and then dividing asymmetrically. Do you find near diploidy often in esophagus as we do in colorectal adenomas? We have found several and also in the literature of Barrett's esophagus the majority in a few reports is near diploidy. Would you comment on that?

Reid: Yes, we find some near diploid but not a lot. You know it is not vanishingly small and it is not huge. It is just if you do kind of a barograph of diploidy versus a number of cases you find that they are less than 2.5N or 2.4N at the same frequency as you find 3N or whatever, in fact, probably a little bit less. So, we do not find an excess of it. I think that you have to be a little careful in Barrett's because, as in some other tumors, a near diploid aneuploidy can result from a surgical misadventure between the tissue and the flow cytometry laboratory in the sense that autolysis of the specimen can give a false positive near diploid. For that reason, we pick up all of our specimens personally in the O.R. and get them into frozen media immediately and the same is true of our endoscopic biopsies. The other thing is that in our near triploid aneuploid cell populations they definitely are p53 inactivated by 17p LOH and mutation.

Livingston: I have three quick questions. First of all, can you enlighten us a little bit on the history of familial Barrett's esophagus? I gather that is a subject of some interest.

Reid: You would predict, from a disease with a rapidly increasing incidence, that it would be all environmental. There have been a series of about six sort of small family studies in the literature where it appears to be inherited. In our epidemiology study, in the first fifty patients, we found better than 10% of them had a first degree family relative with esophageal neoplasia, which is far greater than you would expect in the population. We thought that that was a sampling size problem and so extended it out and the number is holding up. It may be that these, even though there is such a strong appearance of an environmental predisposition, it may be that they are developing in a subset of susceptible individuals, which would be very interesting.

Livingston: Second question: The cells appear to develop and/or migrate north. Why not south?

Reid: You mean up the esophagus? Well, south is the stomach—the stomach has mucus and specialized cells that are resistant to acid. The esophagus relies on a valve between the stomach and the esophagus to protect it. In the patients with Barrett's, that valve breaks down and it is weak and so basically it is like bathing your forearm in acid constantly. When that epithelium is destroyed, I tend to think of it as a protective mechanism that this acid-resistant mucus secretion tends to protect cells against acids. So I think of it as kind of a defense mechanism to replace the epithelium that is being injured, it is just that the defense mechanisms go wrong.

Livingston: In GERD patients, do a substantial fraction have demonstrable histologically evident gastritis, as well?

Reid: Well, they have cardiacgastritis and which is immediately south and that may be the source of the Barrett's epithelium. Some believe that that is where the Barrett's epithelium starts and then grows up and becomes abnormal from there.

Livingston: Third question: In your short term cultures of Barrett's cells, do immortal cells ever grow out?

Reid: We have not seen immortality yet, but what I have to say is that we have been trying to culture cells for years and years and Peter and Corina are the first ones to do it. We do have some cultures that are still in passage after a year without undergoing a crises or an immortality.

Stark: I wonder what you think of the possibility of an infectious agent. You have not really mentioned that explicitly. It seems to me there are a couple of reasons for perhaps suspecting that apart from the precedent of helicobacter. But the epidemiology is so sudden that it really does not correlate very well, I do not think, with predisposition or changes in lifestyle. The other is, I am not sure what it is in this situation that is driving genetic change so rapidly. Is it just the stimulation of cell growth as cells are being destroyed? That happens all over the place without driving genetic change at anything like the rate that you described. It seems to me not impossible that an infectious agent could be driving the genetic change.

Reid: We have actually considered that. I think all the reasons that you articulated make one worry that an infectious agent has got out. It would not be unheard of in this time-frame for a virus to emerge. The only thing I can say is that all of these sort of chronic offenders have been looked at and ruled out. Helicobacter has been ruled out. HPV has been looked at and ruled out. I do not rule it out for all of the reasons that you talked about, but there is not an agent on board and the epidemiology studies at the moment have not concentrated on an infectious approach.

Gray: The question I wanted to ask is coming back to this interesting patchwork typology that you have here. I gather from the one slide that you showed, that you think that this is really just the evolution of one clone and you are seeing stochastic changes throughout that clone to create the patches or do you think that this is multiple independent events?

Reid: Well, at the level of flow cytometry that is of course impossible to prove. What we have shown in two ways, one by sort of classical cytogenetics is that you find common karyotypic abnormalities, common marker chromosomes in the various regions suggesting that they are coming down through a single clone. The second thing that we have done is molecularly showing that the same p53 mutation goes through all of the clones suggesting that they all arose from a common precursor. We do have occasional things that make you wonder about that, such as, two widely separated carcinomas in the same esophagus. We have not had those at a time when we could sequence p53 and see if they rose in the same population.

15

E2F-1 DEGRADATION BY THE UBIQUITIN-PROTEASOME PATHWAY

Francesco Hofmann and David M. Livingston

The Division of Neoplastic Disease Mechanisms
Dana-Farber Cancer Institute and The Harvard Medical School
Boston, Massachusetts 02115

1. INTRODUCTION

E2F-1 is the first cloned member of a family of transcription factors composed of five E2F members and three DP members.[1-4] Individual E2F and DP species form heterodimers and transactivate genes, some of which are required for cell cycle progression or DNA synthesis.[5-11] The product of the retinoblastoma tumor suppressor gene,[12] pRb, binds to some members of the E2F family, including E2F-1, and transforms them into transcriptional repressors.[13-15] In this manner, pRb performs at least part its growth suppressive function.

In late G1, cyclin-dependent kinases phosphorylate pRb leading to its functional inactivation and to the dissociation of the pRb/E2F-1 complex. E2F-1 transactivation function, which depends also on cdk3 kinase activity,[16] then peaks at the G1-S transition and in early S-phase. The deregulated synthesis of E2F-1 leads to neoplastic transformation and tumor formation by E2F-1 transformed cells,[17-19] and its induction in serum-deprived, quiescent cells is mitogenic and leads to apoptosis during or shortly after S-phase.[20-23] Moreover, recent observations, derived from studies on mice homozygous for a nonfunctional E2F-1 allele, strongly suggest that E2F-1 has a tumor suppression function, as well, which might be linked to its apoptosis-inducing activity.[24-26] Taken together, these observations indicate that the tight control of E2F-1 function plays an important role in cellular homeostasis.

E2F-1 synthesis is normally cell cycle-controlled, beginning in G1, as serum-deprived cells emerge from that state. In this regard, we have asked whether the level of E2F-1 is controlled solely at the transcriptional level[27-29] or if other mechanisms like protein stability contributed to its cell cycle regulated function. Several proteins involved in cell cycle regulation are actively degraded at defined time points during the cell cycle. Where studied, degradation is mediated by the ubiquitin-proteasome pathway[30,31] and may be necessary for correct progression through the cell cycle.[32,33] Here we present evidence that E2F-1 is unstable in vivo and that its instability results from its degradation by the ubiquitin-proteasome pathway. Efficient and specific degradation depends upon specific E2F-1 sequence(s) the availability of which to the ubiquitin-proteasome pathway is down-regulated by pRb binding.

2. RESULTS

2.1. The C-Terminus of E2F-1 Is Required for Efficient Degradation

The stability of E2F-1 and of several mutant derivatives was analyzed in pulse-chase experiments using U2OS cells transiently transfected with the corresponding expression plasmids. The E2F-1 mutants either lacked the C-terminal 74 amino acids, (1–363), were defective in DNA binding (E132 and E177), cyclin A-dependent kinase binding (Δ24), or were mildly and more severely impaired for pRb binding, respectively (Δ5 and Δ18). Wild-type E2F-1 and mutants defective for DNA binding, cyclin A-binding or minimally impaired for pRb binding proved to be rather unstable. Indeed, the kinetics of turn-over were indistinguishable among them (Fig. 1A), and the half-life of wild-type E2F-1 was estimated in the 2–3 hr range (Fig. 1B). By contrast, E2F-1-(1–363), which lacks the C-terminal 74 amino acids, was much more stable (half-life > 10 hrs) than wild-type E2F-1 (Fig. 1A and 1C). E2F-1 is therefore an unstable protein, and the data imply that the deletion of the C-terminal region of the protein renders it less susceptible to degradation. Moreover, the results imply that efficient E2F-1 degradation does not depend on the capacity of the protein to bind to DNA, to interact with cyclin A-dependent kinase, or to form a complex with pRb.

2.2. pRb-Binding Protects E2F-1 from Degradation

Given the implication that C-terminal sequences contribute to efficient E2F-1 degra-dation and the knowledge that pRb binds to E2F-1 through a C-terminal E2F-1 sequence,[35–37] we asked whether pRb binding affects the stability of E2F-1. To address this question, an E2F-1 expression plasmid was transfected into U2OS cells, alone and together with a four-fold excess of an expression plasmid encoding a pRb mutant, pRbΔp34, which binds E2F-1 normally, but cannot be phosphorylated in these cells, because a number of its potential phosphorylation sites have been mutated.[38] In pulse-chase experiments a substantial increase in E2F-1 protein stability was observed in the presence of the pRb-phosphomutant (Fig. 2A and C). Most likely the stabilization effect induced by pRb can be ascribed to complex formation between the pRb-phosphomutant and a substantial portion of the overproduced E2F-1 in the transfected culture and not to a state of G1 arrest induced by this pRb mutant, since, under the same experimental condi-tions, the quantity of ectopic pRb was not sufficient to arrest U2OS cells in G1, as deter-mined by FACS analysis (Fig. 2D). In further support of the belief that E2F-1 stabilization was not a consequence of G1 arrest, but rather of pRb binding, we observed that the cyclin-dependent kinase inhibitor, p27, which induces a G1 block in the presence of tran-scriptionally active E2F-1 when both are overproduced in U2OS cells (F. Hofmann and D.M. Livingston unpublished), led to minimal E2F-1 stabilization (Fig. 2A and C) in cells arrested in G1 (Fig. 2D). This effect is, most likely, a result of the binding of a fraction of the available E2F-1 by endogenous, unphosphorylated pRb.

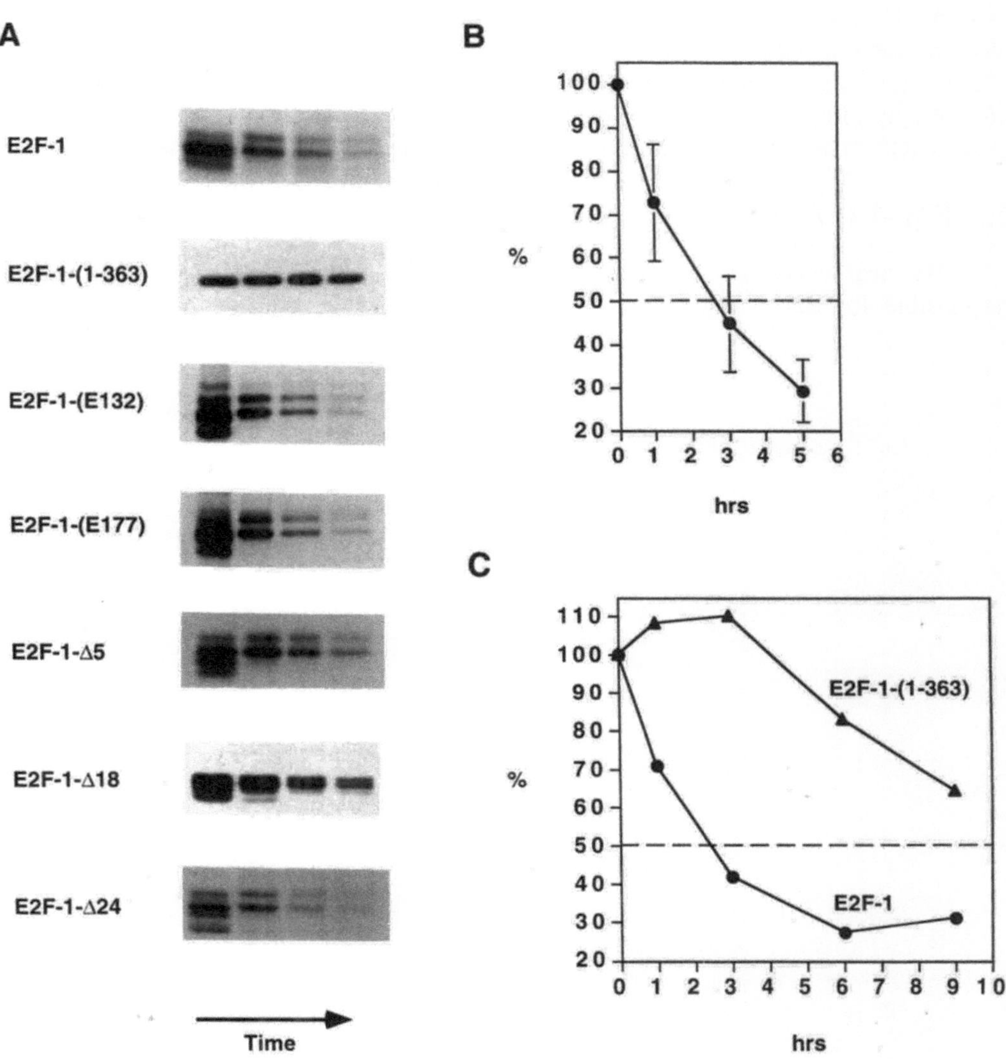

Figure 1. Comparison of wt and mutant E2F-1 protein stability. (A). Expression plasmids for the indicated HA-E2F-1 species (4 μg) were transfected into U2OS cells. Cells were pulse-labelled with ^{35}S-methionine for 30 minutes and chased for 0, 1, 3, and 5 hrs, followed by immunoprecipitation with anti-HA monoclonal antibody (12CA5). Autoradiography reveals the differences in protein turnover among the various E2F species tested. (B). Several pulse-chase experiments for HA-E2F-1 were densitometrically traced and the data were plotted as a function of time. The E2F-1 signal detected at time 0 was set at 100 % and used to normalize the signal at subsequent time points. The vertical bars represent standard deviation. (C). Cells transfected with expression plasmids encoding HA-E2F-1 wild-type (4 μg) or HA-E2F-1-(1–363) (4 μg) were pulse-labelled with ^{35}S-methionine and chased for 0, 1, 3, 6, and 9 hrs to compare their decay over a longer period. The data obtained from the densitometric tracing were plotted as a function of time. The signal at time 0 was set at 100 % and used as a standard for the subsequent time points. (Reprinted, with permission from [34], copyright by Cold Spring Harbor Laboratory Press.)

In an attempt to further test the idea that pRb-induced stabilization is linked to pRb/E2F-1 complex formation, we compared the half-life of E2F-1 wild-type and E2F-1-Δ18 in the presence and absence of pRbΔp34. E2F-1-Δ18, is severely, albeit not completely, impaired in pRb binding, as independently determined under the conditions used in this assay (data not shown). In keeping with their relative abilities to interact with pRb, the same quantity of pRbΔp34 led to a great increase in the half-life of wild-type E2F-1, but only a two fold increase in the half-life of E2F-1-Δ18 (Fig. 3). The stabilization of E2F-1 by pRb binding was also confirmed in an analysis of ectopic wild-type pRb in cells (SAOS-2) lacking both pRb and pRb-specific cyclin-dependent kinase activity.[34]

2.3. E2F-1 Is Degraded by the Ubiquitin Proteasome Pathway

We next performed experiments aimed at identifying the proteolytic pathway responsible for E2F-1 degradation. U2OS cells transfected with expression plasmids

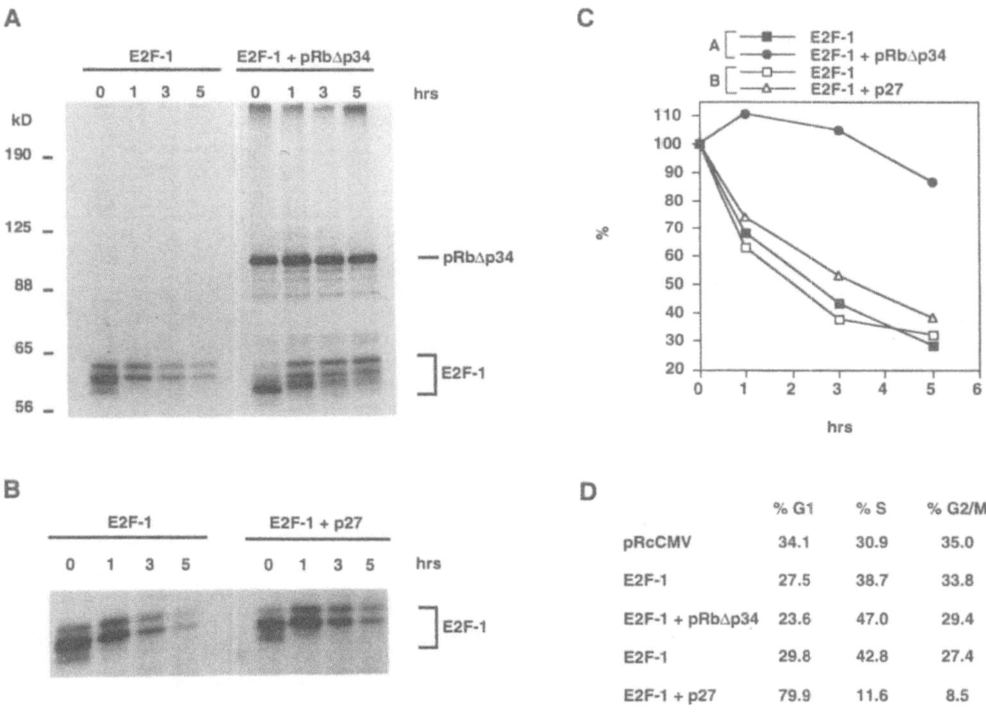

Figure 2. Unphosphorylated pRb stabilizes E2F-1. (A). U2OS cells co-transfected with expression vectors for the surface marker, CD19 (pCMV-CD19, 1 μg), and wild-type HA-E2F-1 (pRcCMV-HA-E2F-1, 4 μg), in the presence and absence of an expression plasmid encoding a pRb-phosphorylation defective mutant (pCMV-Puro-HA-pRbΔp34, 16 μg), were pulsed with ³⁵S-methionine and then chased for 1, 3, and 5 hrs, followed by immunoprecipitation with the anti-HA monoclonal antibody (12CA5). (B). U2OS cells co-transfected with expression vectors for the surface marker CD19 (pCMV-CD19, 1 μg), and wild-type E2F-1 (pRcCMV-HA-E2F-1, 4 μg), in the presence or absence of an expression plasmid encoding the cyclin-dependent kinase inhibitor, p27 (pRcCMV-p27-HA, 5 μg), were subjected to pulse-chase analysis, as noted above. (C). The results of densitometric tracing of the data from the pulse-chase experiments shown in (A) and (B) were plotted as a function of time. The E2F-1 signal at time 0 was set to 100%. (D). Cell cycle distribution of the CD19-expressing cells, obtained by FACS analysis for each of the above-noted pulse-chase experiments and for cells transfected by vector (pRcCMV) plus CD19. (Reprinted, with permission, from [34], copyright by Cold Spring Harbor Laboratory Press.)

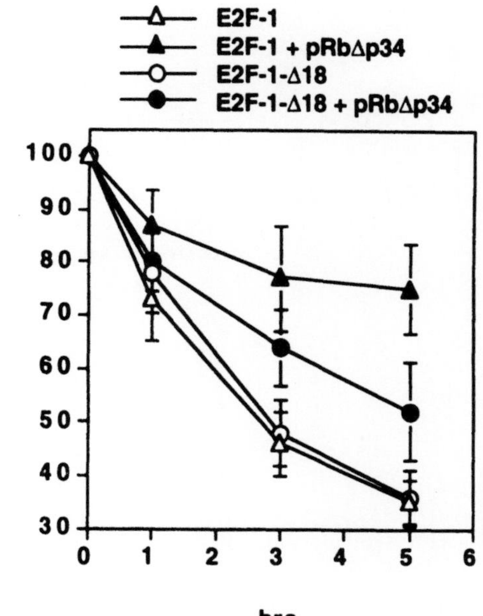

Figure 3. pRb binding stabilizes E2F-1. Pulse-chase experiments were performed on U2OS cells transfected with pRcCMV-HA-E2F-1 (4 µg) or pRcCMV-HA-E2F-1-Δ18 (4 µg) in the presence or absence of an expression vector encoding the pRb-phospho mutant, pCMV-Puro-HA-pRbΔp34 (16 µg). Upon densitometric tracing, the results of five, independent experiments were plotted as a function of time. The signal for E2F-1 or E2F-1-Δ18 at time 0 was set as 100%. Vertical bars represent standard deviation. (Reprinted, with permission, from [34], copyright by Cold Spring Harbor Laboratory Press.)

encoding HA-tagged E2F-1 or E2F-1-(1–363), were treated for 12 hrs with protease inhibitors specific for either lysosomal proteases (E64), calpain (LLM), the proteasome (LLnL) or with the solvent in which the inhibitors were dissolved (DMSO).[39] Subsequent analysis of protein accumulation by western blotting with an anti-HA antibody, showed that E2F-1 accumulated to significantly higher levels in cells treated with LLnL, but was unaffected by the other treatments (Fig. 4A). On the other hand, the accumulation of the more stable, HA-E2F-1-(1–363) was unaffected by any of these compounds, suggesting that wt E2F-1 is efficiently degraded by the ubiquitin-proteasome pathway. In keeping with these results, pulse-chase experiments showed that the proteasome inhibitor led to a substantial increase in the half-life of wt E2F-1. (Fig. 4B and 4C).

In search of additional evidence supporting the idea that E2F-1 is a target of the ubiquitin-proteasome pathway, we performed experiments aimed at identifying ubiquitinated species of the transcription factor. U2OS cells were transfected with an expression plasmid encoding myc epitope-tagged ubiquitin,[40] in the presence and absence of an E2F-1 expression plasmid. After treatment of the cells for 12 hrs with LLnL or with DMSO, equivalent quantities of HA-E2F-1 were immunoprecipitated with a mixture of anti-HA and anti-E2F-1 monoclonal antibodies followed by western blot analysis using an anti-myc polyclonal antibody. Myc-immunoreactive material of high molecular weight was observed only in the combined E2F/HA immunoprecipitate derived from LLnL-treated cells co-transfected with HA-E2F-1 and myc-ubiquitin (Fig. 4D). Thus, E2F-1 is ubiquitinated and appears to be efficiently degraded by the proteasome, unless the latter is inactivated by a specific inhibitor.

3. DISCUSSION

The results reported above indicate that E2F-1 is an unstable protein and a target of the ubiquitin-proteasome degradation pathway. Its degradation does not require the tran-

A

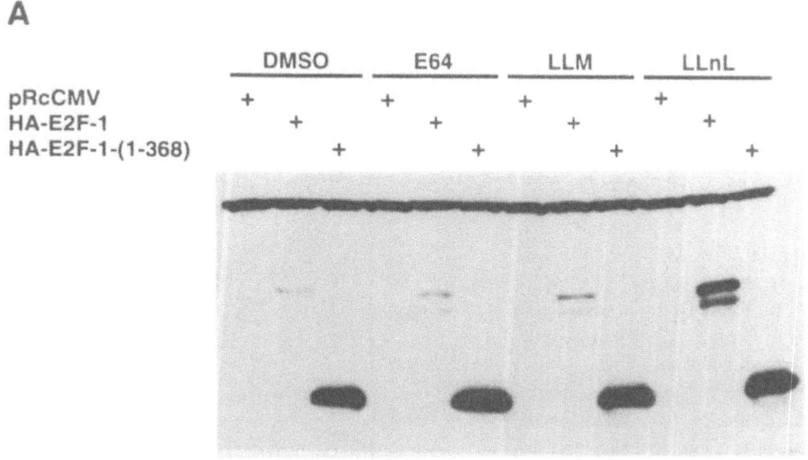

B

C

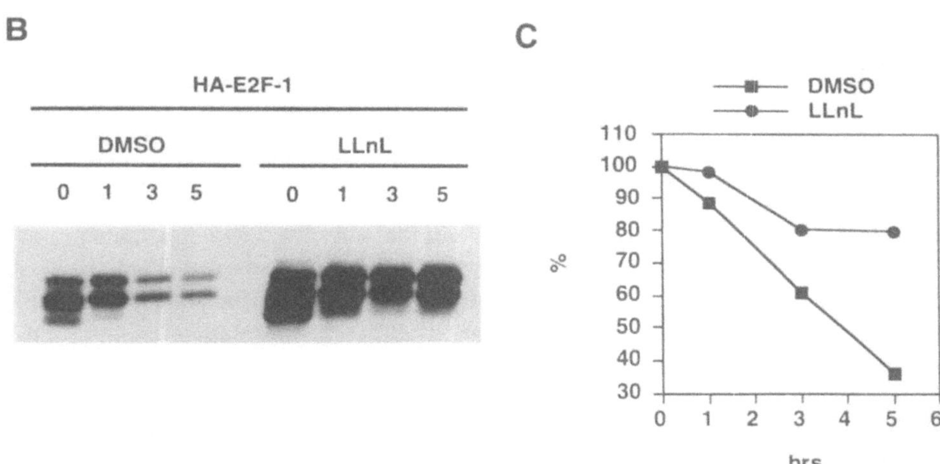

Figure 4. E2F-1 is degraded by the ubiquitin-proteasome pathway. (A). U2OS cells transfected with either pRcCMV (4 µg), pRcCMV-HA-E2F-1 (4 µg) or pRcCMV-HA-E2F-1-(1–363) (4 µg) were treated with either DMSO, a lysosomal protease inhibitor (E64), a calpain inhibitor (LLM), or a proteasome inhibitor (LLnL) for 12 hrs. Cell extracts were prepared, and protein accumulation was analyzed by western blotting using a monoclonal antibody to the HA-tag (12CA5). (B) U2OS cells transfected with pRcCMV-HA-E2F-1 (4 µg) were treated with either DMSO or LLnL for 12 hrs and pulse-chased in the presence of either DMSO or LLnL. (C). The data obtained from densitometric tracing of the pulse-chase experiment shown in (B) were plotted as a function of time. The E2F-1 signal at time 0 in the presence of either DMSO or LLnL was set as 100%. (D). Cell extracts, obtained from U2OS cells transfected with pRBG4-His6-myc-Ub (20 µg) in the presence or absence of pRcCMV-HA-E2F-1 (4 µg) and treated with either DMSO or LLnL for 15 hrs, were first analyzed by western blotting, using a monoclonal antibody to the HA-tag (12CA5) in order to be able to equalize the quantity of E2F-1 to be immunoprecipitated from samples treated with either DMSO or LLnL. Immunoprecipitation with a mixture of monoclonal antibodies to the HA-tag (12CA5) and to E2F-1 (SQ41) was performed, followed by western blot analysis using the anti-myc antibody, A14 as probe. (Reprinted, with permission from [34], copyright by Cold Spring Harbor Laboratory Press.)

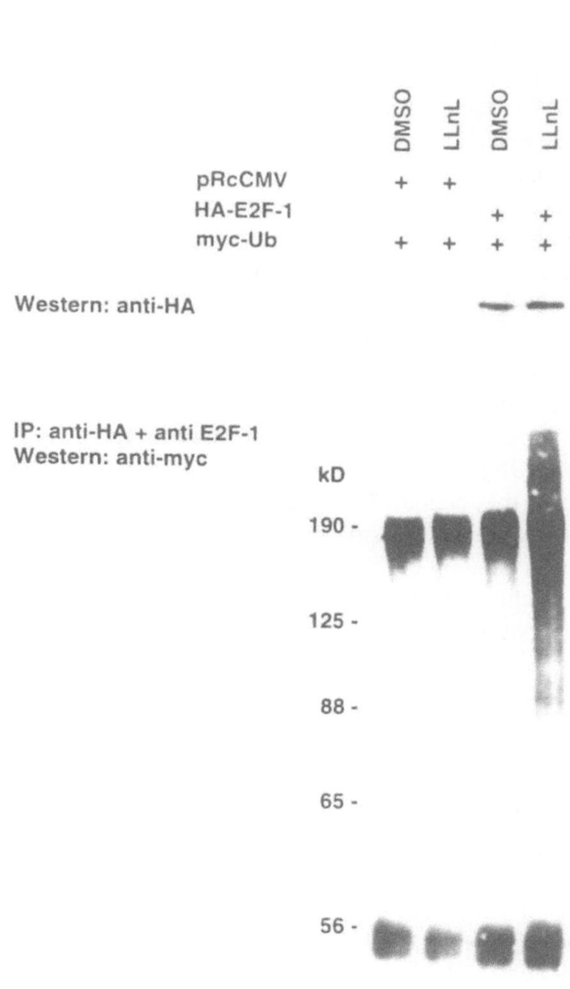

Figure 4D. (*Continued*)

scription factor to bind to DNA or to interact with cyclin A or pRb. However, E2F-1 is protected from degradation when bound to pRb. It appears that efficient degradation depends, at least in part, on particular segments of its primary structure, some of which lie within the C-terminal region of the protein. In keeping with this observation, pRb protection of E2F-1 from degradation is reminiscent of pRb repression of E2F-1 transactivation function. In both cases, the pRb binding site and the sequences responsible for transactivation and for efficient degradation are overlapping or adjacent, suggesting that pRb binding sterically inhibits the relevant E2F-1 property.

Although the biological significance of E2F-1 degradation by the ubiquitin-proteasome pathway is not yet understood, it is conceivable that the abundance of the transcription factor must be tightly controlled. Indeed, E2F-1 can both transform and kill a cell. Evidence in support of the latter was recently provided by the phenotypes of mice lacking both E2F-1 alleles[26] and a clear correlation between level of accumulation of E2F-1 and potential for both cellular killing and transformation had been previously observed.[22] One might, therefore, consider a scenario in which efficient degradation of this protein contributes to normal cellular homeostasis by avoiding a situation wherein the accumulation of E2F-1 can give raise to either a neoplastic phenotype or cell death.

On the other hand, protection of E2F-1 from degradation by pRb can be viewed as an important component of pRb-mediated growth suppression. In fact, during G1, pRb binds to some members of the E2F family (among them, E2F-1) and converts them into transcriptional repressors. This mechanism, at least in part, accounts for the growth suppressive function of pRb.[15,25] Hence, it may be that pRb protection of E2F-1 from degradation contributes to the accumulation of sufficient pRb/E2F-1 complex to sustain a physiological G1 block.

These data indicate that two, distinct mechanisms contribute to the downregulation of E2F-1 function. Specifically, cyclin A-dependent kinase activity has been previously shown to inhibit the DNA binding capacity of the E2F-1/DP-1 heterodimer in S-phase[41] Here we have shown that degradation by the ubiquitin-proteasome pathway also participates in the E2F-1 downregulation process. In this context, it is worth noting that the inhibition of E2F-1 DNA binding capacity by cyclin A-dependent kinase activity represents part of an S-phase checkpoint control system.[42] This, in turn, raises the question of whether E2F-1 degradation is also linked to a checkpoint control function.

4. REFERENCES

1. Lam, E.W.-F. and N.B. La Thangue. 1994. DP and E2F proteins: coordinating transcription with cell cycle progression. *Curr. Op. Cell Biol.* 6: 859–866.
2. Hijmans, E.M., P.M. Voorhoeve, R.L., Beijersbergen, L.J. van't Veer and R. Bernards. 1995. E2F-5, a new E2F family member that interacts with p130 in vivo. *Mol. Cell Biol.* 15: 3082–3089.
3. Sardet, C., M. Vidal, D. Cobrinik, Y. Geng, C. Onufryk, A. Chen and R.A. Weinberg. 1995. E2F-4 and E2F-5, two members of the E2F family, are expressed in the early phase of the cell cycle. *Proc. Natl. Acad. Sci. USA* 92: 2403–2407.
4. Ormondroyd, E., S. de la Luna and N.B. La Thangue. 1995. A new member of the DP family, DP-3, with distinct protein products suggests a regulatory role for alternative splicing in the cell cycle transcription factor DRTF1/E2F. *Oncogene* 11: 1437–1446.
5. Nevins, J.R. 1992. A link between the Rb tumor suppressor protein and viral oncoproteins. *Science* 258: 424–429.
6. De Gregori J., T. Kowalik and J.R. Nevins. 1995. Cellular targets for activation by the E2F-1 transcription factor include DNA Synthesis- and G1/S-regulatory genes. *Mol. Cell Biol.* 15: 4215–4224.
7. Ohtani, K., J. De Gregori and J.R. Nevins. 1995. Regulation of the cyclin E gene by transcription factor E2F-1. *Proc. Natl. Acad. Sci. USA* 92: 12146–12150.
8. Duronio, R.J. and P.H. O'Farrell. 1995. Developmental control of the G1 to S transition in Drosophila: cyclin E is a limiting downstream target of E2F. *Genes & Dev.* 9: 1456–1468.
9. Schulze, A., K. Zerfass, D. Spitovsky, J. Berges, S. Middendrop, P.Jansen-Dürr and B. Henglein. 1995. Cell cycle regulation of cyclin A gene transcription by a variant E2F binding site. *Proc. Natl. Acad. Sci. USA* 92: 11264–11268.
10. Geng, Y., N. Eaton, M. Picon, J.M. Roberts, A.S. Lundberg, A. Gifford, C. Sardet and R.A. Weinberg. 1996. Regulation of cyclin E transcription by E2Fs and retinoblastoma protein. *Oncogene* 12: 1173–1180.
11. Botz, J., K. Zerfass-Thome, D. Spitkovsky, H. Delius, B. Vogt, M. Eilers, A. Hatzigeorgiou and P. Jansen-Dürr. 1996. Cell cycle regulation of the murine cyclin E gene depends on an E2F binding site in the promoter. *Mol. Cell Biol.* 16: 3401–3409.
12. Weinberg, R.A. 1995. The retinoblastoma protein and cell cycle control. *Cell* 81: 323–330.
13. Weintraub, S.J., C.A. Prater and D.C. Dean. 1992. Retinoblastoma protein switches the E2F site from positive to negative element. *Nature* 358: 259–261.

14. Weintraub, S.J., K.N.B. Chow, R.X. Luo, S. Zhang, S. He and D.C. Dean. 1995. Mechanisms of active transcriptional repression by the retinoblastoma protein. *Nature* 375: 812–815.

15. Sellers, W.R., J.W. Rodgers and W.G. Kaelin. 1995. A potent transrepression domain in the retinoblastoma protein induces a cell cycle arrest when bound to E2F sites. *Proc. Natl. Acad. Sci. USA* 92: 11544–11548.

16. Hofmann, F. and D. M Livingston. 1996. Differential effects of cdk2 and cdk3 on the control of pRb and E2F function during G1 exit. *Genes & Dev.* 10: 851–861.

17. Johnson, D.G., W. Cress, L. Jakoi and J.R. Nevins. 1994. Oncogenic capacity of the E2F-1 gene. *Proc. Natl. Acad. Sci. USA* 91: 12823–12827.

18. Singh, P., S. Wong and W. Hong. 1994. Overexpression of E2F-1 in rat embryo fibroblasts leads to neoplastic transformation. *EMBO J.* 13: 3329–3338.

19. Xu, G., D.M. Livingston and W. Krek. 1995. Multiple members of the E2F transcription factor family are the products of oncogenes. *Proc. Natl. Acad. Sci. USA* 92: 1357–1361.

20. Johnson, D.G., J.K. Schwarz, W.D. Cress and J.R. Nevins. 1993. Expression of E2F-1 induces quiescent cells to enter S-phase. *Nature* 365: 349–352.

21. Qin, X.Q., D.M. Livingston, W.G. Kaelin and P. Adams. 1994. Deregulated E2F-1 expression leads to S-phase entry and p53-mediated apoptosis. *Proc. Natl. Acad. Sci. USA* 91: 10918–10922.

22. Shan, B. and W.-H. Lee. 1994. Deregulated expression of E2F-1 induces S-phase entry and leads to apoptosis. *Mol. Cell Biol.* 14: 8166–8173.

23. Wu, X., and A.J. Levine. 1994. p53 and E2F-1 cooperate to mediate apoptosis. *Proc. Natl. Acad. Sci. USA* 91: 3602–3606.

24. Weinberg, R.A. 1996. E2F and cell proliferation: a world turned upside down. *Cell* 85: 457–459.

25. Yamasaki, L., T. Jacks, R. Bromson, E. Goillot, E. Harlow and N.J. Dyson. 1996. Tumor induction and tissue atrophy in mice lacking E2F-1. *Cell* 85: 537–548.

26. Field, S.J., F.-Y. Tsai, F. Kuo, A.M. Zubiaga, W.G. Kaelin, D.M. Livingston, S.H. Orkin and M.E. Greenberg. E2F-1 functions in mice to promote apoptosis and suppress proliferation. *Cell* 85: 549–561.

27. Hsiao, K.-M., S.L. McMahon and P.J. Farnham. 1994. Multiple DNA elements are required for the growth regulation of the mouse E2F-1 promoter. *Genes & Dev.* 8: 1526–1537.

28. Johnson, D. G., K. Ohtani and J.R. Nevins. 1994. Autoregulatory control of E2F-1 expression in response to positive and negative regulators of cell cycle progression. *Genes & Dev.* 8: 1514–1525.

29. Neuman, E., E.K. Flemington, W.R. Sellers and W.G. Kaelin. 1994. transcription of the E2F-1 gene is rendered cell cycle dependent by E2F DNA-binding sites within its promoter. *Mol. Cell Biol.* 14: 6607–6615.

30. Ciechanover, A. 1994. The ubiquitin-proteasome proteolytic pathway. *Cell* 79: 13–21.

31. Hochstrasser, M. 1995. Ubiquitin, proteasomes, and the regulation of intracellular protein degradation. *Curr. Op. Cell Biol.* 7: 215–223.

32. Nurse, P. 1994. Ordering S phase and M phase in the cell cycle. *Cell* 79: 547–550.

33. Wuarin J. and P. Nurse. 1996. Regulating S-phase: cdks, licensing and proteolysis. *Cell* 85: 785–787.

34. Hofmann, F., F. Martelli, D.M. Livingston and Z. Wang. 1996. The retinoblastoma gene product protects E2F-1 from degradation by the ubiquitin-proteasome pathway. *Genes & Dev.* 10: 2949–2959.

35. Helin, K., J.A. Lees, M. Vidal, N. Dyson, E. Harlow and A. Fattaey. 1992. A cDNA encoding a pRb-binding protein with properties of the transcription factor E2F. *Cell* 70: 337–350.

36. Kaelin, W.G.Jr., W. Krek, W.R. Sellers, J.A. DeCaprio, F. Ajchenbaum, C.S. Fuchs, T. Chittenden, Y. Li, P.J. Farnham, M.A. Blanar, D.M. Livingston and E.K. Flemington. 1992. Expression cloning of a cDNA encoding a retinoplastoma-binding protein with E2F-like properties. *Cell* 70, 351–364.

37. Shan, B., X. Zhu, P.-L. Chen, T. Durfee, Y. Yang, D. Sharp and W.-H. Lee. 1992. Molecular cloning of cellular genes encoding retinoblastoma-associated proteins: identification of a gene with properties of the transcription factor E2F. *Mol Cell Biol.* 13: 5620–5631.

38. Hamel, P.A., B.L. Cohen, L.M. Sorce, B.L. Gallie and R.A. Phillips. 1990. Hyperphosphorylation of the retinoblastoma gene product is determined by domains outside the Simian virus 40 large T antigen-binding region. *Mol. Cell Biol.* 10: 6586–6595.

39. Rock, K.L., C. Gramm, L. Rothstein, K. Clark, R. Stein, L. Dick, D. Hwang and A.L. Goldberg. 1994. Inhibitors of the proteasome block the degradation of most cell proteins and the generation of peptides presented on MHC class I molecules. *Cell* 78: 761–777.

40. Ward, C.L., S. Omura and R.R. Kopito. 1995. Degradation of CFTR by the ubiquitin-proteasome pathway. *Cell* 83: 121–127.

41. Krek, W., M.E. Ewen, S. Shirodkar, Z. Arany, W.G. Kaelin and D.M. Livingston. 1994. Negative regulation of the growth-promoting transcription factor E2F-1 by a stably bound cyclin A-dependent protein kinase. *Cell* 78: 161–172.

42. Krek, W., G. Xu and D.M. Livingston. 1995. Cyclin A kinase regulation of E2F-1 DNA binding function underlies suppression of an S phase checkpoint. *Cell* 83: 1149–1158.

DISCUSSION

Bernards: I find your data absolutely beautiful on the ubiquitination and, as you said, many of our data seem to agree with what you say. There is only one sort of extension that I would like to make on what you showed because if you assume this paradigm that you have shown that an E2F pocket protein complex is stable and free E2F is unstable then Adenovirus would have a real problem when it would infect a cell. It would release E2F from the pocket protein leading to a rapid degradation of E2F and no cell cycle entry would result. So there the virus would have to do something to prevent a degradation of E2F. In our hands we see that the E1A protein even though it liberates E2F and should make the protein then unstable. In fact our data suggests that E1A can stabiles E2F to the same extent as RB binding can, increasing the half-life to over eight hours so that would suggest that that Adenovirus has two tricks up its sleeve. One is to release RB from E2F and the second being that it can then sort of inhibit one way or another ubiquitination of the free E2F and stabilize it allowing it then to drive cells into S-phase.

Livingston: I think that is apt. The only comment that might be made is whether after a couple of rounds of DNA replication, under those conditions, the abundance of E2F has something to say about whether or not cells die under normal virus infection conditions, as they should.

Nasmyth: This is really a question for you and René. What are the levels of E2F during the cell cycle in cycling cells? Either in an *in vivo* situation or even in culture?

Livingston: There is so little protein that one could say that it is almost unrecognizable compared to its negative regulator, RB. Now in the case of RB, in many of the cell lines that we have looked at, human or monkey, somewhere between 50 and say, 100,000 or 150,000 molecules per cell are present. In that setting there have to be less than ten percent that many E2F molecules. They are very difficult to detect.

Nasmyth: There are two issues here. One is that there was the increased synthesis in late G1, is that right? E2F regulates itself so you would expect to see it increasing in late G1, E2F1 for example, and then according to what you are saying you would then expect to see it decreasing as cells get into S-phase and in G2. Is that actually what one sees if you measure the levels?

Livingston: Right now, it is an exceedingly difficult experiment to do.

Nasmyth: Could I quickly ask another question. I was at a cell cycle meeting last week where Nevins described, what seemed to me, rather similar experiments to yours, which was claiming that you could inhibit CDK's, cyclin E, cyclin A kinases, by producing very high levels of p21 in the cell and then ectopically expressing E2F and finding that cells still went into S-phase.

Livingston: Yes, I am completely aware of them and they speak for themselves. There are people here who can speak to that, like Michele Pagano for example.

Pagano: I agree that it looks quite strange but we have not done any of these experiments while expressing E2F.

Livingston: In fact others have reported that the experiment that we have shown here, the introduction of E2F in the face of a dominant negative, in the presence of a dominant negative CDK2 allele will override the block and the argument has been that there is no cyclin E or A present under those conditions. The facts are that micro-injection of antibody to cyclin E, as Giulio Draetta showed years ago, stops the show quite well. And we have no reason to suspect or indicate that E and A both are not being activated in these cases. So I think there is a possibility that one might be looking at unscheduled or abonormal DNA synthesis there.

Pagano: Do you know if there is any PEST sequence in the C terminus of E2F?

Livingston: It does not appear to be the case.

Pagano: And do you know if phosphorylation plays any role in signaling ubiquitination of E2F?

Livingston: Good question—I do not know yet. The relevant mutants are being generated for the appropriate experiment.

Pagano: And finally, do you know the half-life of E2F1 in RB deficient cells?

Livingston: Not yet. That is a difficult experiment to do, of course. One would like to look at the endogenous protein. You would like not to have to overproduce it. The ideal experiment is to look at the endogenous E2F1 in RB in plus and minus fibroblasts.

Pagano: And the very last question is about CDK3. Which cells have you used for these experiments? Because we found that CDK3 is particularly expressed at high levels in hematopoietic cells. In other cell lines or cell types we do not find too much CDK3. Is this also your experience?

Livingston: These experiments were done in U20S cells, a human osteogenic sarcoma cell line but they have been repeated in other lines with similar results. There is not much CDK3 in those cells, although, if you collect enough extract, it is possible to see it by immuno-precipitation of S35 labeled extracts. It is not an abundant protein. I agree with you.

Bernards: Concerning the effect of CDK3 dominant negative on E2F1, have you asked whether the cyclin A site is relevant there by using the mutants in the cyclin A site where those are still subject to inhibition by CDK3 dominant negative?

Livingston: They are. It was actually on one of the slides. It is independent of cyclin A binding.

Nasmyth: This is a related question. One explanation for the dominant negative result is you do need CDK3 for some subsequent activation of liberated E2F. Given that you have found that CDK3 does interact, is it possible to rule out the possibility that your dominant negative is going in there and binding to E2F and then just gluing up the works, as it were. That is, what you have is a relative gain of function rather than an interference with the wild type function.

Livingston: What you want to know is whether it is a pharmacologic result. I doubt it because if you do the same experiment by overproducing wild type CDK3, you can detect those complexes as well.

Nasmyth: Yes, but given that it does make those complexes, if you made a form of CDK3 which is now abnormal which you clearly have with the dominant negative, then maybe it cannot release, for example. Because it binds it brings something in, which normally does not. Do you know why it is a dominant negative?

Livingston: No. We can speculate based on the structure. The activity has been measured by others but it is weak. Presumably its catalytic activity is off because it cannot bind substrate, cannot bind ATP. Now the best I can tell you is that overproducing either the wild type or the dominant negative results in complexes and that you do not have to co-transfect either E2F1 or DP1 to detect those complexes. The actual mechanism, is anybody's guess at the moment. The simplest speculation is that it is a direct, CDK3-dependant phosphorylation, but, right now, the mechanism is unknown.

Nasmyth: So, these dominant negative alleles of CDK2 and CDK3 are completely different, they are not, as it were, homologous alleles, they are not the same sort of mutation in different kinases?

Livingston: No, they are the same kind of mutation in different kinases.

Wahl: I would like to ask a kind of physiological question. As I understand it the levels of dioxynucleotides in the cell are just sufficient to allow for about ten minutes of DNA replication. Now pKb is a pretty stable protein. If I understand the data correctly then what you have is a situation where E2F1, let us say, which regulates DHFR synthesis would be available only for a very brief interval to regulate DHFR. What difference is that going to make in terms of cell cycle progression if DHFR is turned on briefly at the G1S boundary in a cycling cell?

Livingston: In a cycling cell, very little as you know from the work of Tom Kelly and others. But in cells coming out of G0 it is like 10 fold or more.

Wahl: So is this mechanism really in place to govern exit from G0?

Livingston: Nobody has yet donethe correct biological experiment in the native state with wild type and mutant alleles. So one can only reason by correlation, where the data imply that there is a role for E2F proteins in promoting exit from G0. That is why we drew the graph that way in the beginning. In cycling cells, I do not know whether E2F function is important.

Wahl: I just wanted to point out that even in light of the conversation that we have just had, where you would think that RB deficiency might not contribute very substantially at all to misregulation of DHFR activity and therefore would not impact substantially on responsive cells with an activated RB function to chemotherapeutic agents, Joe Bertino and we have done some work showing that sarcomas that are RB deficient are intrinsically resistant to FUDR and methotrexate. If you replace RB in those cells, the sensitivity goes back to normal.

Livingston: Well there is actually some data on that point. Peggy Farnham showed a long time ago that in G1, as cells exit from G0, the DHFR promoter is off until the cell reaches a certain point where it turns on. The work of Kaelin and Nevins showed that the E2F1 promoter does not turn on until one goes through mid-late G1 upon exit from G0.

Nasmyth: If I understand it, what you are saying is that this S-phase is a E2F driven program of gene expression, may be much more important physiologically for cells coming out of G0 than it is in cycling cells. That is because a lot of the genes seem to encode proteins that are relatively stable once you are into the cycle. That is largely what is found in the yeast system. I mean a huge number of genes, I did not mention them yesterday, but there are 100 genes or more whose transcription is under the control of SBF and MBF. inety-nine percent of those genes have relatively stable proteins, but there are, of course, one or two of those genes that encode very unstable proteins, like the cyclins, and like CDC6. It seems to me quite likely that E2F will be found to be regulating, say, human CDC6 and it seems to be quite likely that CDC6 will turn out to be an unstable protein. So I think there will be examples where this transcriptional program does play a very important role at every G1 S-phase transition, even in cycling cells. We just do not know what the targets are as yet. I think cyclin E would be one candidate.

Livingston: And cyclin A as well.

Evan: So, it seems to me, that one of the most important differences between the G0-G1 transition and the G1S transition is that you need mitogenic stimulus there from G0 all the way to the restriction points. So in a sense the cell is time-averaging the information it is getting and you can abort from any point up until the restriction point. Indeed that is how the restriction point is defined. So the cell needs this continuous mitogenic stimulation and therefore, there must be at least one unstable system during the whole of this period that needs continuous priming.

Livingston: Yes. In that regard, Gerard went right to the heart of it. There is very little RB in G0 T-cells, and none of it as far as one can tell, is associated with an E2F molecule. But there is another pocket protein, P130, that is very abundant in G0 cells and it is complexed with E2F4. That pocket protein E2F complex is interesting because it may function in G0 entry and/or maintenance. Now, one needs to reveal the signal pathway that tells this complex to disappear.

16

Cdk4-CYCLIN D1 AND Cdk2-CYCLIN E/A PHOSPHORYLATE DIFFERENT SITES IN THE RB PROTEIN

Yoichi Taya,[1] Hai-Kwan Jung,[1] Masako Ikeda,[2] Katsuyuki Tamai,[2] Hideaki Higashi,[3] and Masatoshi Kitagawa[3]

[1]Biology Division, National Cancer Center Research Institute
Tsukiji 5-1-1, Chuo-ku, Tokyo 104, Japan
[2]Ina Laboratories, MBL Co. Ltd.
Oohara Terasawaoka 1063-103
Ina, Nagano 396, Japan
[3]Banyu Tsukuba Research Institute in Collaboration with Merk Research
    Laboratories
Okubo 3, Tsukuba 300-33, Japan

1. INTRODUCTION

The growth-suppressive activity of the retinoblastoma protein (pRB) is regulated by the cell cycle-dependent phosphorylation.[1-3] It has been suggested that Cdk4/6-cyclin D, Cdk2-cyclin E and Cdk2-cyclin A are successively involved in this phosphorylation at G1/S transition.[2,3] However, it was not known why several different Cdks are needed and how their role is distinguished. We have now found that the consensus motif for phosphorylation by Cdk4-cyclin D1 is different from that (Ser/Thr-Pro-X-Arg/Lys) for phosphorylation by Cdk2-cyclin E/A, and that at least one site among 13–14 potential phosphorylation sites of pRB is specifically phosphorylated by Cdk4-cyclin D1 during the cell cycle.[4]

2. PURIFICATION OF Cdk-CYCLIN

Active cyclin-Cdk complexes were produced in Sf9 cells that had been co-infected with recombinant baculoviruses that encoded cDNA for cyclin (A, E or D1) and Cdk(2 or 4). Purification was carried out by column chromatography on Mono Q, hydroxyapatite and MonoS. Purified kinases showed only corresponding cyclin or Cdk bands in silver staining after SDS-polyacrylamide gel electrophoresis. Those bands were subsequently confirmed as such by Western blotting using antibodies against those Cdks or cyclins.

Genomic Instability and Immortality in Cancer
edited by Mihich and Hartwell, Plenum Press, New York, 1997

3. COMPARISON OF THE SUBSTRATE SPECIFICITY USING SYNTHETIC PEPTIDES

First, several synthetic peptides containing potential phosphorylation sites of pRB were tested for phosphorylation by cyclin D1-Cdk4, cyclin A-Cdk2 and cyclin E-Cdk2. Most of these peptides were phosphorylated well by cyclin A-Cdk2 or cyclin E-Cdk2 except a peptide RPPTLSPIPHIPR which was used as a control peptide because this peptide contains Ser-Pro motif but not basic amino acid at +3 position from the Ser residue. Surprisingly, however, this peptide was found to be the best substrate for phosphorylation by cyclin D1-Cdk4. Some other peptides were also phosphorylated by cyclin D1-Cdk4 but others were not, whereas all these peptides were good substrates for Cdk2-cyclin E/A, suggesting that consensus motif for phosphorylation is different between Cdk4-cyclin D1 and Cdk2-cyclin E/A.

It is well known that histone H1 is not phosphorylated by Cdk4-cyclin D1 while it is a good substrate for Cdk2-cyclin E/A or Cdc2-cyclin B.[5-7] In histone H1, a sequence TPKK is repeated several times, and S1 peptide (AKAKKTPKKAKK) containing this sequence is usually used as a substrate for kinase assay. We expected from the previous experiment, using RB-peptides, that S1 peptide will be phosphorylated by Cdk4-cyclin D1 if one amino acid is substituted, and such trial was performed. As was expected, S1 peptide became a good substrate for Cdk4-cyclin D1 when Pro was substituted for Lys at position −2 from the Thr residue. When Lys at position −1 was additionally substituted by Ala, it became even better substrate. On the contrary, extent of phosphorylation by Cdk2-cyclin E/A did not change so much by these amino acid substitution. Thus, consensus motif for phosphorylation by Cdk4-cyclin D1 is clearly different from that for Cdk2-cyclin E/A.

4. PRESENCE OF A Cdk4-CYCLIN D-SPECIFIC PHOSPHORYLATION SITE IN pRB

Peptide experiments suggested that Ser780 of pRB (Fig. 1), which is contained in the sequence RPPTLSPIPHIPR, will be phosphorylated by Cdk4-cyclin D1 but not Cdk2-cyclin E/A. We tested this possibility by constructing a mutant RB in which Ser780 was replaced by Leu. Moreover, an antibody to recognize the phospho-Ser780 was generated using chemically synthesized phosphopeptides as antigen. This antibody detected wild type GST-RB well when phosphorylated by Cdk4-cyclin D1, but not when phosphorylated by Cdk2-cyclin E/A. In contrast, mutant GST-RB was not detected by this antibody after phosphorylation by Cdk4-cyclin D1, suggesting that this antibody indeed recognize phos-

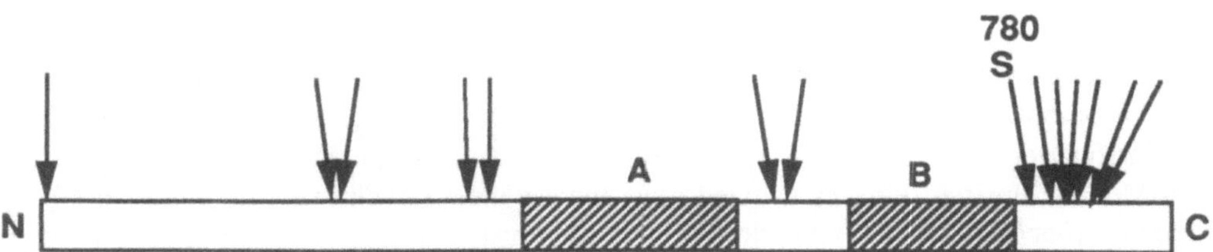

Figure 1. Potential phosphorylation sites in pRB. These sites are indicated by arrows. A and B are E1A-binding pockets.

pho-Ser780 specifically and that Ser780 is phosphorylated by Cdk4-cyclin D1 but not by Cdk2-cyclin E/A *in vitro*.

In order to know whether Ser780 is phosphorylated by Cdk4-cyclin D *in vivo*, human glioblastoma T98G cells were synchronized and cell extracts were subjected to Western blotting by several antibodies. Phospho-Ser780-specific antibody began to detect the pRB band 8 hours after addition of serum. Thus, Ser780 is indeed phosphorylated *in vivo*. Cdk4 band increased after 4hrs, whereas the band of active form of Cdk2 was detected only after 10 hrs, suggesting that Ser780 is phosphorylated by Cdk4 but not Cdk2 *in vivo*.

5. CONCLUSION

We have shown that consensus motif for phosphorylation by Cdk4-cyclin D1 is different from that for phosphorylation by Cdk2-cyclin E/A. In addition, at least one site among 13–14 potential phosphorylation sites in pRB was shown to be specifically phosphorylated by Cdk4-cyclin D1 but not by Cdk2-cyclin E/A. This site is located in the border between pocket B and the C-terminal domain (Fig. 1). Thus, it is conceivable that binding of pRB with pRB-binding proteins is considerably affected by phosphorylation of this site. We are now generating monoclonal antibodies to recognize all 14 potential phosphorylation sites of pRB. These antibodies will be useful to distinguish the physiological role of each sites as well as to identify the Cdk-cyclin species involved in phosphorylation of each site.

REFERENCES

1. Weinberg, R.A. (1991) Science 254, 1138–1146
2. Sherr, C.J. (1994) Trends Cell Biol 4, 15–18
3. Taya, Y. (1995) Mol. Cells 5, 191–195
4. Kitagawa, M., Higashi, H., Jung, H.-K., Suzuki-Takahashi, I., Ikeda, M., Tamai, K., Kato, J., Segawa, K., Yoshida, E., Nishimura, S. and Taya, Y. (1996) EMBO J., in press
5. Matsushime, H., Ewen, M.E., Strom, D.K., Kato, J., Hanks, S.K., Roussel, M.F. and Sherr, C.J. (1992) Cell, 71, 323–334
6. Ewen, M.E., Sluss, H.K., Sherr, C.J., Matsushime, H., Kato, J. and Livingston, D.M. (1993) Cell, 73, 487–497
7. Kato, J., Matsushime, H., Hiebert, S.W., Ewen, M.E. and Sherr, C.J. (1993) Genes & Dev., 7, 331–342

DISCUSSION

Krek: Have you done western blotting with this antibody on extracts collected from S, G2 and M-phase fractions to see whether the site disappears again at later stages?

Taya: I did not do such an experiment but, yes, we want to do it.

Bernards: Do you have evidence that the serine 780 is critical for phosphorylation, for release of E2F?

Taya: We tested the E2F and RB complex formation using the antibody to recognize phosphoserine 780. We found if the serine 780 is phosphorylated E2F is released. After phosphorylation of that site, complex is not present between E2F1 and RB protein. We are testing other kinds of E2F families as well as other RB binding proteins; Elf1, UBF, Brm and other binding proteins that are known to bind RB protein.

17

CELL CYCLE REGULATORY PROTEINS AS TARGETS OF ONCOGENIC EVENTS

Francesca Fiore and Giulio F. Draetta[*]

Department of Experimental Oncology
European Institute of Oncology
435 Via Ripamonti, 20141 Milan, Italy

CELL CYCLE REGULATORY PROTEINS

Many years have passed since the fundamental discoveries of Paul Nurse and Lee Hartwell described the characteristic phenotypes of yeast cell division mutants and the isolation of the first cell cycle regulatory genes (Hartwell et al., 1974; Nurse, 1975). Using very different approaches a large number of investigations have now allowed us to understand how the molecular engine that drives a cell through the division cycle works. At the core of the system are a family of serine/threonine kinases, the cyclin-dependent kinases (Cdks) whose periodic activation leads to the initiation of each cell cycle transition. Their name stems from the fact that they require an associated protein subunit, a cyclin (Evans et al., 1983; Rosenthal et al., 1983) for activity. This biochemical mechanism has been well conserved throughout the evolution, but while in yeast a single Cdk regulates cell cycle progression upon association with distinct cyclins, in higher eukaryotes a family of structurally related proteins exists, each responsible for triggering a certain transition in the division cycle. The various Cdks can associate with different cyclin subunits and form complexes that are activated at distinct times in the cell cycle. In mammals, during the G1 phase, two major protein complexes are activated, the cyclin D and cyclin E associated kinases, while cyclin A complexes are functional throughout S-phase and G2 and cyclin B/Cdc2 is activated at the G2/M-phase transition (see review by Morgan, 1995).

The interaction with the cyclin subunit provides the Cdk with structural stabilization, it controls its subcellular localization and its substrate specificity by providing a protein-protein interaction surface. As an example, the cyclin D1/Cdk4 complex, is localized to the nucleus and it has a unique preference for the retinoblastoma protein (pRb) as a substrate, due to the specific interaction of pRb with cyclin D1.

Cyclin mRNA transcription is periodically activated during the cell cycle and is down regulated in cells that are induced to quiescence. Once a cyclin accumulates in the

---

* Tel.: +39 2 57 48 98 31; fax. + 39 2 57 48 98 51; e-mail: gdraetta@ieo.cilea.it

Genomic Instability and Immortality in Cancer
edited by Mihich and Hartwell, Plenum Press, New York, 1997

cell, it forms a complex with a Cdk subunit. In addition to cyclins several other polypeptides have been shown to bind Cdks, including a small protein called Suc1, PCNA (a subunit of DNA polymerase δ and ε), a protein called p21 (Waf1, Cip1), as well as two novel proteins Skp1 and Skp2 (Draetta et al., 1987; Zhang et al., 1994; Zhang et al., 1993) (Xiong et al., 1992; Xiong et al., 1993). While it is clear that they are not required for the catalytic activity of the enzyme, and the precise role of these subunits in the complexes is not well established at present, yet one could speculate that their association with the Cdks is involved in targeting the complex to various subcellular compartments. It has also recently been found that a protein structurally related to the S. cerevisiae Cdc37 gene product associates with Cdk4 and with the heat shock protein 90, HSP90 (Dai et al., 1996; Lamphere et al., 1996; Stepanova et al., 1996), but the significance of this interaction is still unknown.

Cdk complex activation (Fig.1) is regulated by phosphorylation of residues within two regions of the protein. Phosphorylation of Thr161 in human Cdc2 (and on the equivalent residue in other Cdks) by the Cdk-activating kinase is required for Cdk function. The phosphorylation is located in the large lobe of the Cdk, within the conserved domain VIII of protein kinases (Damagnez et al., 1995; Espinoza et al., 1996; Kaldis et al., 1996; see review by Morgan, 1995; Thuret et al., 1996). Thr161 phosphorylation although not involved in catalysis, is necessary to stabilize the enzyme structure through the formation of ionic interactions with amino acid residues present in both the small and large lobe of the Cdk. Cdks can also be reversibly phosphorylated on a Tyr and a Thr residue located within the ATP-binding loop of the small lobe of the kinase by members of the Wee1 family of protein kinases. This inhibitory phosphorylation is reversed by members of the Cdc25 family of protein phosphatases (see below).

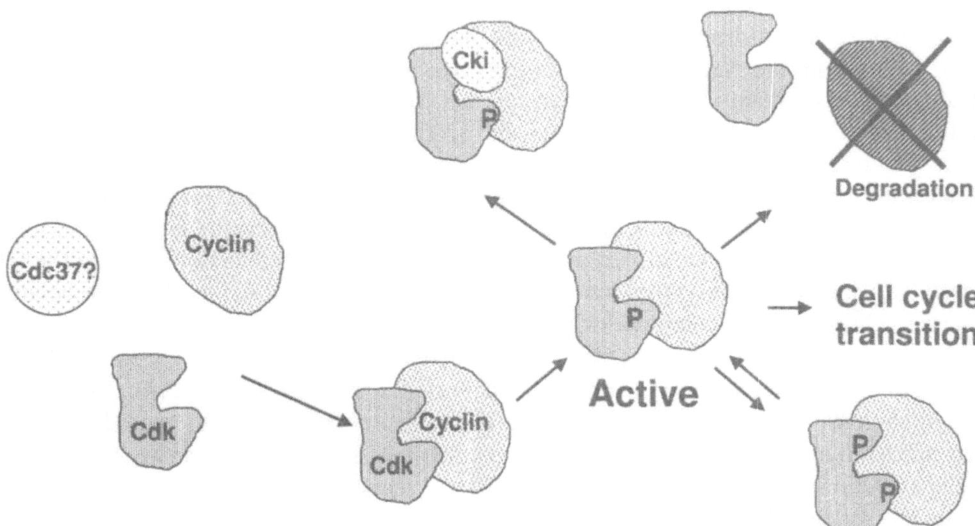

Figure 1. Current view of the biochemical regulation of cyclin D1-dependent kinase. The activation of cyclin-dependent kinases requires their binding to a cyclin and phosphorylation on a threonine residue located within the Cdk large lobe. Inactivation can occur reversibly by phosphorylation by a Wee1 protein kinase family member and subsequent dephosphorylation by Cdc25, or permanently by cyclin degradation or binding to Cdk-inhibitory proteins. Cdc37, a protein which binds monomeric Cdk4 might have a role in directing the folding of Cdk4 to favor its interaction with cyclin.

Cdks complexes can be permanently inactivated by different mechanisms. Once past a given cell cycle transition, as it is well known for mitotic Cdk complexes, cyclins are degraded through a process which involves the ubiquitin pathway (Glotzer et al., 1991; Hershko et al., 1994; King et al., 1995; Sudakin et al., 1995). A distinct mechanism involves the binding of inhibitory proteins, Cdk-inhibitory proteins (Ckis) to Cdk/cyclin complexes. Two family of inhibitors have been described, one specific for Cdk4/6-cyclin D complexes, the prototype member of which is p16, the other able to bind and inhibit all Cdk complexes at least in vitro, and including p21, p27, p57 (see review by Sherr and Roberts, 1995). Interestingly, as also mentioned above, p21 appears to be an intrinsic subunit of Cdk complexes which when present at low stoichiometry does not inhibit the kinase, while at higher level it blocks Cdk function (Zhang et al., 1994).

Since a primary characteristic of tumor cells is their uncontrolled growth, the cell cycle regulatory pathways are very often a primary target of oncogenic events (see reviews by Draetta and Pagano, 1996; Karp and Broder, 1995). The alterations found in tumors affect particularly the ability of the targeted cells to respond to growth-regulatory signals during the G1 phase of the cell cycle (Table 1). These signals feed onto the Cdk4/cyclin D1 and the Cdk2/cyclin E complexes. Once phosphorylated by cyclin D1 first and then by cyclin E kinases, pRb releases the transcriptional factor E2F leading to the expression of genes that will drive the cell through S phase (see review by Weinberg, 1995). In tumors, alterations can either be ascribed to a lack of negative regulators as is the case of p16 and p15 mutations (see review by Elledge and Harper, 1994) or due to overexpression of positive ones (e.g. cyclins).

At present most scientific efforts are directed towards finding links between the activation of Cdk/cyclin complexes and the upstream signaling. The attention is focused mainly on the G1 phase since it is during G1 that cells are sensitive to growth stimulatory or growth inhibitory signals from the external environment converging on the cell cycle machinery (see review by Sherr, 1994).

CYCLIN D1 ONCOGENE

Cyclin D1 was identified as a the product of a cDNA expressed during G1 upon stimulation of hemopoietic cells with growth factors (Matsushime et al., 1991), and for its ability to rescue the phenotype of *S. cerevisiae* strains lacking G1 cyclin genes (Koff et al.,

Table 1. Known alterations of cyclin-dependent
kinase cascades in cancer

- Rb pathway alterations
    - Cyclin D1 overexpression
    - Deletion/mutations of p16, p15
    - Cdk4 mutations that make it insensitive to p16
    - Rb mutations

- Alterations that involve all Cdk controlled pathways
    - Cdc25 A, B overexpression
    - Abrogation of p21 response in p53 mutant tumors
    - Acceleration of p27 degradation

1991; Lew et al., 1991; Xiong et al., 1991). The corresponding gene, PRAD1, was also found rearranged with the parathyroid hormone locus in parathyroid tumors (Motokura et al., 1991). The rearrangement brought the 5′ regulatory region of the hormone to the D1 coding region causing the deregulated expression of the cyclin. Interestingly, the 11q13 region where PRAD1 maps, has been found amplified in 15–20% of human breast and squamous cell carcinomas (Lammie et al., 1991). Furthermore, the cyclin D1 gene has been found amplified in twenty-five percent of a series of squamous esophageal tumors (Jiang et al., 1992) and in carcinomas of the head and neck (Lammie et al., 1991). The cyclin D1 gene is also involved in the BCL-1 translocation identified in B cell lymphomas (see review by Korsmeyer, 1992).

Overexpression of cyclin D1 mRNA has also been reported for a significant number of breast cancer cases in the absence of any increase in gene copy number (Buckley et al., 1993) and in some breast cancers a 3′ truncated version of the transcript is present, that appears to be more stable than the full length one (Lebwohl et al., 1994). Upregulation of cyclin D1 protein expression has also been detected in primary tumor specimen by immunohistochemistry (Gillett et al., 1994).

Recent in situ hybridization studies have demonstrated that overexpression of cyclin D1 mRNA is detected at very high frequency in breast carcinomas suggesting the possibility to use cyclin D1 mRNA detection as a potential tumor diagnostic (Weinstat-Saslow et al., 1995). A fundamental role for cyclin D1 in the development of breast cancer has also been supported by studies showing that cyclin D1 message upregulation occurs as an early response to mitogenic stimulation by estrogen and progestins and it is quickly down regulated in response to the inhibitory effects of anti estrogens (Sutherland et al., 1995). Furthermore, mice carrying a cyclin D1 transgene targeted to the breast display mammary hyperplasia and succumb to breast cancer (Wang et al., 1994). Conversely, mice in which the gene has been inactivated show abnormalities in breast tissue development (Sicinski et al., 1995). Altogether, these findings establish a strong correlation between upregulation of cyclin D1 expression and the occurrence of breast cancer and cyclin D1.

Cdc25 PHOSPHATASES

The Cdc25 protein phosphatases are essential activators of cell cycle progression, since they remove the inhibitory phosphate from Cdks (Draetta and Eckstein, 1996). While in yeast there is only one gene encoding Cdc25 (Millar et al., 1991) (Russell et al., 1989), in humans at least three genes encoding proteins related to yeast Cdc25 have been identified (Galaktionov and Beach, 1991; Nagata et al., 1991; Sadhu et al., 1990). The three phosphatases, known as Cdc25 A, B and C, are involved in controlling different stages of the cell cycle (Galaktionov and Beach, 1991; Hoffmann et al., 1993; Hoffmann et al., 1994; Jinno et al., 1994) and they likely recognize and dephosphorylate different Cdk complexes (Xu and Burke, 1996). Cdc25 A mRNA is mainly expressed in G1, B appears later during G1 and S-phase (Jinno et al., 1994) while C levels are constant in the cell cycle, but the enzyme is activated at the G2/M-phase transition (Sadhu et al., 1990).

Cdk activation can also be controlled by phosphorylation on Thr and Tyr residues located within the ATP binding loop of the small lobe. This phosphorylation, in the case of the Cdc2/cyclin B complex keeps the complex inactive during S-phase and G2. Upon completion of DNA replication, the activation of specific phosphatase of the Cdc25 family dephosphorylates Cdc2 and induces its activation.

Cdc25 A and B interact with the Raf-1 protein kinase, which is a primary target of the Ras oncogene (Galaktionov et al., 1995). In vitro Cdc25 can be phosphorylated and activated by Raf-1 and a fraction of the cellular Cdc25A has been shown to colocalize with Raf-1 to the plasma membrane. As for a potential role of Cdc25 in tumor growth, it was initially shown that Cdc25 B is overexpressed in human cancer cell lines compared to non-transformed cells (Nagata et al., 1991). Recent evidence has proven that Cdc25 A and B can promote oncogenic transformation in vitro and in vivo (Galaktionov et al., 1995). The two proteins Cdc25 A and B cooperate with H Ras in generating focus formation in primary mouse fibroblasts and these transformed cells are able to proliferate in soft agar, and upon injection into nude mice they generate tumors. Using in situ hybridization, Cdc25 B overexpression has been detected in thirty-two percent of a series of 124 breast cancer cases, supporting its role in oncogenesis (Galaktionov et al., 1995).

It has also been recently shown that the transcription of Cdc25 A and B messenger RNAs is regulated by c-myc (Galaktionov et al., 1996). A c-myc binding element has been identified in the promoter region of Cdc25 A and B and it has been shown that activation of a c-myc-estrogen receptor fusion protein results in the induction of Cdc25 A and B in intact cells (Fig. 2). Furthermore, overexpression of Cdc25 A under conditions in which c-myc induces apoptosis causes apoptosis as well, and inhibition of Cdc25 A expression upon c-myc activation blocks apoptosis, suggesting that Cdc25 A is a likely determinant of c-myc induced apoptosis. Thus, both myc and ras converge on a pathway that controls the activation of Cdc25 in G1. These results, although not sufficient to explain the mechanism of oncogene cooperation in its entirety (i.e. Cdc25 is not the sole target of Myc and Ras) nicely demonstrate how these two cellular oncogenes signal onto the cell cycle machinery.

The possibility of utilizing Cdc25 inhibitors as anti cancer agents is supported by the fact that these phosphatases are absolutely needed for cell cycle progression, and that they are likely targets of oncogenic events. In vitro assays have allowed the identification of

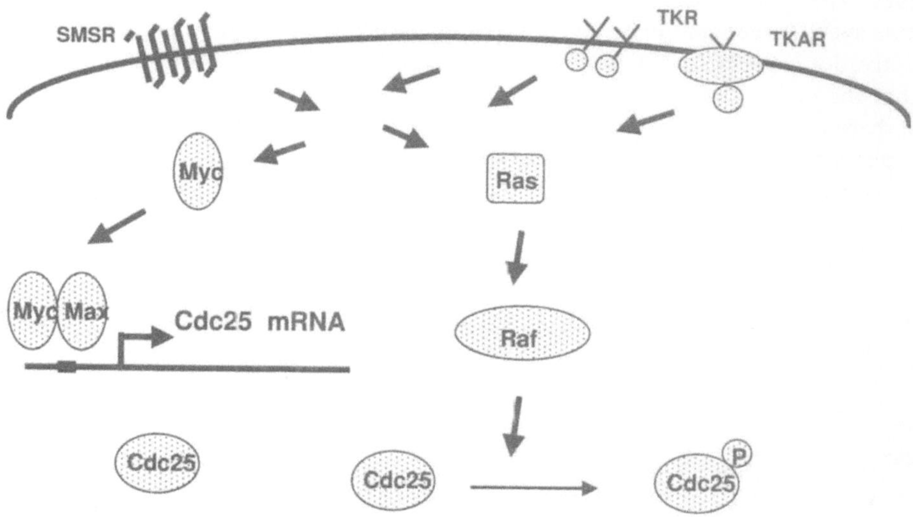

Figure 2. Cdc25 phosphatase activation by Ras and Myc pathways. Cdc25 A and B mRNA transcription is controlled by c-myc. In response to growth factor myc induces Cdc25 A and B. Proper activation of Cdc25 A and B requires the Ras-induced phosphorylation of Cdc25 by Raf-1 kinase. Fully active Cdc25 A or B can then dephosphorylate cyclin-dependent kinases (the identity of the Cdk involved is still unclear).

compounds showing Cdc25 phosphatase inhibitory activity. Interestingly, some of these substances had already been shown to have anti-tumor activity in vivo (Baratte et al., 1992; Horiguchi et al., 1994).

UBIQUITIN PATHWAY

The degradation of intracellular proteins occurs via three major biochemical pathways. One involves the lysosome, through which degradation of exogenous proteins occurs, another requires the activation of site specific proteases which recognize and cleave specific amino acid sequences within certain proteins. A distinct pathway is the one known as the ubiquitin pathway, where proteins are tagged with ubiquitin molecules before being driven to the proteasome, a multi protein complex able to degrade ubiquitin tagged proteins (see reviews by Deshaies, 1995; Hochstrasser, 1996; Hochstrasser, 1995; Jentsch and Schlenker, 1995). This pathway controls the degradation of many cellular regulatory proteins including p53 (Scheffner et al., 1990), cyclin A (Glotzer et al., 1991; Hershko et al., 1994), cyclin B (King et al., 1995; Sudakin et al., 1995), c-jun (Treier et al., 1994), c-myc (Ciechanover et al., 1991) c-Mos (Nishizawa et al., 1992), c-fos (Stancovski et al., 1995), IkB/NFkB (Orian et al., 1995; Palombella et al., 1994; Scherer et al., 1995), p27 (Pagano et al., 1995), Cdc25 (Nefsky and Beach, 1996) and others.

Ubiquitination is a multi step process (see review by Ciechanover, 1994), which involves first an ATP dependent reaction (Fig. 3), during which a thioester is formed between the carboxy terminus of ubiquitin and a Cys residue of a ubiquitin activating-enzyme (E1); the ubiquitin is then transferred from E1 to a Cys in the active site of a ubiquitin-conjugating enzyme (UBC or E2). The ubiquitin molecule can either be directly transferred from the E2 to a target protein, or through an intermediate step first to a additional enzyme (ubiquitin-ligase, E3) and then to the target substrate. The substrate actually binds ubiquitin as an isopeptide through covalent linking to the ε-amino group of lysine residues. The specific targeting of a target molecule by the ubiquitin pathway depends upon its specific recognition by a given UBC and most likely by mechanisms that control the activation of E2s and E3s which are presently unknown. While it is not known whether the activity of E2 and E3 enzymes is regulated by post-translational modifications, there is evidence that protein ubiquitination can be triggered by phosphorylation of the target proteins (see review by Isaksson et al., 1996).

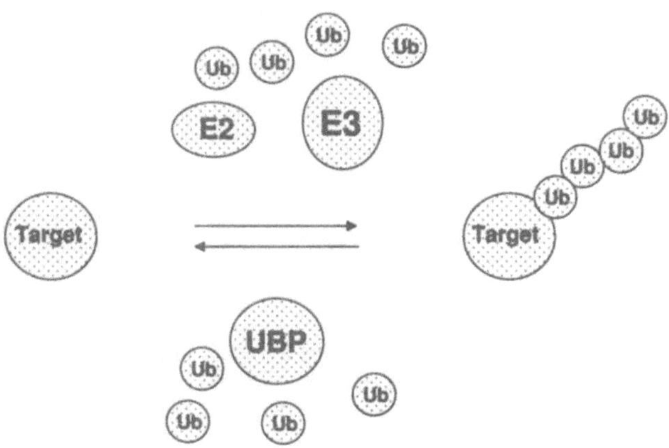

Figure 3. Current view of the regulation of protein degradation by the ubiquitin pathway. The ubiquitin pathway involves the E2, E3 enzyme cascade to generate multi-ubiquitinated proteins and the ubiquitin isopeptidases (UBPs) to counteract this effect.

Members of ubiquitin-conjugating enzyme family share a core region of homology, which by three-dimensional modeling of the known structure of two UBC family members (J. Eckstein and GFD, unpublished), includes a distinct surface that is the likely recognition site for E1 and ubiquitin, and a less well conserved region that is probably responsible for substrate interaction. UBC enzymes have a clear substrate specificity (King et al., 1995; Pagano et al., 1995; Rolfe et al., 1995; Scheffner et al., 1994). The E3 enzymes described so far belong to distinct families one of which has the human papillomavirus E6-associated proteins (E6AP) as the first best characterized member (Huibregtse et al., 1995; Huibregtse et al., 1991; Huibregtse et al., 1993). Another comprises the APC or cyclosome, a large multiprotein complex which contains homologues of the yeast Cdc27 and Cdc16 gene products (King et al., 1995; Sudakin et al., 1995). Additional components involved in targeting G1 cyclins and Cdk-inhibitors have been recently identified (see review by Jackson, 1996).The hypothesis has also been put forward that ubiquitination reaction are in equilibrium with removal of ubiquitin, which is attributed to a family of thiol proteases, known as ubiquitin isopeptidases (UBPs) that would remove the ubiquitin molecules from protein targets (see review by Hochstrasser, 1996; Hochstrasser, 1995).

The tumor suppressor protein p53 is an example of target of the ubiquitin pathway. In growing cells p53 has a short half-life, and its half life increases in response to agents which cause DNA damage, hypoxia, checkpoint alterations or other cellular stresses (see review by Gottlieb and Oren, 1996). This generates an increase in p53 protein level, which activates the transcription of a number of p53 target genes. This can result in either the induction of a block in cell cycle progression via the induction of the p21 Cki, and/or in the induction of apoptosis. Human cells infected with high-risk human papilloma virus (HPV) are unable to activate the p53 pathway in response to DNA damage or checkpoint alterations. This is due to the action of the viral protein E6 which binds to p53 and stimulates its degradation via the ubiquitin pathway. E6 binds thiol-ubiquitinated E6AP and favors the transfer of ubiquitin to p53 (Rolfe et al., 1995; Scheffner et al., 1995). The biochemical mechanisms leading to the degradation of p53 in non-HPV infected cells are still poorly understood. On the other hand, the accumulation of p53 in cells added with inhibitors of the proteasome, or in cells carrying mutations in the gene encoding the E1 enzyme, as well as the recent identification of ubiquitinated forms of p53 in cell extracts (Dietrich et al., 1996; Maki et al., 1996), provides us with sufficient evidence to stimulate further work in this direction.

p27 IN COLORECTAL AND BREAST CANCERS

p27 was initially identified in cells arrested in G1 by TGFβ as an inhibitor of the Cdk2 and Cdk4 protein kinases (Hengst et al., 1994; Polyak et al., 1994). The p27 protein level is higher in quiescent fibroblasts or in resting T-lymphocytes than in growing cells and it has been shown that p27 overexpression causes cell cycle arrest (Firpo et al., 1994; Nourse et al., 1994). The increase in the cellular abundance of p27 upon induction of cell quiescence is primarily due to a decrease in the rate of its degradation (Pagano et al., 1995). It has been demonstrated that p27 is poly -ubiquitinated both in vivo and in vitro and that a lower amount of p27 ubiquitinating activity is present in proliferating cells compared to quiescent cells. Post-translational modifications of p27, as well as changes in the abundance/activity of E2 and E3 enzymes are likely responsible for the observed differences. Contrary to other genes encoding Cdk-inhibitory proteins, the p27 gene has not been found altered in tumors. Recent studies in human colorectal and breast cancers have demonstrated that a subset of these tumors have low or absent levels of p27 as detected by

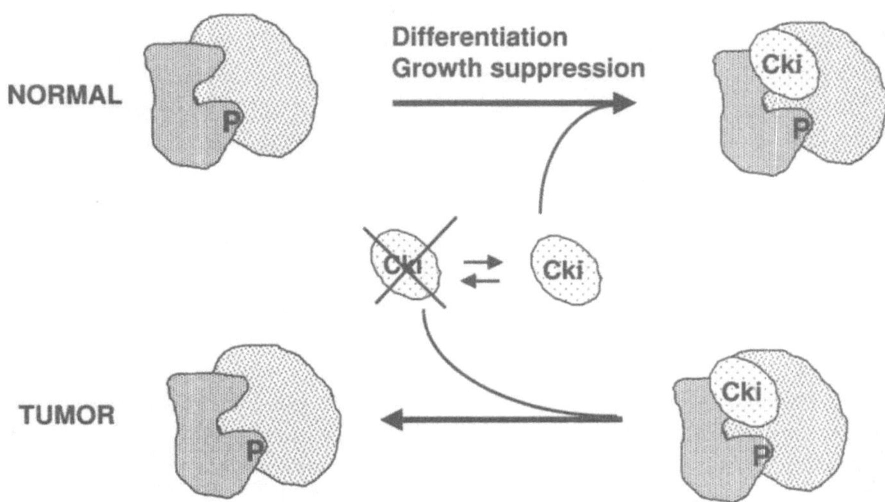

Figure 4. p27 degradation in normal and tumor cells. The p27 levels in growing cells is maintained low due to high level of ubiquitinating activity. Upon growth factor removal, cell–cell contact or removal of adherence to the extracellular matrix, p27 is stabilized. In tumor cells, upon activation of yet unknown pathways, there is an induction of p27 degradation.

immunohistochemistry. In the case of the colon cancer study (Loda et al., 1996) it has also been shown that lack of p27 did not correlate with lack of mRNA expression suggesting that post-transcriptional mechanism of control were altered in those tumors. Indeed extracts from tumors lacking p27 contain high p27 degradation activity, suggesting that the lack of p27 in the tumor tissues is due to an enhancement of its degradation (Fig. 4). These recent findings represent a further demonstration that the ubiquitin pathway is targeted in tumors, and suggest the possibility of targeting the ubiquitination reactions in vivo to restore p27 accumulation.

In conclusion, the discovery of molecular alterations of cell cycle regulatory pathway in tumors justifies the need to continue these studies: a. to extend these observations to other cell cycle regulatory proteins; b. to establish a correlation between Cdk pathway alterations and other known genetic alterations, caused by somatic or germline mutations; c. to start large clinical studies to assess the diagnostic/prognostic value of these alterations; and finally, e. to verify the potential use of these pathways as targets for the discovery of novel anti-proliferative drugs.

ACKNOWLEDGMENTS

Work in our laboratory is supported by grants from AIRC, CNR, Mitotix, Inc. and the Italian Health Ministry.

REFERENCES

Baratte, B., Meiyer, L., Galaktinov, K., and Beach, D. (1992). Screening for antimitotic compounds using the cdc25 tyrosine phosphatase, an activator of the mitosis-inducing p34cdc2/cyclinB protein kinase. Anticancer Research *12*.

Buckley, M. F., Sweeney, K. J. E., Hamilton, J. A., Sini, R. L., Manning, D. L., Nicholson, R. I., deFazio, A., Watts, C. K. W., A., M. E., and Sutherland, R. L. (1993). Expression and amplification of cyclin genes in human breast cancer. Oncogene *8*, 2127–2133.

Ciechanover, A. (1994). The Ubiquitin-Proteasome Proteolytic Pathway. Cell *79*, 13–21.

Ciechanover, A., DiGiuseppe, J., Bercovich, B., Orian, A., Richter, J., Schwartz, A., and Brodeur, G. (1991). Degradation of nuclear oncoproteins by the ubiquitin system *in vitro*. Proc. Natl. Acad. Sci. USA *88*, 139–143.

Dai, K., Kobayashi, R., and Beach, D. (1996). Physical interaction of mammalian CDC37 with CDK4. J. Biol. Chem. *271*, 22030–22034.

Damagnez, V., Makela, T. P., and Cottarel, G. (1995). Schizosaccharomyces pombe Mop1-Mcs2 is related to mammalian CAK. Embo J. *14*, 6164–72.

Deshaies, R. J. (1995). Make it or break it: the role of ubiquitin-dependent proteolysis in cell regulation. Trends in Cell Biol. *10*, 428–434.

Dietrich, C., Bartsch, T., Schanz, F., Oesch, F., and Wieser, R. J. (1996). p53-dependent cell cycle arrest induced by N-acetyl-L-leucinyl-L-leucinyl-L-norleucinal in platelet-derived growth factor-stimulated human fibroblasts. Proc. Natl. Acad. Sci. USA *93*, 10815–10819.

Draetta, G., Brizuela, L., Potashkin, J., and Beach, D. (1987). Identification of p34 and p13, human homologs of the cell cycle regulators of fission yeast encoded by cdc2+ and suc1+. Cell *50*, 319–25.

Draetta, G., and Pagano, M. (1996). Cell cycle control and cancer. In Topics in Biology, W. W. Wong, ed.: Academic Press), pp. 241–248.

Draetta, G. F., and Eckstein, J. W. (1996). Cdc25 phosphatases in cell proliferation. Biochim. Byophys. Acta *in press*.

Elledge, S., and Harper, W. (1994). Cdk inhibitors: on the threshold of checkpoints and development. Curr. Opin. Cell Biol. *6*, 847–852.

Espinoza, F. E., Farrell, A., Erdjument-Bromage, H., Tempts, P., and Morgan, D. O. (1996). A cyclin-dependent kinase-activating kinase in budding yeast unrelated to vertebrate CAK. Science *273*, 1714–1716.

Evans, T., Rosenthal, E. T., Youngblom, J., Distel, D., and Hunt, T. (1983). Cyclin: A protein specified by maternal mRNA in sea urchin eggs that is destroyed at each cleavage division. Cell *33*, 389–396.

Firpo, E. J., Koff, A., Solomon, M., and Roberts, J. (1994). Inactivation of a Cdk2 inhibitor during Interleuki 2-induced proliferation of human T lymphocytes. Mol. Cell Biol. *14*, 4889–4901.

Galaktionov, K., and Beach, D. (1991). Specific activation of cdc25 tyrosine phosphatase by B-type cyclins: evidence for multiple roles of mitotic cyclins. Cell *67*, 1181–1194.

Galaktionov, K., Chen, X., and Beach, D. (1996). Cdc25 cell-cycle phosphatase as a target of c-myc. Nature *382*, 511–517.

Galaktionov, K., Jessus, C., and Beach, D. (1995). Raf1 interaction with cdc25 phosphatase ties mitogenic signal transduction to cell cycle activation. Genes & Develop. *9*, 1046–1058.

Galaktionov, K., Lee, A., Eckstein, J., Draetta, G., Meckler, J., Loda, M., and Beach, D. (1995). Cdc25 phosphatases as potential human oncogenes. Science *269*, 1575–1577.

Gillett, C., Fantl, V., Smith, R., Fisher, C., Bartek, J., Dickson, C., Barnes, D., and Peters, G. (1994). Amplification and overexpression of cyclin D1 in breast cancer detected by immunohistochemical staining. Cancer Res *54*, 1812–7.

Glotzer, M., Murray, A., and Kirschner, M. (1991). Cyclin is degraded by the ubiquitin pathway. Nature *349*, 132–138.

Gottlieb, T. M., and Oren, M. (1996). p53 in growth control and neoplasia. Biochim. Biophys. Acta, Gene Struct. Expr. *1287*, 77–102.

Hartwell, L. H., Culotti, J., Pringle, J. R., and Reid, B. J. (1974). Genetic control of the cell division cycle in yeast. Science *183*, 46–51.

Hengst, L., Dulic, V., Slingerland, J., Lees, E., and Reed, S. (1994). A cell cycle regulated inhibitor of cyclin-dependent kinases. Proc. Natl. Acad. Sci. USA *91*, 5291–5295.

Hershko, A., Ganoth, D., Sudakin, V., Dahan, A., Cohen, L., Luca, F., Ruderman, J., and Eytan, E. (1994). Components of a System That Ligates Cyclin to Ubiquitin and Their Regulation by the Protein Kinase cdc2. J. Biol. Chem. *269*, 4940–4946.

Hochstrasser, M. (1996). Protein degradation or regulation: Ub the judge. Cell *84*, 813–815.

Hochstrasser, M. (1995). Ubiquitin, proteasomes, and the regulation of intracellular protein degradation. Current Opinion in Cell Biology *7*, 215–223.

Hoffmann, I., Clarke, P. R., Marcote, M. J., Karsenti, E., and Draetta., G. (1993). Phosphorylation and activation of human cdc25-C by cdc2-cyclin B and its involvement in the self-amplification of MPF at mitosis. EMBO J. *12*, 53–63.

Hoffmann, I., Draetta, G., and Karsenti, E. (1994). Activation of the phosphatase activity of human cdc25A by a cdk2-cyclin E dependent phosphorylation at the G1/S transition. Embo J *13*, 4302–4310.

Horiguchi, T., Nishi, K., Hakoda, S., Tanida, S., Nagata, A., and Okayama, H. (1994). Dnacin A1 and dnacin B1 are antitumor antibiotics that inhibit cdc25B phosphatase activity. Biochem Pharmacol 48, 2139–2141.

Huibregtse, J., Scheffner, M., Beaudenon, S., and Howley, P. (1995). A family of proteins structurally and functionally related to E6-AP ubiquitin-protein ligase. Proc. Natl. Acad. Sci. USA 92, 2563–2567.

Huibregtse, J., Scheffner, M., and Howley, P. (1991). A cellular protein mediates association of p53 with the E6 oncoprotein of human papillomavirus types 16 or 18. EMBO J. 10, 4129–4135.

Huibregtse, J., Scheffner, M., and Howley, P. (1993). Cloning and Expression of the cDNA for E6-AP, a Protein That Mediates the Interaction of the Human Papillomavirus E6 Oncoprotein with p53. Mol. Cell. Biol. 13, 775–784.

Isaksson, A., Musti, A. M., and Bohmann, D. (1996). Ubiquitin in signal transduction and cell transformation. Biochim. Byophis. Acta 1288, F21-F29.

Jackson, P. K. (1996). Cull and destroy. Curr. Biol. 6, 1209–1212.

Jentsch, S., and Schlenker, S. (1995). Selective protein degradation: a journey's end within the proteasome. Cell 82, 881–884.

Jiang, W., Kahan, S., Tomita, N., Zhang, Y., Lu, S., and Weinstein, B. (1992). Amplification and expression of the human cyclin D gene in esophageal cancer. Cancer Research 52, 2980–2983.

Jinno, S., Suto, K., Nagata, A., Igarashi, M., Kanaoka, Y., Nojima, H., and Okayama, H. (1994). Cdc25A is a novel phosphatase functioning early in the cell cycle. Embo J 13, 1549–56.

Jinno, S., Suto, K., Nagata, A., Igarashi, M., Kanaoka, Y., Nojima, H., and Okayama, H. (1994). Cdc25A is a novel phosphatase functioning early in the cell cycle. EMBO J. 13, 1549–1556.

Kaldis, P., Sutton, A., and Solomon, M. J. (1996). The Cdk-activating kinase (CAK) from budding yeast. Cell 86, 553–564.

Karp, J. E., and Broder, S. (1995). Molecular foundations of cancer: new targets for intervention. Nature Medicine 1, 309–320.

King, R., Peters, J., Tugendreich, S., Rolfe, M., Hieter, P., and Kirschner, M. (1995). A 20S complex containing Cdc27 and Cdc16 catalyzes the mitosis-specific conjugation of ubiquitin to Cyclin B. Cell 81, 279–288.

Koff, A., Cross, F., Fisher, A., Schumacher, J., Leguellec, K., Philippe, M., and Roberts, J. M. (1991). Human cyclin E, a new cyclin that interacts with two members of the CDC2 gene family. Cell 66, 1217–1228.

Korsmeyer, S. J. (1992). Chromosomal translocations in lymphoid malignancies reveal novel proto- oncogenes. Annu Rev Immunol 10, 785–807.

Lammie, G., Fantl, V., Smith, R., Schuuring, E., Brookes, S., Michalides, R., Dickson, C., Arnold, A., and Peters, G. (1991). D11S287, a putative oncogene on chromosome 11q13, is amplified and expressed in squamous cell and mammary carcinimas and linked to BCL-1. Oncogene 6, 439–444.

Lamphere, L., Fiore, F., Xu, X., Brizuela, L., Keezer, S., Sardet, C., Draetta, G. F., and Gyuris, J. (1996). Interaction between Cdc37 and Cdk4 in human cells. Oncogene submitted.

Lebwohl, D. E., Muise-Helmericks, R., Sepp-Lorenzino, L., Serve, S., Timaul, M., Bol, R., Borgen, P., and Rosen, N. (1994). A truncated cyclin D1 gene encodes a stable mRNA in a human breast cancer cell line. Oncogene 9, 1925–9.

Lew, D. J., Dulic, V., and Reed, S. I. (1991). Isolation of three novel human cyclins by rescue of G1 cyclin (cln) function in yeast. Cell 66, 1197–1206.

Loda, M., Cukon, B., Tam, S., Lavin, P., Fiorentino, M., Draetta, G., Jessup, J., and Pagano, M. (1996). Increased proteasome-dependent degradation of the cyclin-dependent kinase inhibitor p27 in aggressive colorectal carcinomas. submitted.

Maki, C., Hulbregtse, J., and Howley, P. (1996). In vivo ubiquitination targets wild-type p53 for degradation by the proteasome. Cancer Res. in press.

Matsushime, H., Roussel, M., Ashmun, R., and Sherr, C. J. (1991). Colony-stimulating Factor 1 regulates novel cyclins during the G1 phase of the cell cycle. Cell 65, 701–713.

Millar, J., McGowan, C. H., Lenaers, G., Jones, R., and Russell, P. (1991). p80cdc25 mitotic inducer is the tyrosine phosphatase that activates p34cdc2 kinase in fission yeast. EMBO J. 10, 4301–4309.

Morgan, D. O. (1995). Principles of CDK regulation. Nature 374, 131–134.

Motokura, T., Bloom, T., Kim, Y. G., Jueppner, H., Ruderman, J., Kronenberg, H., and Arnold, A. (1991). A novel cyclin encoded by a bcl1-linked candidate oncogene. Nature 350, 512–515.

Nagata, A., Igarashi, M., Jinno, S., Suto, K., and Okayama, H. (1991). An additional homolog of the fission yeast cdc25 gene occurs in humans and is highly expressed in some cancer cells. New Biol. 3, 959–967.

Nefsky, B., and Beach, D. (1996). Pub1 acts as an E6-AP-like protein ubiquitiin ligase in the degradation of cdc25. Embo J. 15, 1301–1312.

Nishizawa, M., Okazaki, K., Furuno, N., Watanabe, N., and Sagata, N. (1992). The 'second-codon rule' and autophosphorylation govern the stability and activity of Mos during the meiotic cell cycle in Xenopus oocytes. Embo J 11, 2433–46.

Nourse, J., Firpo, E., Flanagan, M., Coats, S., Polyak, C., Lee, M., Massague, J., Crabtree, G., and Roberts, J. (1994). Interleukin-2-mediated elimination of p27Kip1 cyclin-dependent kinase inhibitor prevented by rapamycin. Nature *372*, 570–573.

Nurse, P. (1975). Genetic control of cell size at division in yeast. Nature *256*, 547–551.

Orian, A., Whitedside, S., Isral, A., Stancovski, I., Schwartz, A., and Ciechanover, A. (1995). Ubiquitin-mediated processingof NF-kB transcreptional activator precursor p105. J. of Biol. Chem. *270*, 21707–21714.

Pagano, M., Tam, S. W., Theodoras, A. M., Romero-Beer, P., Del Sal, G., Chau, V., Yew, R., Draetta, G., and Rolfe, M. (1995). Role of the Ubiquitin-Proteasome pathway in regulating aboundance of the Cyclin-dependent kinase inhibitor p27. Science *269*, 682–685.

Pagano, M., Tam, S. W., Theodoras, A. M., Romero-Beer, P., Del Sal, G., Chau, V., Yew, R., Draetta, G., and Rolfe, M. (1995). Role of the Ubiquitin-Proteasome pathway in regulating abundance of the Cyclin-dependent kinase inhibitor p27. Science *269*, 682–685.

Palombella, V., Rando, O., Goldberg, A., and Maniatis, T. (1994). The Ubiquitin-Proteasome Pathway Is Required for Processing the NF-κB1 Precursor Protein and the Activation of NF-kB. Cell *78*, 773–785.

Polyak, K., Kato, M., Solomon, M. J., Sherr, C. J., Massague, J., Roberts, J. M., and Koff, A. (1994). p27[Kip1]and Cyclin D-Cdk4 are interacting regulators of Cdk2, and link TGF-β and contact inhibition to cell cycle arrest. Genes & Dev. *8*, 9–22.

Rolfe, M., Romero, P., Glass, S., Eckstein, J., Berdo, I., Theodoras, A., Pagano, M., and Draetta, G. (1995). Reconstitution of p53-ubiquitinylation reaction from purified components: the role of human UBC4 and E6AP. Proc. Natl. Acad. Sci. USA *92*, 3264–3268.

Rolfe, M., Romero, P., Glass, S., Eckstein, J., Berdo, I., Theodoras, A., Pagano, M., and Draetta, G. (1995). Reconstitution of p53-ubiquitinylation reaction from purified components: the role of human UBC4 and E6AP. Proc. Natl. Acad. Sci. USA *92*, 3264–3268.

Rosenthal, E. T., Tansey, T. R., and Ruderman, J. V. (1983). Sequence-specific adenylations and deadenylations accompany changes in the translation of maternal messenger RNA after fertilization of Spisula oocytes. J. Mol. Biol. *166*, 309–327.

Russell, P., Moreno, S., and Reed, S. I. (1989). Conservation of mitotic controls in fission and budding yeasts. Cell *57*, 295–303.

Sadhu, K., Reed, S. I., Richardson, H., and Russell, P. (1990). Human homolog of fission yeast cdc25 is predominantly expressed in G1. Proc. Natl. Acad. Sci. USA *87*, 5139–5143.

Scheffner, M., Huibregtse, J., and Howley, P. (1994). Identification of a human ubiquitin-conjugating enzyme that mediates the E6-AP-dependent ubiquitination of p53. Proc. Natl. Acad. Sci. USA *91*, 8797–8801.

Scheffner, M., Nuber, U., and Huibregtse, J. (1995). Protein ubiquitination involving an E1-E2-E3 enzyme ubiquitin thioester cascade. Nature *373*, 81–83.

Scheffner, M., Werness, B., Huibregtse, J., Levine, A., and Howley, P. (1990). The E6 Oncoprotein Encoded by Human Papillomavirus Types 16 and 18 Promotes the Degradation of p53. Cell *63*, 1129–1136.

Scherer, D., Brockman, J., Chen, Z., Maniatis, T., and Ballard, D. (1995). Signal-induced degradation of IkBa requires site-specific ubiquitination. Proc. Natl. Acad. Sci. USA *92*, 11259–11263.

Sherr, C. (1994). G1 phase progression: cycling on cue. Cell *79*, 551–555.

Sherr, C., and Roberts, J. (1995). Inhibitors of mammalian G1 cyclin-dependent kinases. Genes& dev. *9*, 1149–1163.

Sicinski, P., Donaher, J. L., Parker, S. B., Li, T., Fazeli, A., Gardner, H., Haslam, S. Z., Bronson, R. T., Elledge, S. J., and Weinberg, R. A. (1995). Cyclin D1 provides a link between development and oncogenesis in the retina and breast. Cell *82*, 621–630.

Stancovski, I., Gonen, H., Orian, A., Schwartz, A., and Ciechanover, A. (1995). Degradation of the proto-oncogene c-Fos by the ubiquitin proteolytic system in vivo and in vitro: Identification and characterization of the conjugating enzymes. Mol. Cell Biol. *15*, 7106–7116.

Stepanova, L., Leng, X., Parker, S. B., and Harper, J. W. (1996). Mammalian p50Cdc37 is a protein kinase-targeting subunit of Hsp90 that binds and stabilizes Cdk4. Genes Dev. *10*, 1491–1502.

Sudakin, V., Ganoth, D., Dahan, A., Heller, H., Hershko, J., Luca, F., Ruderman, J., and Hershko, A. (1995). The cyclosome, a large complex containing cyclin selective ubiquitin ligase activity, targets cyclins for destruction at the end of mitosis. Mol. Biol. Cell *6*, 185–198.

Sutherland, R. L., Hamilton, J. A., Sweeney, K. J., Watts, C. K., and Musgrove, E. A. (1995). Steroidal regulation of cell cycle progression. Ciba Found. Symp. *191*, 218–228.

Thuret, J. Y., Valay, J. G., faye, G., and Mann, C. (1996). Civ1 (CAK in vivo), a novel Cdk-activating kinase. Cell *86*, 565–576.

Treier, M., Staszewski, L., and Bohmann, D. (1994). Ubiquitin-Dependent c-Jun Degradation In Vivo Is Mediated by the δ Domain. Cell *78*, 787–798.

Wang, T. C., Cardiff, R. D., Zukerberg, L., Lees, E., Arnold, A., and Schmidt, E. V. (1994). Mammary hyperplasia and carcinoma in MMTV-cyclin D1 transgenic mice. Nature *369*, 669–671.

Weinberg, R. A. (1995). The retinoblastoma protein and cell cycle control. Cell *81*, 323–330.

Weinstat-Saslow, D., Merino, M. J., Manrow, R. E., Lawrence, J. A., Bluth, R. F., Wittenbel, K. D., Simpson, J. F., Page, D. L., and Steeg, P. S. (1995). Overexpression of cyclin D mRNA distinguishes invasive and in situ breast carcinomas from non-malignant lesions. Nat Med *1*, 1257–1260.

Xiong, Y., Connolly, T., Futcher, B., and Beach, D. (1991). Human D-type cyclin. Cell *65*, 691–699.

Xiong, Y., Zhang, H., and Beach, D. (1992). D Type cyclins associate with multiple protein kinases and the DNA replication and repair factor PCNA. Cell *71*, 504–514.

Xiong, Y., Zhang, H., and Beach, D. (1993). Subunit rearrangement of the cyclin dependent kinases is associated with cellular transformation. Genes & Dev. *7*, 1572–1583.

Xu, X., and Burke, S. P. (1996). Roles of active site residues and the NH2-terminal domain in the catalysis and substrate binding of human Cdc25. J. Biol. Chem. *271*, 5118–5124.

Zhang, H., Hannon, G., and Beach, D. (1994). p21-containing Cyclin kinases exist in both active and inactive states. Genes & Dev. *8*, 1750–1758.

Zhang, H., Xiong, Y., and Beach, D. (1993). PCNA and p21 are universal elements of the cell cycle kinase family. Mol. Biol. Cell *4*, 897–906.

DISCUSSION

Bernards: Do you have evidence for a cell cycle regulated phosphorylation of p27 as a means to explain its destabilization during the cell cycle?

Draetta: I do not think we do have evidence for p27 phosphorylation in the cell cycle that might deregulate stabilization.

Pagano: I do not know if the book of the last Pezcoller symposium is out or not, but in that book we presented a figure showing that p27 is not very much phosphorylated in G0 cells. When during G1 phosphorylation goes up p27 starts to be degraded, but that is the only correlation. I know that at least two labs mutated the threonine present in the PEST sequence, but I do not know if they looked at protein stability, what are the results and so on. We will know soon.

Hartwell: Many years ago gross alterations in CDK cyclin behavior were reported in tumors. It seems like a very high proportion, either overexpressing things or, what seems really strange was expressing CDK cyclin combinations at the wrong times in the cycle. I do not see much more of that. What is your sense of whether that sort of misregulation is common in tumors?

Draetta: If we let aside all the observations done on cell lines, there is clearly, as you say, altered expression of these elements in tumors. Well, what still surprises me is the ability of the tumor cells to be held quiescent so you have the priming of the cell cycle, you see the appearance of the fraction that Dr. Reid was talking about this morning. You have that priming, you have, from what Jim Roberts was telling me, expression of cyclin E in 80% of the breast cancers and very little proliferative activity. What would be interesting to see, whether most of these cancers are held back from growing very fast by the presence of CDK inhibitors? Or what is preventing them from being constantly in S-phase and what is preventing all these cells from killing themselves in doing so. That would also lead to the idea of intervening at the checkpoint level by perhaps blocking the expression of p27 induced cells. Everybody is telling me that they find cyclin E expression, they find cyclin D expression, so why are they quiescent, what is holding them back? This needs large studies to be able to correlate large numbers of cyclin E to p27 expression, p53 and so on.

Ferrari: My impression is that in several cases of colon cancer tumors the biological problem is not in increased proliferation activity but a differentiation block. In acute leukemias, for example, we have a block of differentiation and this is the real problem from the therapeutic point of view, since you have to induce differentiation not to block proliferation. So, can you comment on the differentiation markers expressed in these tumors? The expression of p27 can be important for this aspect. So, the tumor is not only due to increased proliferation but in several cases to a block of differentiation and, therefore, of apoptosis.

Draetta: I do not know because, I think that there is no evidence from the studies done so far, there is correlation between p27 expression and the real differentiation of the tumors. So that does not seem to be an absolute requirement for p27 in the differentiation pathway of these tumors. These tumors also do have p21 and if you do p21 staining, in the normal *mucosa,* you find that p21 has exactly the same profile

Ferrari: Are these tumors G1 arrested, because from G1, the cells can go in differentiation or apoptosis. Do you have data on study of the distribution of the tumor cells in the cell cycle?

Draetta: I do not know. I think that proliferative index in colon is much higher than it is in breast. And definitely it is true that cells have to die very frequently and quickly of apoptosis, because otherwise they would go into a very big size in a very short time.

Nasmyth: You have now very good antibodies against p27 that allow you to see it, as far as I can tell, mainly in the nucleus. Using those antibodies which permit an *in situ* analysis of when p27 disappears, such an analysis obviously gives you much more information than looking in cultures whose synchrony is usually very poor. Can you come to conclusions, at present, concerning when the p27 dips in the cell cycle relative to say, the onset of DNA replication as measured by the bromodeoxyuridine incorporation. Is there a clear linkage there?

Draetta: No, we have not done it yet. But it is clearly worth doing. By the way, there are commercially available monoclonals from transduction lines, if any of you is interested. It is a very good antibody. I like very much the idea of thinking about the control of G1 S as we think about the control of G2 and that there are forces opposing each other and there is an issue of threshold. Opposing forces, p27 being there, and slowly disappearing through this mechanism, and the accumulation of cyclin E occurs until you reach a certain threshold in p27. At that point, the cell will enter S.

INDEX

MIX
Papier aus verantwortungsvollen Quellen
Paper from responsible sources
FSC® C105338

If you have any concerns about our products,
you can contact us on
ProductSafety@springernature.com

In case Publisher is established outside the EU,
the EU authorized representative is:
Springer Nature Customer Service Center GmbH
Europaplatz 3, 69115 Heidelberg, Germany

Printed by Libri Plureos GmbH
in Hamburg, Germany